Neural and Fuzzy Systems

THE EMERGING SCIENCE OF INTELLIGENT COMPUTING

SPIE Institutes for Advanced Optical Technologies

Transformations in Optical Signal Processing
William T. Rhodes, James R. Fienup, Bahaa E. A. Saleh, eds., 1984, SPIE Vol. 373 (out of print)

Optical and Hybrid Computing
Harold H. Szu, ed., 1987, SPIE Vol. 634

Photonics: High-Bandwidth Analog Applications
James Chang, ed., 1987, SPIE Vol. 648

Large-Area Chromogenics: Materials and Devices for Transmittance Control
Carl M. Lampert, Claes G. Granqvist, eds., 1990, Vol. IS 4

Dosimetry of Laser Radiation in Medicine and Biology
Gerhard J. Müller, David H. Sliney, eds., 1989, Vol. IS 5

Future Directions and Applications in Photodynamic Therapy
Charles J. Gomer, ed., 1990, Vol. IS 6

Automatic Object Recognition
Hatem N. Nasr, ed., 1991, Vol. IS 7

Holography
Pál Greguss, Tung H. Jeong, eds., 1991, Vol. IS 8

Invisible Connections: Instruments, Institutions, and Science
Robert Bud, Susan E. Cozzens, eds., 1992, Volume IS 9

Technology of Proximal Probe Lithography
Christie R. K. Marrian, ed., 1993, Vol. IS 10

Medical Optical Tomography: Functional Imaging and Monitoring
Gerhard J. Müller et al., eds., 1993, Vol. IS 11

Neural and Fuzzy Systems: The Emerging Science of Intelligent Computing
Sunanda Mitra, Madan M. Gupta, Wolfgang Kraske, eds., 1994, Vol IS 12

SPIE INSTITUTES

Volume IS 12

Institutes for
Advanced Optical
Technologies

Neural and Fuzzy Systems

THE EMERGING SCIENCE OF INTELLIGENT COMPUTING

Sunanda Mitra
Computer Vision and Image Analysis Laboratory
Texas Tech University

Madan M. Gupta
Intelligent Systems Research Laboratory
University of Saskatchewan

Wolfgang F. Kraske
Northrop Corporation

Editors

SPIE OPTICAL ENGINEERING PRESS

A publication of SPIE—The International Society for Optical Engineering
Bellingham, Washington USA

Library of Congress Cataloging-in-Publication Data

Neural and fuzzy systems : the emerging science of intelligent
computing / editors, Sunanda Mitra, Madan M. Gupta, Wolfgang Kraske.
p. cm. – (SPIE Institutes for advanced optical technologies
v. IS 12)
Includes bibliographical references.
ISBN 0-8194-1566-9. – ISBN 0-8194-1565-0 (pbk.)
1. Neural networks (Computer science) 2. Fuzzy systems.
I. Mitra, Sunanda. II. Gupta, Madan M. III. Kraske, Wolfgang.
IV. Series.
QA76.87.N472 1994 94-1640
006.3–dc20 CIP

Published by
SPIE—The International Society for Optical Engineering
PO Box 10
Bellingham, Washington 98227-0010

Printed in the United States of America.

This volume is dedicated to

Professor Lotfi A. Zadeh

for providing us with a mathematical tool of graded membership.

This tool has revolutionized the field of mathematical logic and computing methods, and has given rise to the birth of cognitive robots. Also, the researchers have faced new challenges in the study of chaotic and fuzzy neural systems, the subject of the present volume.

CONTENTS

SERIES INTRODUCTION

Neural and Fuzzy Systems: The Emerging Science of Intelligent Computing, twelfth in the SPIE Institute Series, is representative of SPIE's primary goal for this series: to provide, in a timely manner, authoritative overall introductions and reviews of emerging or critical technologies based on or related to optics and optoelectronics.

The eleven contributions present topics in what one might term the "science of reasoning," discussing, as they do, the principles and algorithms for neural network and autonomous robot systems. Matters of control and decision in systems having nonlinear dynamic characteristics are at the core of modern research in these areas. Again, as demonstrated by previous Institute volumes, the time lapse between fundamental research and technological applications in this field is narrowing significantly. Developments on one side of the R&D ledger lead to, and are balanced by, new concepts and approaches in the other. The Institute Series hopes to encourage and abet such synergism.

It is expected that this book will be of interest to those bodies of scientists and engineers involved in the R&D of a variety of technologies—computer vision, image processing, artificial intelligence, neural networks, and robotic command and control. That there are other technologies that might have an interest in and benefit from developments in this field of neural and fuzzy systems, I have no doubt.

Roy F. Potter
General Editor
SPIE Institutes for Advanced Optical Technologies

PREFACE

The concept of adaptive learning of system parameters in nonlinear systems has been addressed by system theorists for quite some time. The independent use of neural networks and fuzzy logic to facilitate such adaptive learning is, however, very recent. An even more recent trend has been to combine the strength of adaptive learning in neural networks with the fuzzy membership values of input/output data for designing better identification, recognition, and control systems.

The idea of publishing an SPIE Institute Series volume on the emerging science of intelligent computing was first discussed with Roy F. Potter, the series editor, while planning a workshop on Neuro-Control Systems that was held during the SPIE Symposium on Aerospace Sensing, 20–24 April 1992 in Orlando. This workshop was made possible by the sponsorship of Northrop Corporation through a symposium grant awarded to Texas Tech University. Keynote speakers at the workshop included Lotfi Zadeh (Univ. of California/Berkeley), James Bezdek (Univ. of West Florida), Madan Gupta (Univ. of Saskatchewan), James Keller (Univ. of Missouri/Columbia), and Robert Lea (NASA/JSC). Wolfgang Kraske (Northrop Corp.) and Sunanda Mitra (Texas Tech Univ.) were among the other speakers.

This twelfth volume in the SPIE Institute Series, *Neural and Fuzzy Systems*, presents some of the ideas and applications involving neural networks and/or fuzzy logic for better representation and operation of dynamical systems. Eleven chapters from researchers in six countries have been included. Chapter 1 describes the general behavior of nonlinear dynamical systems that are often chaotic. During the evolution of a dynamical system, information content may undergo degradation under iteration of usual fuzzified chaotic mappings. This chapter investigates more general classes of fuzzy chaos based on t-norms/conorms. Chapter 2 discusses the implementation of fuzzy rules for approximate case-based reasoning in neural network units. Chapter 3 presents a hierarchical structure of intelligent servo control for robotic motion, including fuzzy logic combined with neural networks. Chapter 4 describes a neural model for constrained inverse kinematic transformations for two- and three-linked robots. Chapter 5 presents a fuzzy controller model of highly complex and nonlinear dynamics of tether operations in space. Chapters 6 through 8 investigate different approaches for fuzzy and neuro-fuzzy techniques as applied to problems in computer vision, image processing, and pattern recognition. Chapters 9 through 11 discuss fuzzy and fuzzy morphological operators and algorithms used in analyzing higher-dimensional data sets. Chapters 6 and 8 present new neuro-fuzzy and possibilistic approaches to clustering techniques, while Chapter 11 reviews current fuzzy morphological filtering techniques for clustering noisy multidimensional data.

Clustering is a well-known technique that has been used primarily for partitioning complex data sets of unknown data distribution. However, interest has grown recently in using clustering for identification of system parameters in nonlinear dynamical systems.

I would like to thank Madan Gupta for his encouragement in organizing the workshop in Orlando, as well as the other keynote speakers, Lotfi Zadeh, James Bezdek, James Keller, and Robert Lea, for their valuable help in making the workshop successful. I am grateful also to Wolfgang Kraske, without whose support the workshop and book would not have materialized, and to Roy Potter, editor of the SPIE Institute Series. Finally I would like to thank SPIE staff Mary Horan, Anne Noteboom, and Eric Pepper, and all the authors, for their valuable contributions and efforts in making this volume possible.

Sunanda Mitra
Texas Tech University
February 1994

CONTRIBUTORS

Phil Diamond
University of Queensland
Mathematics Department
Queensland QLD 4072
Australia

Liya Ding
National University of Singapore
Institute of Systems
Heng Mui King Terrance
Kent Ridge 0511
Singapore

Edward R. Dougherty
Rochester Institute of Technology
Center for Imaging Science
One Lomb Dr.
Rochester, NY 14623

Toshio Fukuda
Nagoya University
Department of Mechano-Informatics and Systems
1 Furo-cho, Chikusa-ku
Nagoya 464-01
Japan

Ashish Ghosh
Indian Statistical Institute
Machine Intelligence Unit
203 B. T. Road
Calcutta 700 035
India

Madan M. Gupta
University of Saskatchewan
Intelligent Systems Research Laboratory
College of Engineering
Saskatoon SK S7N 0W0
Canada

Yashvant Jani
Togai InfraLogic, Inc.
17000 El Camino Real, Ste.204C
Houston, TX 77058-2636

James M. Keller
University of Missouri/Columbia
Department of Electrical and Computer Science
217 Electrical Eng. Building
Columbia, MO 65211

Alok Kher
Texas Tech University
Computer Vision and Image Analysis Laboratory
MS 3102, Department of Electrical Engineering
P.O. Box 43102
Lubbock, TX 79409-3102

Yong Soo Kim
Texas Tech University
Computer Vision and Image Analysis Laboratory
Department of Electrical Engineering
P.O. Box MS 3102
Lubbock, TX 79409-3102

Wolfgang F. Kraske
Northrop Corporation
Aircraft Group
B-2 Div., Department of Advanced Special Projects
8900 E Washington Blvd., W921
Pico Rivera, CA 90660-3765

Raghu J. Krishnapuram
University of Missouri/Columbia
221 EE Bldg., E&CE Department
Columbia, MO 65211-0001

Robert N. Lea
NASA/Johnson Space Center
Software Technology Group, PT4
Houston, TX 77058

Sunanda Mitra
Texas Tech University
Computer Vision and Image Analysis Laboratory
MS 3102, Department of EE
P.O. Box 43102
Lubbock, TX 79409-3102

Sankar K. Pal
Indian Statistical Institute
Machine Intelligence Unit
203 B. T. Road
Calcutta 700 035
India

Suryalakshmi Pemmaraju
Texas Tech University
Computer Vision and Image Analysis Laboratory
Department of Electrical Engineering
P.O. Box MS 3102
Lubbock, TX 79409-3102

Dandina H. Rao
807-107, Cumberland
Saskatoon SK S7N 0W0
Canada

Zuliang Shen
National University of Singapore
Institute of Systems
Heng Mui Keng Terrace
Kent Ridge 0511
Singapore

Takanori Shibata
MITI
Biorobotics Division
Mechanical Engineering Laboratory
Namiki 1-2
Tsukuba-shi Ibaraki 305
Japan

Divyendu Sinha
CUNY/College of Staten Island
Computer Science Department
2800 Victory Blvd.
Staten Island, NY 10314-6609

Purnendu Sinha
Stevens Institute of Technology
Electrical Engineering and Computer Science Department
Castle Point Station
Hoboken, NJ 07030

Neural and Fuzzy Systems

THE EMERGING SCIENCE OF INTELLIGENT COMPUTING

Chaos and Information Loss in Fuzzy Dynamical Systems*

Phil Diamond

Mathematics Department
University of Queensland
QLD 4072 Australia

ABSTRACT

That dynamical systems can produce chaotic and erratic behavior is now an accepted feature of systems modeling and controller design. This behavior is not necessarily something to be avoided at all costs, but is an aspect which should be recognized and taken into account when dealing with a control system.

Despite the intense interest in chaotic systems, there is no universally accepted mathematical definition of chaos. Common descriptions are in terms of positive topological entropy, Li-Yorke scrambled sets, sensitive dependence of initial conditions and positive Liapunov exponents. The precise relationships between these competing signatures is unclear and some notions are not even independent.

Fuzzy dynamical systems occur in many areas of interest and it is important to understand the repertoire of possible behaviors of fuzzy systems. This review gives an overview of chaos in fuzzy dynamical systems, studying

- Dynamical systems and definitions of chaos
- Metric space of fuzzy sets ($\mathcal{D}^n$, d_∞).
- Chaotic mappings on $\mathcal{D}^n$ and possible loss of information due to chaotic dynamics.
- A general definition of fuzzification and level set, based on *t*-norms/conorms and their diagonal functions, and the effect of fuzzifying chaotic mappings.
- A relationship between a very simple criterion, positive topological entropy and Li-Yorke chaos.
- Examples of fuzzy chaos

1. INTRODUCTION

That dynamical systems can produce chaotic and erratic behavior is now an accepted feature of systems modeling and controller design. This behavior is not necessarily something to be avoided at all costs, but it is an aspect which should be recognized and taken into account when dealing with a control system.

---

* This research was funded under Australian Research Council grants A 490 894 and A 8913 2609.

Despite the intense interest in chaotic systems, there is no universally accepted mathematical definition of chaos. Common descriptions are in terms of positive topological entropy,[3] Li-Yorke scrambled sets,[18] sensitive dependence of initial conditions,[6,24] and positive Liapunov exponents.[11] The precise relationships between these competing signatures is unclear and some notions are not even independent.[1]

Even granted that there is no unequivocal concept of chaos, it is still inadequate to describe complicated, erratic behavior as chaotic without further criteria. Numerous authors have given sufficient conditions for types of chaos to appear. Among many we mention Li and Yorke,[18] Kloeden,[16] Marotto,[19] Wiggins.[24] A very simple criterion was given in[7] but it only produces periodic sets of arbitrary periods rather than periodic points.

The situation becomes still more complicated with fuzzy controllers. Here we have systems which, while still dynamical systems, operate with functions (fuzzy sets) rather than with points in a Euclidean space. Nevertheless, mathematical descriptions of chaos can be formulated in much the same way and there has been some progress in giving sufficient conditions for the behavior.[17,10] Some specific examples can be found in.[8,23] It has been shown that if $f : \Omega \to \Omega$ is chaotic, so also is its fuzzification

$$\tilde{f}: g(\Omega) \to g(\Omega),$$

$$\left(\tilde{f}u\right)(y) = \sup\left\{u(x): x \in f^{-1}(y)\right\} \tag{1}$$

defined on the space $\mathcal{D}(\Omega)$ of fuzzy sets on a set Ω. However, this chaos is in a sense degenerate, because the iterates are asymptotically crisp and, ultimately, one sees chaos of a mapping of ordinary sets rather than of fuzzy output.[8] Apparently, the usual fuzzification and sup-min composition are inadequate to describe all of the complexities which may arise in fuzzy control. This is not completely unexpected since other t-norm/conorm operations are being used[13] and more general extension principles have been developed.[12]

In this paper we draw some of these threads together. Chaos is studied in two distinct metric spaces of fuzzy sets. We examine the way in which a measure of information content changes, during the evolution of a dynamical system. More general fuzzifications are also considered. These do not necessarily degenerate under chaotic iteration. It is shown that if the fuzzification operation is reasonably well-behaved then chaotic maps on Ω induce chaos in the corresponding fuzzy system. A relationship between Li-Yorke chaos, positive topological entropy and the simple criterion of[7] is demonstrated. We give a theorem which generalizes and improves the chaotic theorem for set valued mappings of Benhabib and Day.[2]

Below, Section 2 gives definitions and some background. Section 3 gives an overview of dynamical systems and chaos, while Section 4 introduces a measure of information content of fuzzy sets. Section 5 discusses how information can degrade under iteration of fuzzified chaotic mappings when the usual fuzzification is used. More general classes of Γ - fuzzifications are considered in Section 6 and a general theorem linking various types of chaos to a simple sufficient condition is proved in Section 7. An example of a chaotic fuzzified mapping with non degenerate periodicities is given in Section 8, and a number of examples of fuzzy chaos are studied in the last section. A brief appendix defining topological entropy concludes the paper.

2. PRELIMINARIES

Let Ω be a compact metric space with metric ρ. Denote by $\mathcal{D}(\Omega)$ the totality of fuzzy sets

$$u : \Omega \to [0,1] = I$$

which are *upper semicontinous*. That is, for each x_0 in the domain of u,

$$u(x_0) \geq \overline{\lim}_{x \to x_0} u(x).$$

Note that the *characteristic function* χ_A of set $A \subset \Omega$, defined by

$$\chi_A(x) = \begin{cases} 1 \text{ if } x \in A, \\ 0 \ \text{ otherwise,} \end{cases}$$

is a fuzzy set.

Let A, B be any two nonempty closed subsets of a metric space (Þ , $\overline{\rho}$) The *Hausdorff separation* of A from B is defined to be

$$\overline{\rho}_H(A,B) = \sup_{x \in A} \overline{\rho}(x,B),$$

where $\overline{\rho}(x,B) = \inf_{y \in B} \overline{\rho}(x,y)$ The separation $\overline{\rho}_H$ is not a metric because it is not symmetric (See Fig. 2.1) It may also be shown that $\overline{\rho}_H(A,B) = 0$ if and only if $A \subseteq B$. The quantity

$$d_H(A,B) = \max\{\overline{\rho}_H(A,B), \overline{\rho}_H(B,A)\}$$

is said to be the *Hausdorff distance* between A and B. It is a metric on the space S(Þ) of closed nonempty subsets of Ω. The metric space (S(Þ), d_H) is compact provided Þ is compact.[5]

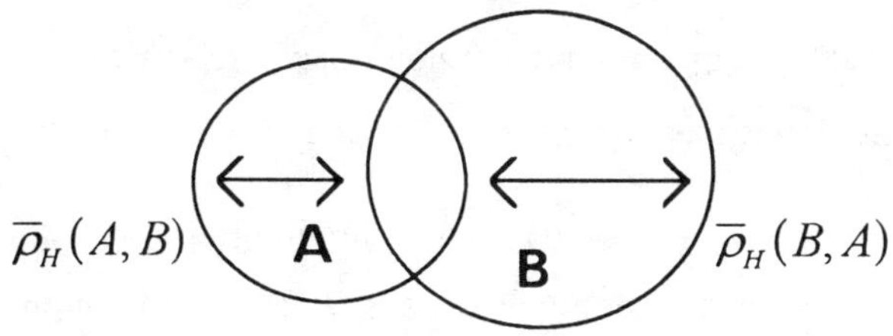

Fig. 2.1: The Hausdorff separations of A and B.

For any $u \in \mathcal{D}(\Omega)$ define

$$\text{send}\ (u) = \{(x,\alpha) \in \Omega \times I : u(x) \geq \alpha\}$$

Since u is upper semicontinous, send(u) is closed in $Þ = \Omega \times I$ with respect to the natural product metric $\overline{\rho}((x,\alpha),(y,\beta)) = \rho(x,y) + |\alpha - \beta|$. Consequently we can define a distance function on $\mathcal{D}(\Omega)$ by

$$H(u,v) = d_H(\text{send}(u), \text{send}(v)),\ u,\ v \in \mathcal{D}(\Omega).$$

This is indeed a metric on $\mathcal{D}(\Omega)$ and it is straightforward but lengthy to show

Theorem 2.1[15] *The metric space* ($\mathcal{D}(\Omega)$, H) *is compact.*

For any $\delta \in (0,1]$, define $\mathcal{D}_\delta(\Omega)$ to be class of all $u \in \mathcal{D}(\Omega)$ satisfying

$$\max_{x \in \Omega} u(x) \geq \delta.$$

Observe that the class of all normal fuzzy sets on Ω is just $\mathcal{D}_1(\Omega)$ and so the $\mathcal{D}_\delta(\Omega)$ can be thought of as generalizing the idea of normality.

Yet another metric may be defined on the class of normal upper semicontinous fuzzy sets $\mathcal{D}_1(\Omega)$. This metric is especially important when levels of possibility are part of the context, and is defined in terms of level sets $[u]^\alpha = \{x \in \Omega : u(x) \geq \alpha\}$, $0 < \alpha \leq 1$. We put

$$d_\infty(u,v) = \sup_{0<\alpha\leq 1} d_H\left([u]^\alpha,[v]^\alpha\right)$$

Theorem 2.2. [21] *The metric space* $\mathcal{D}(\Omega)$, d_∞) *is compact.*

It may be shown that convergence in the d_∞ metric implies convergence in the sendograph metric H, but not conversely. Hence the topology of H is weaker than $_\infty$.

If the fuzzy sets are defined on a space which is not compact, for example $\Re^n$ where only bounded and closed sets are compact spaces, it is more natural to define the space $\mathcal{D}^n \stackrel{def}{=} \mathcal{D}(\Re^n)$ of upper semicontinous fuzzy sets with *compact support*. The metrics H, d_∞ are defined as above and the space is *complete* with respect to each.

Theorem 2.3. *The metric spaces* ($\mathcal{D}^n$, H) *and* ($\mathcal{D}^n$, d_∞) *are complete.*

Let $C(\Omega)$ be the set of all continuous maps f: $\Omega \to \Omega$. Suppose that there is a correspondence between each $f \in C(\Omega)$ and a mapping

$$\mathcal{F}(f): \mathcal{D}(\Omega) \to \mathcal{D}(\Omega)$$

Such a correspondence is called a *fuzzification*. The most common fuzzification is $\bar{f}$ defined by

$$(\bar{f}u)(y) = \sup\{u(x) : x \in f^{-1}(y)\} \tag{2}$$

for all $y \in f(\text{supp}(u))$ and $(\bar{f}u(y) = 0$ for all other $y \in \Omega$. Further fuzzifications are discussed in Section 6.

3. DYNAMICAL SYSTEMS AND CHAOS

A dynamical system in a space $\mathcal{H}$ describes the way quantities in the space change with the passage of time. Usually, the space $\mathcal{H}$ is a space of states of a physical system. A differential equation generates dynamics for such a system, as does a difference equation. But so also does the interaction of a practitioner with an expert system, or a fuzzy controller, or a decision maker working with vague and imperfect information. However, it is all a matter of what space (or universe) $\mathcal{H}$ that one is working in, and all these systems have common features. Broadly speaking, each is a smooth mapping $\phi_t : \mathcal{H} \to \mathcal{H}$, More precisely, a *dynamical system* is a continuously differentiable map $\phi : P \times \mathcal{H} \to \mathcal{H}$, where $P = \Re$ of $\mathbb{Z}$, written as $\phi_t : \mathcal{H} \to \mathcal{H}$, $t \in P$, and where ϕ_t satisfies

- ϕ_o is the identity on $\mathcal{H}$, $\phi_o(x) = x$, for all $x \in \mathcal{H}$;
- The composition $\phi_s \, o \, \phi_t = \phi_{s+t}$ for all $s, t \in P$, that is $\phi_s(\phi_t(x)) = \phi_{s+t}(x)$ for every $x \in \mathcal{H}$.

Typically, a differential equation $x' = f(t, x)$, $x(t_o) = x_0$ generates a mapping $\phi_t : \Re^n \to \Re^n$, with $t \in \Re$ (a *continuos* dynamical system), defined by $\phi_t(x_0) = X(t, t_o ; x_o)$, where this denotes the solution of the initial value problem at time t. Similarly, a difference equation $x_{k+1} = f_k(x_k)$ generates $\phi_k : \mathcal{H} \to \mathcal{H}$, $k \in \mathbb{N}$ (a *discrete* dynamical system or *iterated function system*), defined by

$$\phi_k(x) = f_k \circ f_{k-1} \circ \ldots \circ f_1(x)$$

The discrete systems are described by iterations of one or more functions. We shall restrict our discussion to systems consisting of iterates $f^k(x)$ of a single continuous function f, where x is an element of a metric space $\mathcal{H}$. These can arise as solutions of the difference equation $x_{k+1} = f(x_k)$.

The object of study in such systems are the *orbits* $\{x, f(x), f^2(x), \ldots\}$ of points $x \in \mathcal{H}$. *Periodic points* of period k correspond to k-cycles, that is to orbits of finite length k, for then $f^k(x) = x$.

Discrete dynamical systems are capable of enormously complicated behavior, and certain aspects are said to be "chaotic". One definition of chaos is due to Devaney[6] and uses the criteria of

1. f is *transitive*; for all disjoint open sets $U, V \subset \mathcal{H}$, $f^k(U)$ meets V eventually for some k.

2. Periodic points of f are dense in $\mathcal{H}$.

3. *Sensitive dependence on initial conditions*; there is a sensitivity constant δ such that for any unequal x, y, eventually (for some k) $f^k(x)$ and $f^k(y)$ are more than a distance δ apart.

It has recently been shown that (1) + (2) $\Rightarrow$ (3) [1]

A closely related definition, that is easier to handle in many respects, involves the notion of a *scrambled set.*[15,18] The criteria here are that

(LY1) There exist k-cycles, for all $k \geq K$.

(LY2) There exists a set $S \subset \mathcal{H}$, containing no cycles, such that

(a) $f(S) \subset S$;
(b) for every $x, y \in S, x \neq y$,

$$\limsup_{k\to\infty} \left\| f^k(x) - f^k(y) \right\| > 0.$$

(c) for every $x \in S$ and cyclic point p,

$$\limsup_{k\to\infty} \left\| f^k(x) - f^k(p) \right\| > 0.$$

(LY3) There exist an uncountable subset $S_0 \subset S$ such that for all $x, y \in S_0$,

$$\liminf_{k\to\infty} \left\| f^k(x) - f^k(y) \right\| = 0.$$

Definition 3.1. A mapping ϕ of a metric space $\mathcal{H}$ to itself is said to be *LY-chaotic* if it satisfies (LY1) and (LY2) above.

Definition 3.2. The map $\phi : \mathcal{H} \to \mathcal{H}$ is said to be *E-chaotic* if it has positive topological entropy $E(\phi)$. See the appendix for a brief description of topological entropy.

Now suppose that we have an iterative system on a space $\mathcal{H}$,

$$x_{k+1} = f(x_k), \quad k = 0, 1, 2, \ldots,$$

where f is a continuous mapping from $\mathcal{H}$ into itself. Numerous sufficient conditions for chaos have been found: see,[7,15,18,19] to list just a few. When $\mathcal{H} = \mathfrak{R}$, the "period three is chaos" result of Li and Yorke is especially evocative.

Definition 3.3. A map $f \in C(\Omega)$ is said to be *erratic* if there exists a nonempty compact set $Y \subset \mathfrak{R}^n$ satisfying

(C1) $Y \cap f(Y) = 0$;

(C2) $Y \cup f(Y) \subset f^2(Y)$.

It should be noted that if the set Y is also specified to be convex, erratic maps have much the same behavior as the Kloeden conditions[15] and are considerably simpler to verify. When $\Omega = \mathfrak{R}$ then Y is a compact interval and these erratic conditions include the period three criterion. However, all these conditions have similar structure. There is first an expansion in one direction, often accompanied by shrinkage in others. This is followed by a wraparound step, which not only keeps the iterates in a bounded region but also jumbles them up.

In Section 7 we will show that erratic mappings induce both E-chaos and LY-like chaos on $(\mathcal{D}(\Omega), H)$, provided that the fuzzification of the erratic map is well-behaved in the sense that it preserves possibility levels.

4. INFORMATION CONTENT OF FUZZY SETS

If one thinks of greyscale images as fuzzy sets, it is clear that some fuzzy sets convey more information than others. A crisp set A, given by the characteristic function $\mathcal{X}_A$ (also a fuzzy set), is totally white on the greyscale and we would want to say that it had less information content than an image which included various shades and gradations of grey.

Information has most commonly been defined in a probabilistic context. Recently, the theory has been extended to uncertainty based on discrete *possibility distributions.*[14] Continuos information measures were introduced for elements of $\mathcal{D}^1$ by Ramer[22] and involve the concept of a *rearrangement* compatible with the possibility levels of fuzzy sets. The rearrangement of functions of a single variable is a nonincreasing function which generalizes the idea of a permutation in decreasing order. Rearrangements of functions defined on $\Re^n$ do not have such a natural interpretation, because this domain is only partially ordered for n > 1, nor is the rearrangement a function on $\Re^n$.

Denote by $m(A)$ the measure or content of a set $A \subset \Re^n$, and let $u \in \mathcal{D}^1$. Write $P_u(\beta) = m(\{x: u(x) \geq \beta\}) = m([u]^\beta)$. Then $P_u(\cdot)$ is a non-decreasing function and has a generalized inverse P_u^{-1}, and so

$$\tilde{u}(\xi) = P_u^{-1}(\xi), m([u]^1) \leq \xi \leq m([u]^0)$$

is well-defined and $0 \leq \tilde{u}(\xi) \leq 1$. In $\mathcal{D}^1$, Ramer proposes the measure

$$I(u) = \int_{m([u]^1)}^{m([u]^0)} \frac{1-\tilde{u}(\xi)}{\xi} d\xi \tag{3}$$

This measure is additive on joint distributions of independent possibility assignments, subadditive on marginal assignments, symmetric and expansible (see Reference 22 for an explanation of these concepts).

Note that the definition of $P_u(\beta)$ depends only on the measure of the β-level sets of u. This suggests that the same construction for a rearrangement $\tilde{u}$ of $u \in \mathcal{D}^n$ can be carried out, even though the function $\tilde{u}$ is a function of a single variable. In particular, it follows that the measure of the set in which $\tilde{u}(\xi) \geq \beta$ is $P_u(\beta)$, that is

$$m([\tilde{u}]^{\beta}) = m([u]^{\beta})$$

even though $[\tilde{u}]^{\beta} \subset \Re$, while $[u]^{\beta} \subset \Re^n$. Here, the measure $m(\cdot)$ denotes length in $\Re$ and n-dimensional measure in $\Re^n$ respectively, according to context.

Unfortunately, the integral (1) may not exist, as can be seen for

$$u(x) = 1 + (\ln x)^{-1},\ 0 < x < 1/e$$

Here, $[u]^{\beta} = (0, e^{-1/(1-\beta)})$, $P_u(\beta)$and $= e^{-1(1-\beta)}$ and $\tilde{u}(\xi) = 1 + 1/\ln\xi$. Consequently the integral

$$I(u) = \int_0^{1/e} \frac{-1}{\xi \ln \xi} d\xi$$

does not exist because it diverges at the origin. However, there is a further reason for seeking another measure for studying the behavior of information in chaotic systems. The measure I is somewhat unintuitive in so far as the information content of *every* triangular fuzzy number in $\mathcal{D}^1$ is 1. To see this, note that $P_u(\beta) = m([u]^{\beta}) = (1-\beta)\, m([u]^0)$ and so $\tilde{u}(\xi) = 1 - \xi/m([u]^0)$, giving $I(u) = 1$.

For discussing triangular functions, and keeping in mind the difficulty in defining $\tilde{u}$ for functions u of more than one variable, it is more suitable to define a measure of information directly in terms of the level sets, using the function P_u by

$$\Im(u) = \int_0^1 \frac{P_u(0) - P_u(\beta)}{\beta} d\beta \qquad (4)$$

Both definitions imply that $I(\chi_A) = \Im(\chi_A) = 0$ or crisp connected subsets of $\Re^n$. For triangular fuzzy numbers u,

$$\Im(u) = \int_0^1 \frac{m([u]^0) - (1-\beta) m([u]^0)}{\beta} d\beta = m([u]^0)$$

The integral in (4) also need not exist, for example if

$$u(x) = \exp\left(1 + \frac{1}{x-1}\right), 0 \le x \le 1$$

The measure $\Im$ has virtually all of the desirable properties (additivity is replaced by a form of weighted subadditivity) and will suffice for our purposes here.

5. CHAOTIC MAPPINGS ON $(\mathcal{D}^n, d_\infty)$

Let f be a continuous mapping of $(\mathcal{D}^n, d_\infty)$ into itself. Such mappings can be induced from mappings on the base space $\Re^n$ using the classical extension principle.
For $T : \Re^n \to \Re$ and $u \in \mathcal{D}^n$, define $Tu \in \mathcal{D}^n$ by

$$(Tu)(x) = \begin{cases} \sup_{x \in T^{-1}(y)} u(x) & \text{if } T^{-1}(y) \neq \emptyset, \\ 0 & \text{otherwise,} \end{cases}$$

where $x \in T^{-1}(y)$ iff $y = T(x)$. In terms of level sets, this means that

$$[Tu]^\beta = T([u]^\beta) \text{ and so } [T^k u]^\beta = T^k([u]^\beta),\ k = 1, 2, \qquad (5)$$

However, mappings from $\mathcal{D}^n$ to itself can be constructed which are not induced in this way, as is the case with the first and last examples of Section 9.

Consequently, if the mapping T satisfies some conditions such as (C1), (C2) of Section 3, chaotic iterates will be generated by the difference scheme. This chaos can lead to loss of information. To see how this might occur, suppose that the iteration has the effect of increasing the relative measure of $[u]^\beta$ in $[u]^0$, so that

$$m([Tu]^0) - m([Tu]^\beta) \leq m([u]^0) - m([u]^\beta),\ 0 \leq \beta \leq 1,$$

Then $\Im(Tu) \leq \Im(u)$

Another way that information might be lost is when the iterative systems have an attractor Λ which is a set (for example, a strange attractor). So, under iteration, the level sets $[T^k u]^\beta = T^k([u]^\beta)$ of $T^k u$ will approach the same limit set Λ for each $0 \leq \beta \leq 1$. Hence,

$$\Im(T^k u) = \int_0^1 \frac{P_{T^k u}(0) - P_{T^k u}(\beta)}{\beta} d\xi \to \Im(\chi\Lambda) = 0$$

as $k \to \infty$. Thus, there are a number of mechanisms whereby the information content can decay along the orbit

6. Γ - FUZZIFICATION

A *t -norm* is an associative binary operation $T(\cdot,\cdot)$ on I that satisfies

(i) T is non-decreasing in each place,

$$T(x, y) \leq T(x', y')$$

for all $x, x', y, y' \in I$ such that $x \leq x'$ and $y \leq y'$.

(ii) 1 is an identity of T,

$$T(x, 1) = T(1, x) = x, \ \forall x \in I.$$

(iii) T is commutative,

$$T(x,y) = T(y, x)$$

As a simple consequence of (i) and (ii) we have

$$T(x, x) \leq x, \ x \in I \tag{6}$$

Some common examples are $M(x,y) = \min(x,y)$, $\Pi(x,y) = xy$ and $W(x,y) = \max(x + y - 1,0)$. Define the corresponding *t-conorm* by

$$T^*(x, y) = 1 - T(1 - x, 1 - y), \ \forall x, y \in I.$$

A function Γ is a *t*-conorm if there is a *t*-norm T such that $\Gamma = T^*$. The *t* - conorms corresponding to those above are $M^*(x,y) = \max(x,y)$, $\Pi^*(x,y) = x + y - xy$, and $W^*(x, y) = \min(x + y, 1)$. It follows that

$$W < \Pi < M < M^* < \Pi^* < W^*$$

Further, property (ii) is replaced by

$$T^*(x, 0) = T^*(0, x) = x \forall \ x \ \in I,$$

and equation (6) now becomes

$$T^*(x, x) \geq x, \ x \in I. \tag{7}$$

If Γ is any *t* - norm or any *t*-conorm, define the *diagonal of* Γ, $\Delta_\Gamma : I \rightarrow I$ by

$$\Delta_\Gamma(x) = \Gamma(x, x).$$

Let Γ be any *t* - norm or *t* - conorm. A mapping f: $\Omega \rightarrow \Omega$ has a Γ - *fuzzification* : $\tilde{f}_\Gamma$: $\mathcal{D}(\Omega) \rightarrow \mathcal{D}(\Omega)$ defined by

$$\left(\tilde{f}_\Gamma u\right)(y) = \sup_{x,x' \in f^{-1}(y)} \left\{\Gamma\left(u(x), u(x')\right)\right\}$$

for $y \in \Omega$ and $u \in \mathcal{D}(\Omega)$. By virtue of property (9), this can be rewritten in terms of the diagonal as

$$\left(\tilde{f}_\Gamma u\right)(y) = \sup_{x \in f^{-1}(y)} \left\{\Delta_\Gamma\left(u(x)\right)\right\} \tag{8}$$

An α - *level set with respect to* Γ is defined as

$$[u]_\Gamma^\alpha = \{x \in \mathrm{supp}(u) : \Delta_\Gamma(u(x)) \geq \alpha\}$$

for all $u \in \mathcal{D}(\Omega)$ and $0 \leq \alpha \leq 1$.
Example. For $\Gamma = \Pi$, $\Delta_\Pi(u) = u^2$ and so

$$[u]_\Pi^\alpha = \{x : u(x)^2 \geq \alpha\} = [u]^{\sqrt{\alpha}}$$

and similarly, for Π^*, $\Delta_{\Pi^*}(u) = 2u - u^2$ and so

$$[u]_{\Pi^*}^\alpha = [u]^{1-\sqrt{1-\alpha}}$$

Lemma 6.1 *Let T be any t-norm and let T^* be any t-conorm and suppose that $f \in C(\Omega)$. Then for any $u \in \mathcal{D}(\Omega)$ and all $\alpha \in I$*

$$[u]_T^\alpha \subseteq [u]^\alpha \tag{9}$$

$$[u]_{T^*}^\alpha \supseteq [u]^\alpha \tag{10}$$

$$f([u]_\Gamma^\alpha) \subseteq [\tilde{f}_\Gamma u]^\alpha, \Gamma = T \text{ or } T^* \tag{11}$$

Proof Let $x \in [u]_\Gamma^\alpha$. That is, $\Delta_\Gamma(u(x)) \geq \alpha$. But then (6) gives

$$u(x) \geq \Delta_\Gamma(x) \geq \alpha$$

and the first inclusion follows. If $x \in [u]^\alpha$, then by (7)

$$\alpha \leq u(x) \leq \Delta_T{}^* u(x)$$

and the second inclusion is true For $x \in [u]_\Gamma^\alpha$, if $y = f(x) \in f(x \in [u]_\Gamma^\alpha)$, then

$$\begin{aligned}(\tilde{f}_\Gamma u)(y) &= \sup\{\Delta_\Gamma(u(\xi)) : y = f(\xi)\} \\ &\geq \Delta_\Gamma(u(x)) \geq \alpha\end{aligned}$$

and the last inclusion is proved. □

Theorem 6.2 *Let Γ be any t - norm or t - conorm. If $f \in C(\Omega)$ and $u \in \mathcal{D}(\Omega)$, then a necessary and sufficient condition for*

$$f([u]_\Gamma^\alpha) = [\tilde{f}_\Gamma u]^\alpha, \alpha \in (0,1] \tag{12}$$

is that for every $y \in f(supp(u))$,

$$\sup_{\xi \in f^{-1}(y)} \left\{ \Delta_\Gamma(u(\xi)) \right\}$$

is attained

Proof. Necessity: let $f\left([u]_\Gamma^\alpha\right) = \left[\tilde{f}_\Gamma^\alpha\right]$. Choose y such that

$$\sup\left\{ \Delta_\Gamma(u(\xi)) : y = f(\xi) \right\} = \alpha$$

Then $y \in [\tilde{f}_\Gamma u]^\alpha$ and so, by hypothesis, $y \in f\left([u]_\Gamma^\alpha\right)$. But then $y = f(x')$ for some $x' \in [u]_\Gamma^\alpha$ which means that $\Gamma(u(x'), u(x')) \geq \alpha$. But

$$\begin{aligned} \alpha &= \sup_{\xi \in f^{-1}(y)} \left\{ \Delta_\Gamma(u(\xi)) \right\} \\ &\geq \Gamma(u(x'), u(x')) \\ &\Rightarrow \Gamma(u(x'), u(x')) = \alpha \end{aligned}$$

and the supremum is thus attained. Conversely, suppose that the supremum is attained. Take $y \in [\tilde{f}_\Gamma u]^\alpha$. That is,

$$(\tilde{f}_\Gamma u)(y) = \sup_{\xi \in f^{-1}(y)} \left\{ \Delta_\Gamma(u(\xi)) \right\} \geq \alpha$$

Since this is attained, there is x' such that

$$y = f(x') \textit{ and } \Delta_\Gamma(u(x')) \geq \alpha.$$

Hence $x' \in [u]_\Gamma^\alpha$ which means that $y \in f([u]_\Gamma^\alpha)$ and (12) now follows from (11). □

As with (5), the relation (12) provides a procedure for computing the iterates of the fuzzified mapping, in this case of $\tilde{f}_\Gamma$ through the leve; sets induced by Γ. Since Δ_Γ is non decreasing, there is a generalized inverse Δ_Γ^{-1} and

$$[u]_\Gamma^\alpha = [u]^{\Delta_\Gamma^{-1}(\alpha)}$$

From this it follows that

Proposition 6.3. *Let Γ be any t-norm or t-conorm. If $f \in C(\Omega)$ and $u \in \mathcal{D}(\Omega)$ satisfy the condition of Theorem 6.2, then for each positive integer k*

$$\left[\tilde{f}_\Gamma^k u\right]^\alpha = f^k\left([u]^{\Delta_\Gamma^{-k}(\alpha)}\right)$$

Proof. Since

$$\left[\tilde{f}_\Gamma^{k+1}u\right]^\alpha = \left[\tilde{f}_\Gamma\left(\tilde{f}_\Gamma^{k}u\right)\right]^\alpha = f\left(\left[\tilde{f}_\Gamma^{k}u\right]^{\Delta_\Gamma^{-1}(\alpha)}\right) = f\left(f^k\left(\left[u\right]^{\Delta_\Gamma^{-k}\left(\Delta_\Gamma^{-1}(\alpha)\right)}\right)\right) = f^{k+1}\left(\left[u\right]^{\Delta_\Gamma^{k+1}(\alpha)}\right)$$

the result is proved by induction.

7. CHAOS AND FUZZIFICATION

A fuzzification F is said to be *H-continuos* if every mapping $\mathcal{F}(f)$, $f \in \Omega$, is continuous in $\mathcal{D}(\Omega)$ with respect to the sendograph metric H. The H-continuous fuzzification $\mathcal{F}$ is said to be δ *- normal* if the following properties are satisfied: For any $u \in \mathcal{D}(\Omega)$ and any $f \in C(\Omega)$

$$\text{supp}\,(\mathcal{F}\,(f)u) \subseteq f(\text{supp}(u)),$$

and any function of the form

$$\mathcal{F}\,(f)u,\; u \in \mathcal{D}_\delta(\Omega),\; f \in C(\Omega)$$

belongs to the set $\mathcal{D}_\delta(\Omega)$.

Denote by $\mathcal{B}$(N) the set of all binary strings

$$w = \{w_1, w_2, \ldots, w_N\},\; w_i \in \{0, 1\},\; 1 \le i \le N$$

of length N that do not contain two adjacent zeros and begin with $w_1 = 1$. Write the cardinality of $\mathcal{B}(N)$ as $\#\mathcal{B}(N)$. Clearly

$$2^{(N-1)/2} \le \#\mathcal{B}(N) \le 2^{N-1}$$

Consequently,

$$\gamma = \lim_{N\to\infty}\left(N^{-1}\ln(\#\mathcal{B}(N))\right)$$

exists and

$$\gamma \ge \frac{\ln 2}{2}$$

Our principal result is

Theorem 7.1 *Let f be erratic on the set* $A \subset \Omega$. *Let* $\mathcal{F}$ *be a* δ *- normal fuzzification. Then the following statements hold:*

(I) *The map* $\mathcal{F}\,(f)$ *is E-chaotic and its topological entropy satisfies*

$$E\,(\mathcal{F}\,(f)\,) \ge \gamma \tag{13}$$

(II) *The map* $\mathcal{F}(f)$ *is LY-chaotic and, for each natural number k, it has at least* $\#\mathcal{B}(N)$ *distinct periodic points in* $\mathcal{D}_\delta(\Omega)$.

Remark . This theorem shows that the situation is somewhat different from that for erratic maps on $\Re^n$, as described in.[7] In $\Re^n$ rather than periodic points it is only possible to show that *periodic sets* occur in general. The difference arises because compact sets of fuzzy sets have stronger fixed point properties than compact sets in $\Re^n$.[9] Theorem 7.1 also strengthens the theorem of[2] in the same manner, as well as generalizing it from the special case of mappings of crisp compact sets, whose characteristic functions are specific types of fuzzy sets.

The proof proceeds through several lemmas. All statements below assume that f is erratic. Denote by $\mathcal{B} = \lim_{N\to\infty} \mathcal{B}(N)$ Write

$$Ao = A \,,\, A_1 = f(A)\,.$$

For each $w \in \mathcal{B}$, denote by $\mathcal{R}(w)$ the set of all $x \in \Omega$ satisfying

$$x \in A, f^{i}(x) \in A_{wi}\,,\; i = 1,2, \ldots.$$

Lemma 7.2 .

(a) *For each* $w \in \mathcal{B}$, *the set* $\mathcal{R}(w)$ *is nonempty and closed in* Ω

(b) *Let the word* $w \in \mathcal{B}$ *be k - periodic. Then the sequence of sets*

$$f^{i}(\mathcal{R}(w)),\; i = 1,2, \ldots$$

is also k - periodic. If w *and* w^* *are two distinct words in* $\mathcal{B}$ *then the sets* $\mathcal{R}(w)$ *and* $\mathcal{R}(w)^*$ *are disjoint.*

Proof . Statement (a) follows immediately from the defined properties of a δ-normal fuzzification, since supp($\mathcal{F}$u) and supp($\mathcal{F}$v) are disjoint and compact. Now, (b) will follow from the same properties and (b) of Lemma 7.2. □

Let us return to the proof of the theorem. To prove (I) it is sufficient to prove (11), that is $E(\mathcal{F}(f)) \geq \gamma$. Let U_0 and U_1 respectively be neighborhoods in $\mathcal{D}(\Omega)$ of the sets $\mathcal{D}_\delta(A_0)$ and $\mathcal{D}_\delta(A_1)$. Since $A_0 \cap A_1 = \varnothing$ and are compact, the sets U_0, U_1 can be chosen to be disjoint. Introduce the open cover $\mathcal{U}$ with elements U_0, U_1 and

$$U_2 = \mathcal{D}(\Omega) \setminus (\mathcal{D}_\delta(A_0) \cup \mathcal{D}_\delta(A1))$$

Let w be a fixed word in $\mathcal{B}(N)$ and let u_w be a member of $\mathcal{D}(\mathcal{R}(w))$ From the δ-normal properties it follows that

$$\mathcal{F}(f)^{i} u_w \in \mathcal{D}_\delta(A_{wi})\,,\; i = 1,2,\ldots,N\,.$$

Hence the unique element of the cover $\mathcal{U}(\mathcal{F}(f), N)$ which contains u_W is the element

$$U_0 \cap \mathcal{F}(f)^{-1} U_{w1} \cap \ldots \cap \mathcal{F}(f)^{N-1} U_{wn}$$

So $\# \mathcal{U}(\mathcal{F}(f), N) \leq \# \mathcal{B}(N)$ and (11) follows.
To prove (II), let us first construct a set S that satisfies (i) and (ii) of (LY2) Let $\mathcal{Q}$ be an uncountable set of words from $\mathcal{B}$ such that no elements of $\mathcal{Q}$ are periodic, and each pair of distinct words from $\mathcal{Q}$ differ on an infinite number of letters. Let $\mathcal{Q}^*$ be the set of elements $u_W \in \mathcal{D}(\Omega)$ each of which belongs to the disjoint sets $\mathcal{D}_\delta(\mathcal{R}(w))$:

$$\mathcal{Q}^* = \{u_W \in \mathcal{D}_\delta(\mathcal{R}(w)) : w \in \mathcal{Q}\}$$

Then $\mathcal{Q}^*$ satisfies (i) and (ii) of (LY2) because of the property (a) of Lemma 7.3. The property (iii) can be shown to follow by a similar construction.
It remains to show the existence of k - periodic points of the map $\mathcal{F}(f)$. Let k be fixed. From (b) of Lemma 7.3 there exist at least $\#\mathcal{B}(k)$ disjoint invariant sets of the form $\mathcal{D}_\delta(N(w))$. But sets of this type have the fixed point property[9] (see also the appendix) and consequently have fixed points of $\mathcal{F}(f)^k$. The theorem is proved. □

8. NON-DEGENERATE PERIODICITIES AND CHAOS

Consider the t - norm $W(x,y) = \max(x + y - 1, 0)$. The diagonal map is $\Delta_W(u) = \max(2u - 1, 0)$. Hence $[\tilde{f}_W u]^\alpha = f([u]^{(1+\alpha)/2})$, and for all positive integers n

$$\left[\tilde{f}_W^n u\right]^\alpha = f\left([u]^{\alpha_n}\right)$$

where $\alpha_n = \Delta_W^{-n}(\alpha) = 1 - (1-\alpha)/2^n$, since

$$\alpha_{n+1} = \frac{1+\alpha_n}{2} = 1 - \frac{(1-\alpha)}{2^n}$$

Observe that the fuzzification associated with W is not δ-normal because, if $u \in \mathcal{D}_\delta(\mathcal{D})$ we can in general only infer that $\tilde{f}_W u \in \mathcal{D}_{\delta'}(\Omega)$, where $\delta' = \Delta_W(\delta)$. Nevertheless, we will use this fuzzification to construct a type of non degenerate chaos on $\mathcal{D}_1(\Omega)$. That is, not all orbits asymptotically become crisp sets.
Let Ω be the unit interval I and let

$$f(x) = \begin{cases} 2x, & if\ 0 \leq x \leq 1/2, \\ 2 - 2x, & if\ 1/2 < x < 1 \end{cases}$$

If we take $A = \mathcal{D}_1([5/18,4/9])$, then $\tilde{f}_W(A) = \mathcal{D}_1([5/9, 8/9])$, $\tilde{f}_W^2(A) = g_1([2/9,8/9])$ and each of these sets is compact in $\mathcal{D}_1(I)$. So, arguing as in,[7] $\tilde{f}_W$ satisfies (LY2), and has *periodic sets* of any order. We show that in fact $\tilde{f}_W$ has periodic points, that is, periodic fuzzy sets, of any order and is consequently non-degenerate. Let k be any positive integer and let

$$p_k = \frac{2}{1+2^k}$$

so that $f^k(p_k) = p_k$. Consider the isosceles triangular fuzzy number u_k with "feet" at $x = p_k \pm \lambda$ where λ is chosen so that $[p_k - \lambda, p_k + \lambda] \subset [0, 1/2]$. In addition, choose λ so small that

$$f^{k-2}\left(p_k + \lambda / 2^k\right) = \frac{2^{k-1}}{1+2^k} + \frac{\lambda}{2^2} < \frac{1}{2}$$

Note that the α_k - level sets of u are

$$[u]^{\alpha_k} = [p_k - \lambda + \lambda\alpha_k, p_k + \lambda - \lambda\alpha_k]$$

Write $\beta = \lambda(1 - \alpha)$. Then

$$\begin{aligned}
\left[\tilde{f}_W^k u_k\right]^\alpha &= f^k\left(\left[p_k - \beta / 2^k, p_k + \beta / 2^k\right]\right) \\
&= \left[f^k\left(p_k - \beta / 2^k\right), f^k\left(p_k + \beta / 2^k\right)\right] \\
&= [p_k - \beta, p_k + \beta] \\
&= [u_k]_W^\alpha
\end{aligned}$$

if $f^k(u_k) = u_k$. This situation is in marked contrast to the case when the Γ = min fuzzification is used, where the only k - periodic points are singleton sets of the form $\{p_k\}$.

In this example there are just a countable number of periodic fuzzy sets - a lower bound of the actual number is given by (II) of Theorem 7.1.

9. EXAMPLES OF FUZZY CHAOS

9.1 Example 1.

This is due to Kloeden.[17] Consider the space $\mathcal{D}^1([0, 1])$, the set of all $u \in \mathcal{D}^1$ with support $[u]^0 \subseteq [0,1]$. For each such u, there exist functions a, b: $[0,1] \to \Re$ such that the β - level sets of u are the intervals $[u]^\beta = [a(\beta), b(\beta)]$. Denote by $\mathcal{T}$ the subset

Example 9.1: Tent Map.

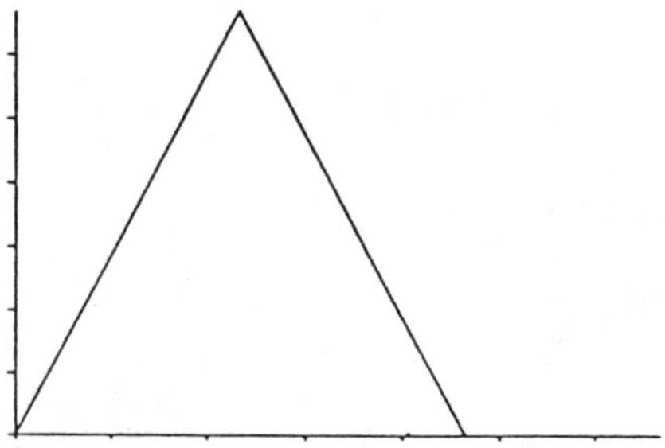

Fig. 9.1.1: Initial fuzzy value, $b = 0.7$.

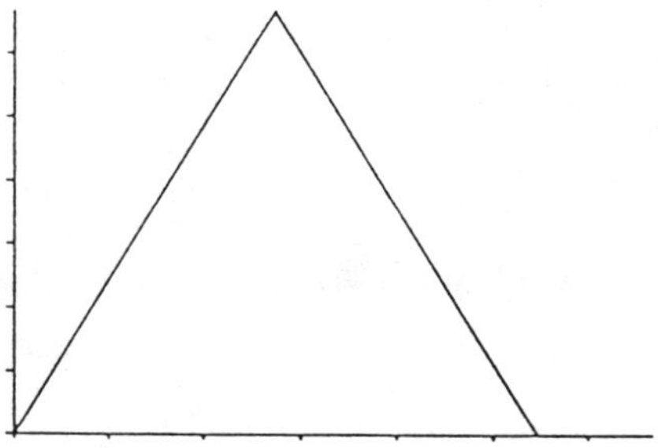

'ig. 9.1.2: First iterate under Tent Map.

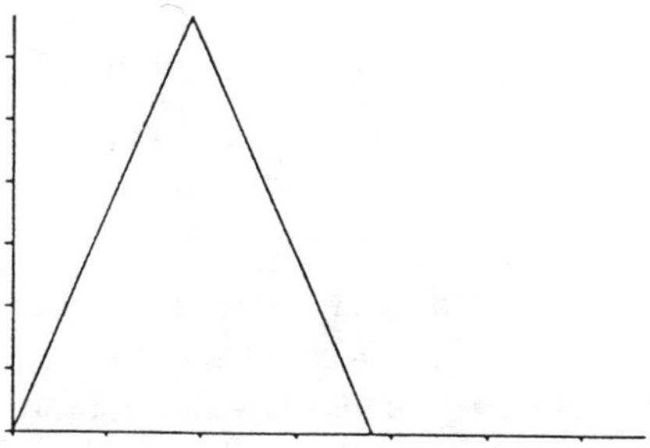

Fig. 9.1.3: $n = 2$ iterate.

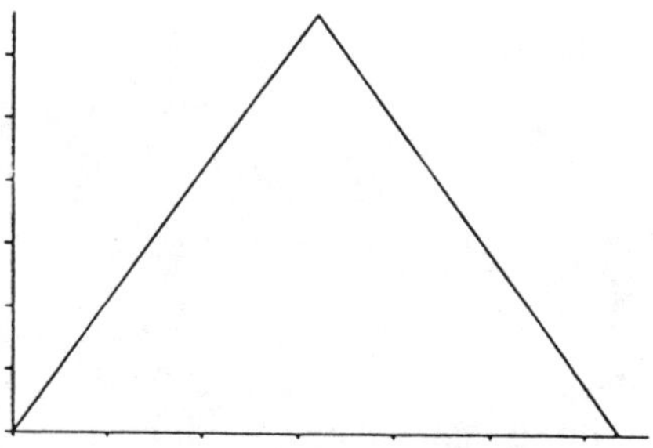

Fig. 9.1.4: $n = 3$ iterate.

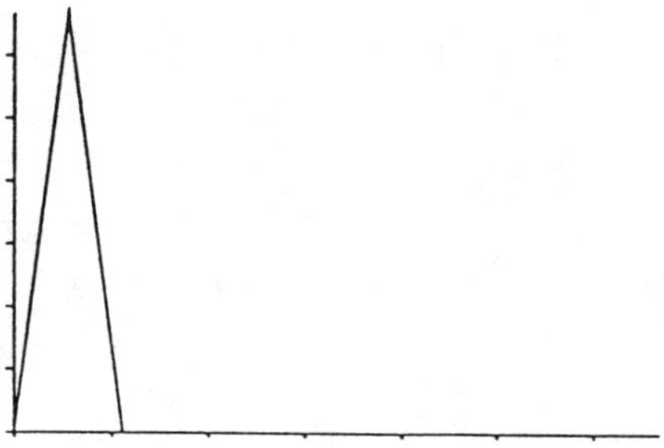

Fig. 9.1.5: $n = 4$ iterate.

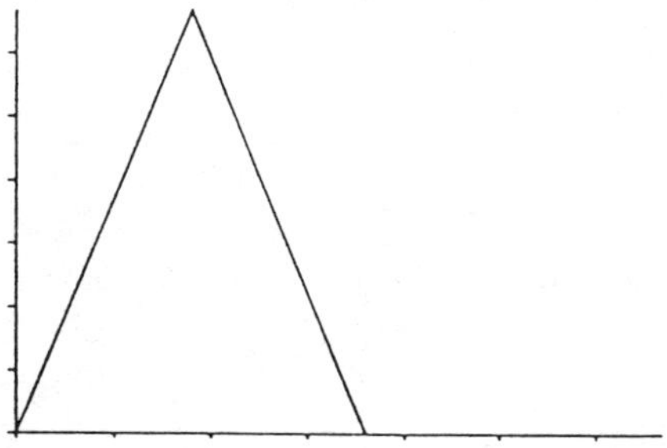

Fig. 9.1.6: $n = 5$ iterate.

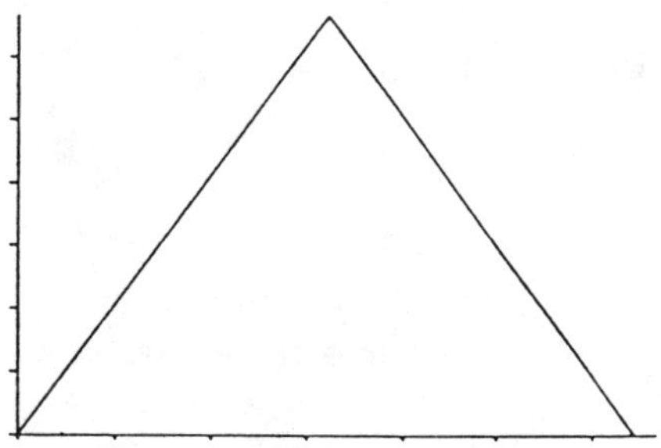

Fig. 9.1.7: $n = 6$ iterate.

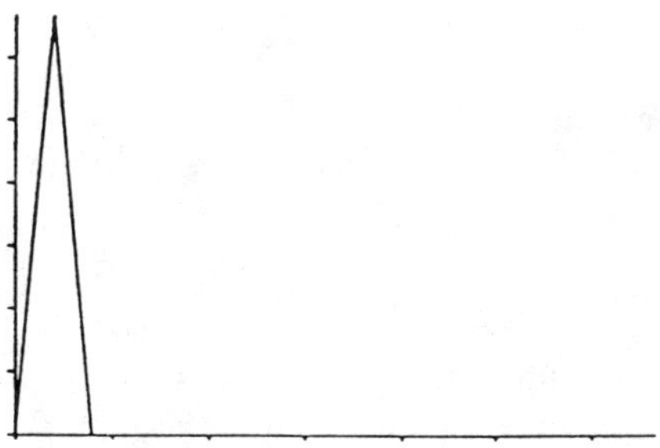

Fig. 9.1.8: $n = 7$ iterate.

$$\mathcal{T} = \{u \in \mathcal{D}^1 : a(\beta) = \beta\, b(0)/2, b(\beta) = (1 - \beta/2) b(0)\}$$

of $\mathcal{D}^1$. Define the mapping $T : \mathcal{T} \to \mathcal{T}$ by

$$[Tu]^\beta = [g\,(b(0))\,a(\beta),\ g\,(b(0))\,b(\beta)],$$

where $xg(x) : [0,1] \to [0,1]$ is the tent function θ,

$$\theta(x) = xg(x) = \begin{cases} 2x & \text{if } 0 \le x \le 1/2 \\ 2 - 2x & \text{if } 1/2 \le x \le 1 \end{cases}$$

The map T is obviously a H - continuous and δ - normal fuzzification of θ. Clearly, any $u \in \mathcal{T}$ is completely determined by the value of $b(0)$, written b, and will be denoted by u_b. Then $T(u_b) = u_{\theta(b)}$.

Let $Y = \{b\colon 1/4 \le b \le 4/9\}$ and note that

$$\theta(Y) = \{b{:}1/2 \le b \le 8/9\}$$

$$\theta^2(Y) = \{b{:}2/9 \le b \le 1\}$$

so that $Y \cap T(Y) = \varnothing$ and $T^2(Y) \supseteq Y \cup T(Y)$, and the mapping T is erratic. Now, for $u_b \in \mathcal{T}$, $P_{u_b}(\beta) = (1 - \beta)b$ and so

$$\Im(u_b) = \int_0^1 \frac{b - b(1-\beta)}{\beta}\, d\beta = b$$

Hence, the information of $T^k u_b$,

$$\Im\left(T^k u_b\right) = \Im\left(u_{\theta^k(b)}\right) = \theta^k(b)$$

and the information in the fuzzy iterates $T^k(u)$ does not degrade. The information content varies in a chaotic fashion as $k \to \infty$, because the trajectory $\{\theta^k(b) : k = 1,2 \ldots\}$ is chaotic for every $0 < b < 1$, but it does not continually decrease. Observe that, if the measure $I(\cdot)$ were used here, then $I(T^k u_b) = 1$ for $k = 1, 2, \ldots$.

We note finally that if $\mathcal{Y} = \{u_b \in \mathcal{T}\colon 1/4 \le b \le 4/9\}$, then also $Y \cap T(Y) = \varnothing$ and $T^2(Y) \supseteq Y \cup T(Y)$.

9.2 Example 2

Let f: $[0,1] \to [0,1]$ be given by $f(x)=3.9x(1-x)$. Define $F : \mathcal{D}^1([0,1]) \to \mathcal{D}^1([0,1])$ by

$$[Tu]^\beta = f([u]^\beta)$$

It is not difficult to show that the whole of the interval $\Lambda = [.0950625, .975]$ is an attractor for the iterative system f^k on $[0,1]$. It follows that for any subinterval $J \subset [0,1]$, no matter how small, the images $f^k(J)$ expand to cover Λ as $k \to \infty$. Now

Example 9.2: sup–Induced Chaos.

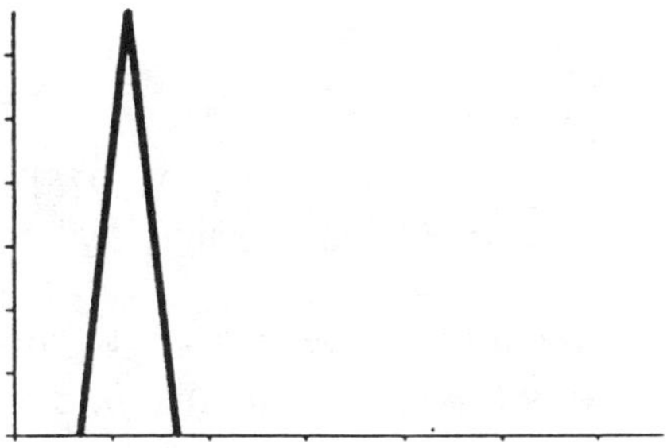
Fig. 9.2.1: Initial fuzzy value.

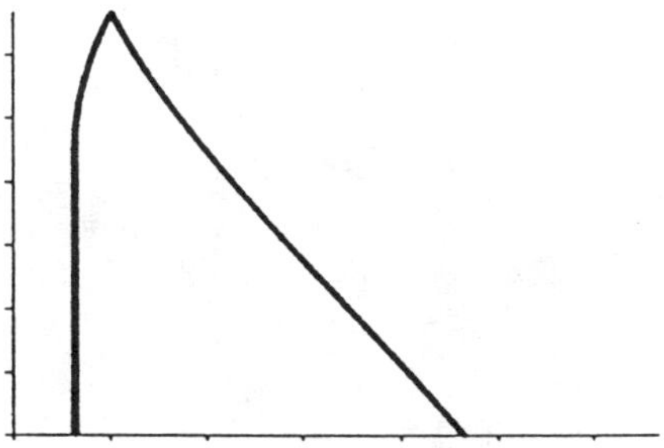
Fig. 9.2.2: $n = 3$ iterate.

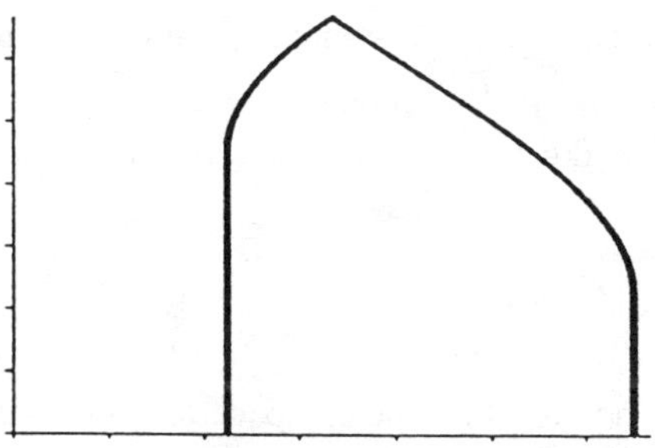
Fig. 9.2.3: $n = 4$ iterate.

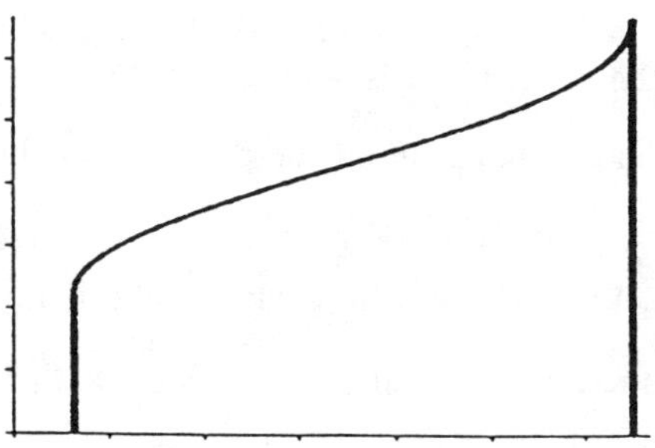
Fig. 9.2.4: $n = 5$ iterate.

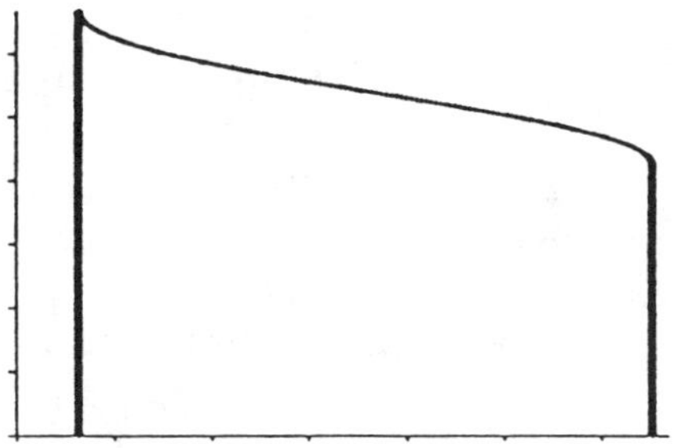
Fig. 9.2.5: $n = 6$ iterate.

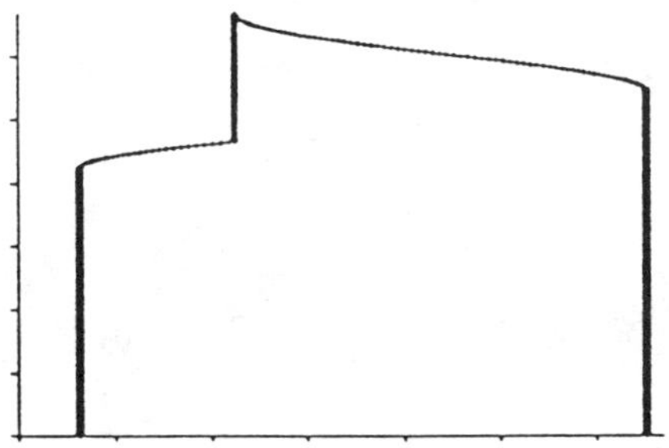
Fig. 9.2.6: $n = 7$ iterate.

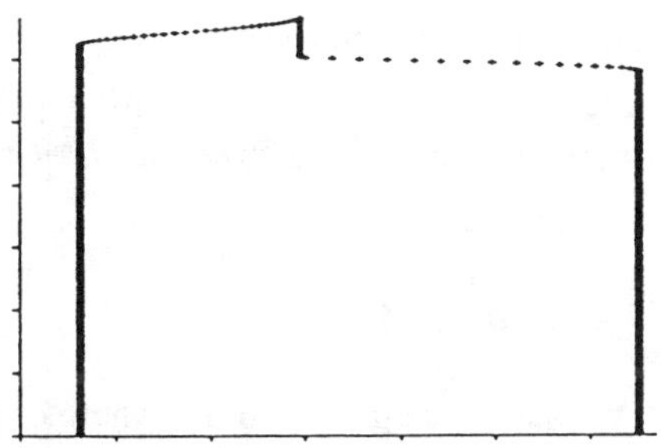
Fig. 9.2.7: $n = 9$ iterate.

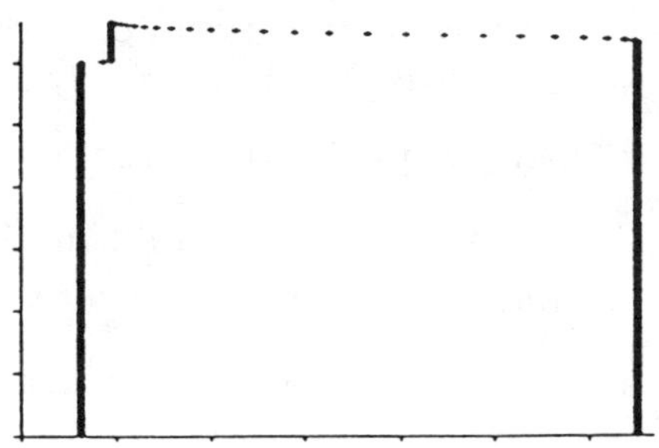
Fig. 9.2.8: $n = 11$ iterate.

consider a fuzzy set $u \in \mathcal{D}^1$, and observe that for $0 \leq \alpha \leq \beta \leq 1$ we have $[u]^0 \supseteq [u]^\alpha \supseteq [u]^\beta$. From the continuity of T, $[Tu]^0 \supseteq [Tu]^\alpha \supseteq [Tu]^\beta$.

Let

$$Y = \mathcal{D}^1([0.15, 0.45]) = \{u \in \mathcal{D}^1 : [u]^0 \subseteq [0.15, 0.45]\}$$

Then $T(\mathcal{D}^1([.15, .45])) = \mathcal{D}^1([.497, .965])$ and $T^2(\mathcal{D}^1([.15, .45])) = \mathcal{D}^1([.131,.975])$. Consequently, the sufficient conditions (C1), (C2) are satisfied and a chaotic regime exists.

However, for this system information is lost as the system evolves with increasing k. To see this, for any $u \in \mathcal{D}^1$ choose any β - level set, $[u]^\beta$, $0 \leq \beta \leq 1$. Then for k sufficiently large, $T^k([u]^\beta) = \Lambda$, and $\Im(T^k u) = 0$.

9.3 Example 3.

Let h: $\Re^2 \rightarrow \Re^2$ be the Hénon map

$$\begin{aligned} x_{k+1} &= 1.4 + 0.3y_k - x^2_k \\ y_{k+1} &= x_k \end{aligned}$$

Define $H : \mathcal{D}^2(K) \rightarrow \mathcal{D}^2(K)$ by $[Hu]^\beta = h([u]^\beta)$, where

$$K = \{(x, y) \in \Re^2 : |x|, |y| \leq 1.904\}$$

It is well known that h maps K into itself, so the map H is well-defined. It seems certain(but is still an open question) that the Hénon mapping has a strange attractor Λ, and that each level set of $u \in \mathcal{D}^2(K)$ is attracted to Λ. Consequently, information degrades as the chaotic iteration proceeds, since $h^k([u]^\beta)$ becomes close to Λ as k increases, that is the fuzzy set $H^k u$ becomes very like the characteristic function χ_Λ, whose information content is zero.

9.4 Example 4.

We take an example from,[8] where chaos was defined on the Banach space of functions in which fuzzy sets were embedded. We restrict ourselves to the subspace $\mathcal{E}^2_L \subset \mathcal{E}^2$ consisting of fuzzy convex members of $\mathcal{D}^2$ such that the map $\beta \rightarrow [u]^\beta$ is Lipschitz with respect to the Hausdorff distance, $d_H([u]^\alpha, [u]^\beta) \leq M|\alpha - \beta|$ for all $0 < \alpha, \beta \leq 1$ and some fixed M depending only on u. Then there exists an isometric embedding j: $\mathcal{E}^2_L \rightarrow C([0,1] \times S^1)$, as a cone. Now, if θ is a chaotic mapping on the embedded cone in that function space, then T defined $j \circ T = \theta \circ j$ is a chaotic mapping on $\mathcal{E}^2_L$. We construct such a mapping θ on the positive cone of $C(\{\beta\} \times S^1)$ for each $0 \leq \beta \leq 1$. For a fixed $\beta \in [0,1]$, the domain of each support function $s_{[u]^\beta}$ can be identified with the

Example 9.3: Hénon Map.

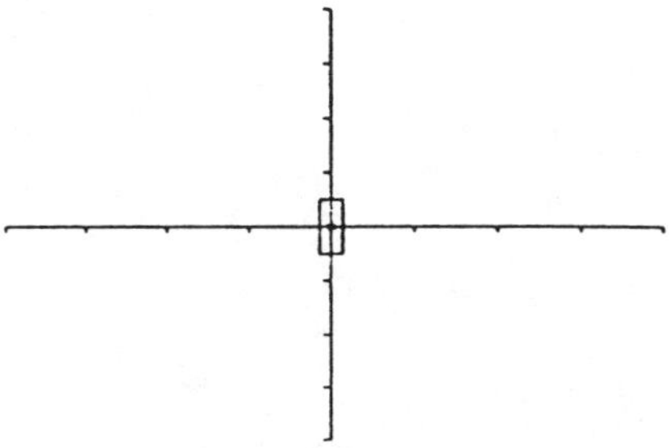

Fig. 9.3.1: Initial cross–section.

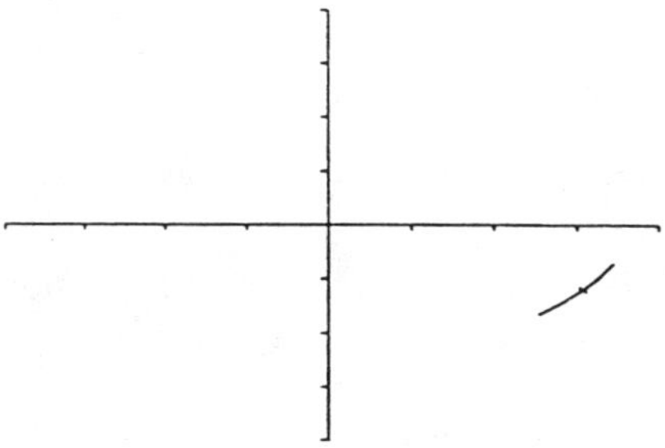

Fig. 9.3.2: $n = 3$ iterate.

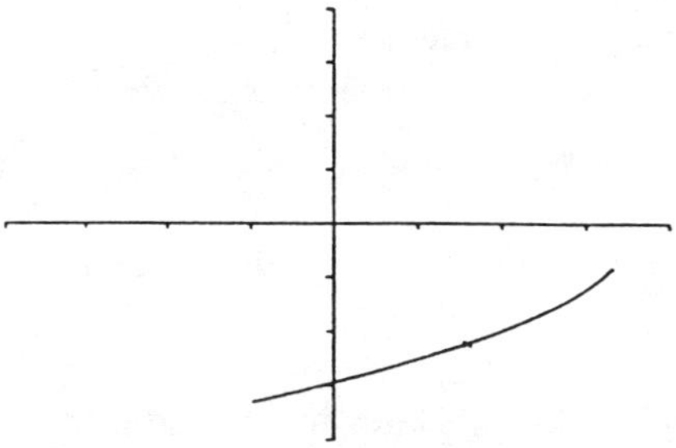

Fig. 9.3.3: $n = 5$ iterate.

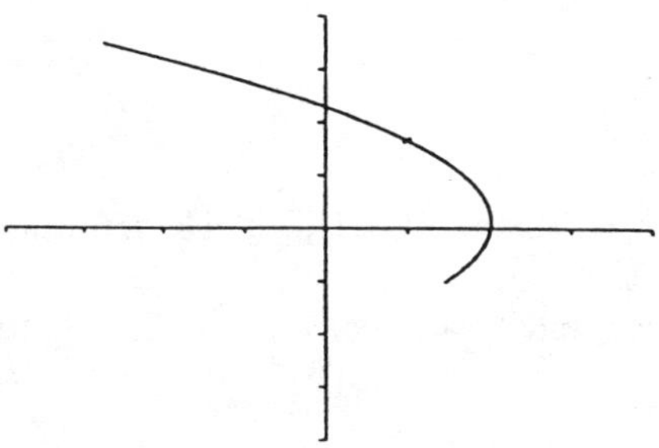

Fig. 9.3.4: $n = 6$ iterate.

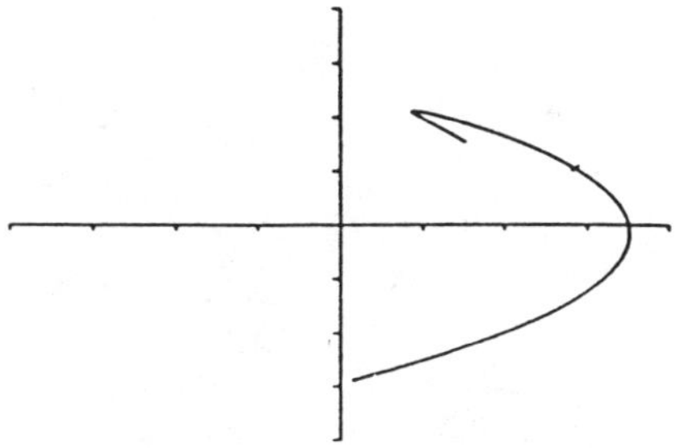

Fig. 9.3.5: $n = 7$ iterate.

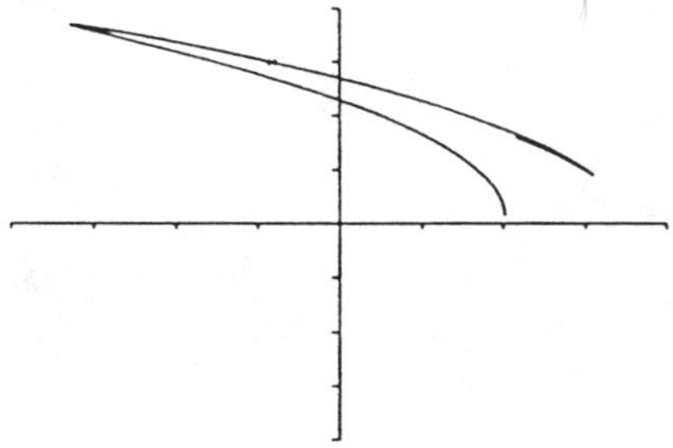

Fig. 9.3.6: $n = 8$ iterate.

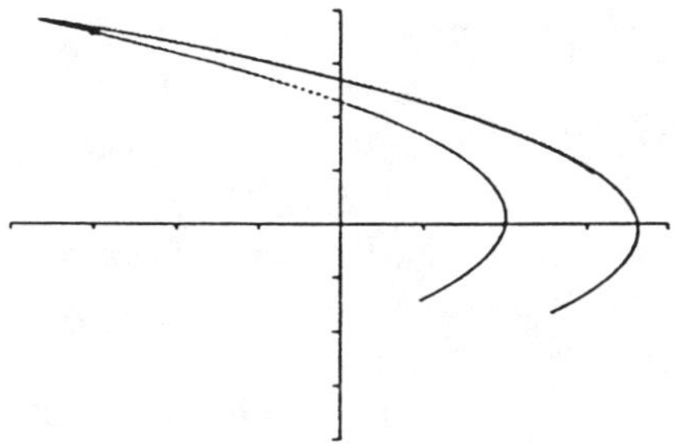

Fig. 9.3.7: $n = 10$ iterate.

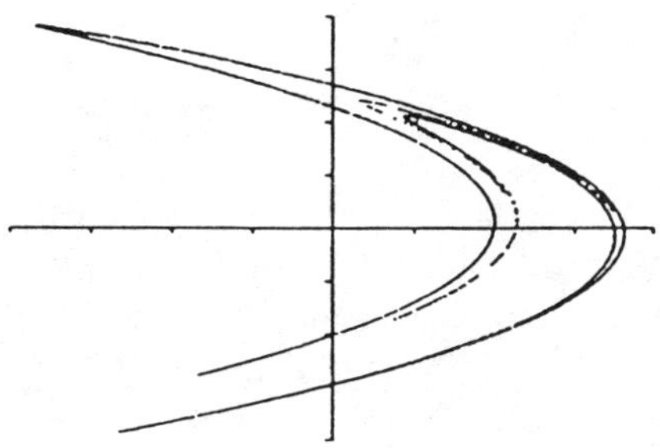

Fig. 9.3.8: $n = 20$ iterate.

interval $[0, 2\pi)$, and we are essentially looking at the 2π - periodic functions. The support function s_u of u is given explicitly by, for each $p = (cost, sint) \in S^1$,

$$\tilde{s}_u(\beta,t) \stackrel{\Delta}{=} s_u(\beta,p) \stackrel{\Delta}{=} s_{[u]^\beta}(p) = \max_{x \in [u]^\beta}(x\cos t + y\sin t)$$

for $0 \leq t < 2\pi$. For a given kernel $V(t, \tau)$, periodic in t, define θ by

$$(\theta\tilde{s}_u)(\beta,t) = 3.9(1-\tilde{s}_u(\beta,t))\int_0^{2\pi} V(t,\tau)\tilde{s}_u(\beta,\tau)d\tau$$

It is possible to show that, for suitable kernels V, the mapping θ^2 is chaotic on the unit ball of the function space. Hence, $T: \mathcal{E}_L^2 \rightarrow \mathcal{E}_L^2$, defined by

$$[j \circ T(u)]_\beta = j \circ T([u]^\beta) = \theta^2(s_u(\beta, \cdot))$$

is chaotic on $\mathcal{E}_L^2$

10 CONCLUSION

- Fuzzy dynamical systems can be generated by either fuzzification of functions on the underlying base space or by intrinsic mappings on the various spaces of fuzzy sets.

- Quite different fuzzifications are produced by different t - norms and t - conorms and their general properties in terms of level sets were investigated.

- Chaos can arise from all these types of maps, provided certain conditions are met. Fuzzy chaotic mappings are most richly described in a *metric space* (like $\mathcal{D}^n, d_\infty$))

- Several definitions of fuzzy chaos were shown to be linked and could each be produced from simple sufficient conditions.

- A measure of information content of fuzzy dynamical states was introduced and discussed.

- Chaos may or may not result in degradation of information along orbits, and examples were given of both circumstance.

REFERENCES

1. J. Banks, J. Brooks, G. Cairns, G. Davis and P. Stacey, "On Devaney's definition of chaos," *Amer. Math. Monthly*, 99 (1992), 332-334.
2. J. Benhabib and R. H. Day, "Rational choice and erratic behavior," *Rev. Econ. Stud.* 48 (1981), 459 - 471.

3. M. L. Blank, "Small perturbations of chaotic dynamical systems," *Russian Math. Surveys*, 44:6 (1989), 1-33.
4. R. Bowen, "Topological entropy and Axiom A," *Proceedings of Symposia in Pure Mathematics*, Volume XIV, American Mathematical Society, Providence, Rhode Island, 1970.
5. G. Debreu, "Integration of correspondences," Proc. *Fifth Berkeley Symp. Math. Statist. Probability*, Vol. II (1967), 351-372.
6. R. L. Devaney, "An Introduction to Chaotic Dynamical Systems", Addison-Wesley, 1989.
7. P. Diamond, "Chaotic behavior of systems of difference equations," *Intern. J. System Sci.*, 7 (1976), 953-956.
8. P. Diamond, "Chaos and fuzzy representations of dynamical systems," *Proc. 2nd International Conference on Fuzzy Logic & Neural Networks* (Iizuka, Japan, July 17-22, 1992), 51-58, FLSI, Iizuka, Japan, 1992.
9. P. Diamond, P. Kloeden and A. Pokrovskii, "Absolute retracts and a general fixed point theorem for fuzzy sets," submitted for publication, 1992.
10. P. Diamond and A. Pokrovskii, "Chaos, entropy and a generalized extension principle," submitted for publication, 1992.
11. J.-P. Eckmann and D. Ruelle, "Ergodic theory of chaos and strange attractors," *Rev. Mod. Phys.*, 57 (1985), 617-656.
12. R. Fuller and T. Keresztfalvi, "On generalization of Nguyen's theorem," *Fuzzy Sets and Systems*, 41 (1991), 371-374.
13. M. M. Gupta and J. Qi, "Design of fuzzy logic controllers based on generalized T - operators," *Fuzzy Sets and Systems*, 40 (1991), 473-489.
14. M. Higashi and G. Klir, "Measures of uncertainty and information based on possibility distributions," *Int. J. Gen. Systems*, 9, (1983), 43-58.
15. P. Kloeden, "Compact supported endographs and fuzzy sets," *Fuzzy Sets and Systems*, 4 (1980), 193-201.
16. P. Kloeden, "Chaotic difference equations in $\Re^n$," *J. Australian Math. Soc. A*, 31 (1981), 217-225.
17. P. Kloeden, "Chaotic iterations of fuzzy sets," *Fuzzy Sets and Systems,* 42 (1991), 37-42.
18. T. Y. Li and J. A. Yorke, "Period three implies chaos," *Amer. Math. monthly*, 82 (1975), 985-992.
19. F. R. Marotto, "Snap-back repellers imply chaos in $\Re^n$," *J. Math. Anal. Appl.*, 63 (1978), 199-223.
20. H. T. Nguyen, "A note on the extension principle for fuzzy sets," *J. Math. Anal. Appl.*, 64 (1978), 369 -380.
21. M. L. Puri and D. A. Ralescu, "The concept of normality for fuzzy random variables," *Ann. Probab.*, 13 (1985), 1373-1379.
22. A. Ramer, "Concepts of fuzzy information measures on continuous domains," *Int. J. Gen. Systems,* 17 (1990), 241-248.
23. H. N. Teodorescu, "Chaos in fuzzy systems and signals," *Proc. 2nd International Conference on Fuzzy Logic & Neural Networks* (Iizuka, Japan, July 17-22, 1992), 21-50, FLSI, Iizuka, Japan, 1992.
24. S. Wiggins, "Global Bifurcations and Chaos", Springer-Verlag, 1988.

1 1 APPENDIX.

Topological Entropy

Roughly speaking, a system is chaotic if the trajectories are mixing. Essentially, the Li-Yorke definition says that such mixing occurs, while topological entropy measures the speed of such mixing. So the system is chaotic if the topological entropy is positive.

Let p be a fixed positive integer and let Ω be a compact metric space. Suppose that

$$\mathcal{U} = \{U_0, U_1, \dots, U_N\}$$

be an open cover of Ω. Denote by $\mathcal{U}_{\min}$ the subcover of $\mathcal{U}$ which contains a minimum number of elements of $\mathcal{U}$ covering Ω. Write $\#\mathcal{U}_{\min}$ for the cardinality of the minimal subcover.

Definition A1 . The quantity

$$E(\mathcal{U}) = \ln(\#\mathcal{U})$$

is called the *topological entropy* of the cover $\mathcal{U}$.
Let $f\colon \Omega \to \Omega$ generate a dynamical system and suppose that $\mathcal{U}$ as above is a fixed open cover of Ω. Then for any positive integer k, the class

$$\mathcal{U}_{f,k} = \{f^{-k}(U_0), \dots, f^{-k}(U_N)\}$$

is also an open cover of Ω. Define the open cover

$$\mathrm{U}(f,K) = \left\{ \bigcap_{k=1}^{K} V_k : V_k \in \mathrm{U}_{f,k}, 1 \le k \le K \right\}$$

Lemma A. 1 . *Let K, L be positive integers. Then*

$$\#\,\mathcal{U}(f, K)\,\#\,\mathcal{U}(f, L) \ge \#\,\mathcal{U}(f, K+L)\,.$$

From this it follows that
Theorem A.2. *The following limit exists:*

$$E(f;\mathrm{x}) = \lim_{K\to\infty} K^{-1} E(\mathrm{x}(f,K))$$

Definition A.2 . The topological entropy of f is defined to be

$$E(f) = \sup_{\mathcal{U}} E(f;\mathcal{U})$$

It can easily be shown that if $\mathcal{U}_i$ is a sequence of open covers of Ω with diam $\mathcal{U}_i \to 0$, then $E(f;\mathcal{U}_i) \to E(f)$

There are relationships between the number of k - periodic points of a mapping and its entropy, especially if additional conditions are placed on f. One typical example is

Theorem A.3. *If f is an expansive homeomorphism of a compact metric space, then*

$$E(f) \geq \overline{\lim}_k \frac{1}{k} \ln P_k(f)$$

where $P_k(f)$ is the number of fixed points of f^k and the limit is finite.
Further details can be found in Reference 4.

The Fixed Point Property

Definition B.1. A metric space (Π, ρ) is said to have the fixed point property (f.p.p.) if any continuous mapping

$$f : \Pi \to \Pi$$

has at least one fixed point.

It was shown in Referencer 9 that quite general classes of fuzzy sets have the f.p.p., provided that the topology of the sendograph metric is used. The conditions of fuzzy convexity and normality can be dropped and the totality $\mathcal{D}(\Omega)$ of upper semicontinous fuzzy sets on a compact metric space (Ω, ρ) can be shown to have the f.p.p. This is so even when Ω is not connected, in which case Ω itself does not even have the f.p.p.

A subset of $\mathcal{H}$ of Π is said to be a *retract* of Π if there exists a continuous mapping, called a *retraction*, $f : \Pi \to \mathcal{H}$, such that $f(x) = x$ for all $x \in \mathcal{H}$. The set $\mathcal{H}$ is called an *absolute retract* if for any pair $\Pi_0 \subseteq \Pi$ II of metric spaces and each continuous map ϕ: $\Pi_0 \to \mathcal{H}$, there exists a continuous map ϕ_*: $\Pi \to \mathcal{H}$ such that $\phi = \phi_*|_{\Pi 0}$.
The connection between retracts and the f.p.p. is given by the following two well-known theorems.
Theorem B.2. *If Π has the f.p.p. and $\mathcal{H}$ is a retract of Π, then $\mathcal{H}$ also has the f.p.p*
Theorem B.3. *Every compact absolute retract has the f.p.p.*
The principal result of Reference 9 is
Theorem B.4. *The metric spaces* $(\mathcal{D}(\Omega), H)$ *and* $(\mathcal{D}_\delta(Q), H)$, for $0 < \delta \leq 1$, *each have the f.p.p.*
This may be proved directly, or by showing that $(\mathcal{D}(\Omega), H)$, $(\mathcal{D}_\delta(\Omega), H)$ are each an absolute retract. By virtue of Theorem B.3, this also gives a proof that each has the f.p.p.

Proofs, references and further details may be found in Reference 9.

Neural Network Implementation of Fuzzy Inference for Approximate Case - based Reasoning

Liya Ding and Zuliang Shen

Institute of Systems Science, National University of Singapore
Heng Mui Keng Terrace, Kent Ridge, Singapore 0511
Phone: 7756666, Fax: 7782571
E-mail: liya@iss.nus.sg zuliang@iss.nus.sg

ABSTRACT

In this chapter, we point out the symbolical-numerical duality of fuzzy logic and the rule-case duality of fuzzy rules in approximate reasoning. By the former, we can use a node of a neural network to represent a fuzzy proposition for the symbolic and the value passing the node (input or output) for the numeric. By the latter, we can construct a neural network structure for describing the relations of fuzzy rules and modify the weights of the neural network to realize learning from cases (examples). The concept of the so-called approximate case-based reasoning[1] *and its neural network implementation is set up on the above understanding.*

We first give the basic mechanism of approximate case-based reasoning and the neural network implementation, then extend it to more general and complex cases by several examples.

Keywords: Approximate Case-based Reasoning, Revision Principle, Fuzzy Inference, Neural Networks, Fuzzy Logic.

1. INTRODUCTION

One of the assumptions in binary logic is that any proposition can only take truth value 1 or 0, where 1 means *true* and 0 means *false* for positive logic (for negative logic, the situation is the opposite). Usually, we write a true proposition as itself, such as P, and a false proposition as the negation of it, such as ¬P. The logical reasoning in binary logic can be always processed as symbolic processing because all the propositions take the *unique truth value* 1. In contrast there are infinitely many truth values used in fuzzy logic, there may be between 0 and 1 (fuzzy valued logic) or any linguistic truth value in [0, 1][2] (fuzzy linguistic valued logic). So fuzzy logic introduces a numeric aspect into the pure symbolic system by using those fuzzy truth values and linguistic truth values.

However, any fuzzy proposition in either fuzzy valued logic or fuzzy linguistic valued logic, can be represented by a true proposition (for instance, in fuzzy valued logic, it can be considered as truth value 1) with some kind of degree. Hence, the true proposition considered as a symbolical term and the degree is processed as a numerical value. This feature is called the *symbolical-numerical duality* of fuzzy logic.

Another assumption in binary logic is that a rule is always considered to be correct, or in other words, it is a *universally true rule.* Fuzzy logic, unlikely binary logic, causes approximate reasoning, where the matching law is not always satisfied. It means that a rule P → Q can be used with a fact P', an approximation of P, to get Q', an approximation of Q. Obviously, P → Q does not include the case for the fact P' that is not equal to P. Since the rules in approximate reasoning are not universally true, we prefer to consider them as examples or case-rules rather than as rules. In a sense, approximate reasoning done by using those fuzzy rules is the *approximate case-based reasoning.*[1] This feature is called the *rule-case duality* of fuzzy rules.

This chapter will investigate fuzzy rules on approximate case-based reasoning. A fuzzy rule P → Q is considered as an example or a case for describing a relation from a universe of discourse X to another universe of discourse Y, where $P \subseteq X$ and $Q \subseteq Y$ are fuzzy sets or fuzzy truth values. In general, we often need to do inference by such a fuzzy rule and a given fact P', an approximation of P. In order to get an approximate consequence Q', the fuzzy rule is applied as an approximate knowledge reference for the reasoning from X to Y. The inference relation between the premise P and the consequence Q of a fuzzy rule P → Q can be expressed by some kind of semantics of the rule. The semantics of a fuzzy rule can be retrieved as a weight of the rule. By representing the weights of fuzzy rules to the weights of a corresponding neural network unit, it is possible to implement fuzzy inference on neural networks.[2] The revision principle based on the semantic revising method[3] can be used on the neural networks for approximate case-based reasoning. Some existing learning approaches[4,5] in neural networks can also be used for modifying the weightage of the corresponding neural network.

The discussion is divided into several sections: Section 2 discusses the basic definitions of fuzzy truth types. The features of fuzzy truth and fuzzy rules are analyzed in section 3. Section 4 discusses the logical meaning of approximate case-based reasoning. The basic mechanism of approximate case-based reasoning is given in section 5. Section 6 reviews several models of neural networks for fuzzy reasoning. Section 7 proposes a basic model for implementing approximate case-based reasoning on neural network. In section 8, the basic model is extended for applications. Conclusions are given in the last section.

2. DEFINITIONS ON FUZZY TRUTH TYPES

When we extend binary logic to fuzzy logic, it is necessary to decide the logical meaning of truth values that are some kind of extension of {0,1}. In this section, we define fuzzy truths in fuzzy valued logic and fuzzy linguistic valued logic.[2,6] By using these basic truth types and their combinations, more complex fuzziness can be represented.

Fuzzy valued logic can be defined as an algebraic system $< \{ [0, 1], F, T, \underline{\text{unknown}}, \underline{\text{undefined}}\}, \wedge, \vee, \neg >$. Where in the truth set {[0, 1], F, T, unknown, undefined}, the closed interval [0, 1] is a set of truth values, the $F \subset [0, 1]$ is a set of values on degree of false, the $T \subset [0, 1]$ is a set of values on degree of true, the $\underline{\text{unknown}} \subset [0, 1]$ is a special set represented by only one point where both F and T are unknown, and the set undefined does not have any definition in [0, 1]. The logical operations AND($\wedge$), OR($\vee$) and NOT($\neg$) in the algebraic system are defined as follows:

$$A \wedge B = \min(A, B), \quad (2.1)$$
$$A \vee B = \max(A, B), \quad (2.2)$$
$$\neg A \text{ (or not (A))} = 1 - A, \qquad A, B \in [0, 1]. \quad (2.3)$$

Definition 2.1: Let the kind of truth be *truth I* in fuzzy valued logic if and only if in the truth set, F = undefined. T = (0, 1], and unknown = 0. (Fig. 2.1)

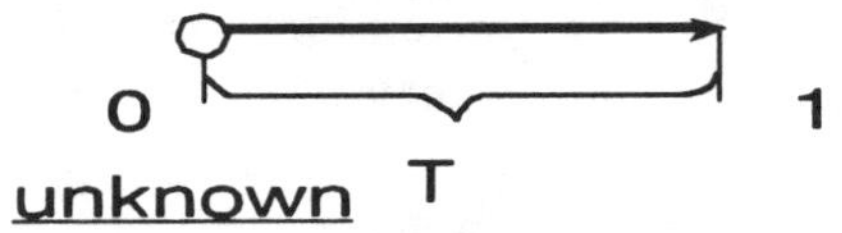

Fig. 2.1. *Truth I* in fuzzy valued logic

Definition 2.2: Let the kind of truth be *truth II* in fuzzy valued logic if and only if in the truth set, F = [0, 0.5), T = (0.5, 1], and unknown = 0.5. (Fig. 2.2)
unknown

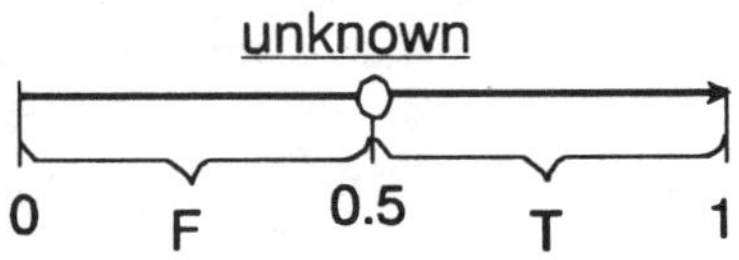

Fig. 2.2. *Truth II* in fuzzy valued logic

According to the above definition, under *truth II*, if a is an element of V = [0, 1], then a = 0.5 is said to be *unknown*. That is, in the set of truth value [0,1], 0 and 1 have different definite information and the ambiguity reaches its maximum at 0.5. But, under *truth I*, it is considered that 1 has definite information and the ambiguity reaches its maximum at 0.

Thereby, we can introduce the following partially ordered relation ($\gg$) to the set [0,1] under *truth II* and the 'less or equal to' relation ($\leq$) to the set [0,1] under *truth I*, which designates some kinds of ambiguity, respectively.

Definition 2.3: Let a, b be elements of $V = [0,1]$ under *truth II*. Then, $a \gg b$ if and only if $0.5 \geq a \geq b$ or $0.5 \leq a \leq b$.

Under *truth I*, when $a \leq b$ holds, the ambiguity of a is larger than or equal to the ambiguity of b (Fig. 2.3(a)). And under *truth II*, when $a \gg b$ holds, the ambiguity of a is larger than or equal to the ambiguity of b (Fig. 2.3(b)).

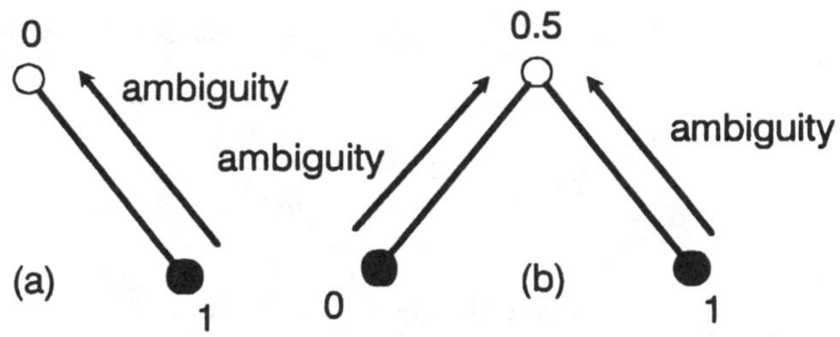

Fig. 2.3. The ambiguity under *truth I* and *truth II*

Definition 2.4: Let a be an element of V = [0, 1]. Then, the *confidence* of a is defined as follows:

C(a) = a	C(a) ∈ [0,1] and a ∈ [0,1] under *truth I*;	(2.4)
C(a) = (a - 0.5) x 2	C(a) ∈ [-1,1] and a ∈ [0,1] under *truth II*;	(2.5)

and the absolute value |C(a)| will designate the non-ambiguity of a

By Definition 2.4, fuzzy valued logic under *truth II* can be redefined by the concept of confidence as the algebraic system < { [-1,1], F, T, unknown, undefined } , ∧, ∨, ¬ >, where the closed interval [-1, 1] is a set of confidences, F is represented by negative confidences, T is represented by positive confidences, unknown = 0, and undefined has no definition in [-1, 1]. Obviously, the *truth I* can be looked as the positive part of *truth II* by using the closed interval [0, 1] instead of [-1, 1].

Clearly, for the partially ordered relation ≫ about the ambiguity, the elements between [0, 0.5) and (0.5, 1] are not comparable to each other. But the absolute values of their confidences are comparable, because in the two half-intervals, one is more ambiguous when it is closer to 0.5.

Example 2.1: If under *truth II*, a = 0.8, b = 0.2 then C(a) = (0.8 - 0.5) x 2 = 0.6 and C(b) = (0.2 - 0.5) x 2 = -0.6, respectively. Therefore, |C(a)| = |C(b)| = 0.6 which means they have the same absolute values of confidence and ambiguity in the T part and F part, respectively.

A truth value which is a point in [0, 1], e.g. T(a) = 0.8, will be referred to as a *numerical* truth value in fuzzy valued logic. The numerical truth values play the role of the values of the base variable for the linguistic variable *Truth*. The linguistic values of the *Truth* will be referred to as the *linguistic truth values*, and by which the so-called *fuzzy linguistic valued* logic can be defined.

Fuzzy linguistic valued logic can be defined as an algebraic system < {T1, T2, **F, T, unknown, undefined**}, ∧, ∨, ¬ >, where in the *truth set* { T1, T2, **F, T, unknown, undefined**}, T1 and T2 are the truth sets in the fuzzy valued logic, and they are the universe of discourse of truth and the membership degree of the truth, respectively, and any linguistic truth value A ∈(**F,T**) is a fuzzy subset $A = \{\mu_A(\tau)/\tau \mid \tau \in T1, \mu_A(\tau) \in T2\}$, the **F** means a set of linguistic truth values of false, the **T** means a set of linguistic values of true, the **unknown** is a set defined by $\int 0/\tau$, **undefined** has no definition in $[0, 1]^2$. The logical operations **AND**($\wedge$), **OR**($\vee$) and **NOT**($\neg$) are defined as follows:

$$A \wedge B = \int \min(\mu_A(\tau), \mu_B(\tau)) / \tau \tag{2.6}$$

$$A \vee B = \int \max(\mu_A(\tau), \mu_B(\tau)) / \tau \tag{2.7}$$

$$\neg A(\text{or not}(A)) = \int (1 - \mu_A(\tau)) / \tau \tag{2.8}$$

where $\int$ means union over τ, and A, B $\subset$ { **F, T**), $\tau \in [0, 1] = T1$; $\mu_A(\tau)$, $\mu_B(\tau) \in [0,1] = T2$.

Definition 2.5: Let the type of truth be *truth I-I* in fuzzy linguistic valued logic if and only if in the truth set, T1 and T2 are all *truth I* in fuzzy valued logic.

Depending on Definition 2.5, one kind of *truth I-I* of fuzzy linguistic valued logic systems may be defined as follows(Fig. 2.4):

$$\text{true} = \int \mu_{\text{true}}(\tau) / \tau, \qquad \text{for } \mu_{\text{true}}(\tau) = \tau, \tau \in [0, 1] \tag{2.9}$$

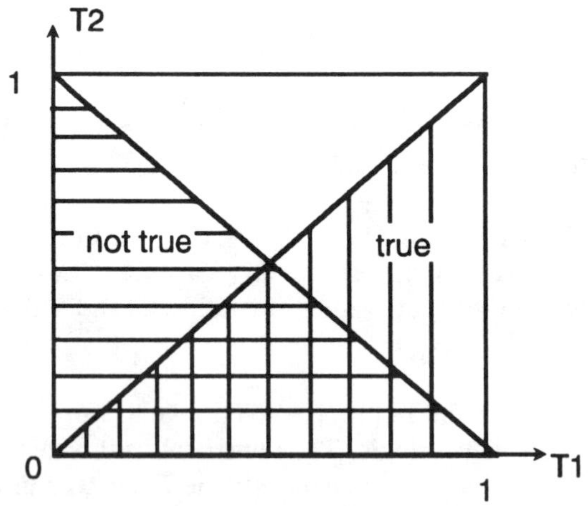

Fig. 2.4. The *truth I-I* in fuzzy linguistic valued logic

Definition 2.6: Let the kind of truth be *truth II-I* in fuzzy linguistic valued logic if and only if in the truth set, T1 is *truth II* and T2 is *truth I* in fuzzy valued logic.

Depending on Definition 2.6, one kind of *truth II-I* of fuzzy linguistic valued logic systems may be defined as follows(Fig. 2.5):

$$\text{true} = \int \mu_{\text{true}}(\tau)/\tau, \quad \text{for } \mu_{\text{true}}(\tau) = 0, \tau \in [0, 0.5] \\ \mu_{\text{true}}(\tau) = (\tau - 0.5) \times 2, \tau \in [0.5, 1] \tag{2.10}$$

and

$$\text{false} = \int \mu_{\text{false}}(\tau)/\tau, \quad \text{for } \mu_{\text{false}}(\tau) = (0.5 - \tau) \times 2, \tau \in [0, 0.5] \\ \mu_{\text{false}}(\tau) = 0, \tau \in [0.5, 1] \tag{2.11}$$

In real applications, the types of linguistic *truth I-I*[7] and *II-I*[8] are often used.

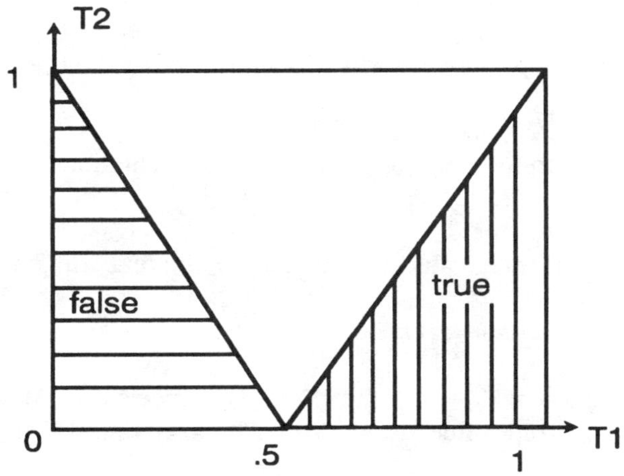

Fig. 2.5. The *truth 11-1* in fuzzy linguistic valued logic

Generally, we can use some kind of operators on the primary terms *true* and *false* to produce a lot of fuzzy linguistic truth values, such as *very true*, *more or less true*, *not true*, *very (not true)*, *essentially true*, etc. which are all produced from the primary term *true* and belong to the set of linguistic truth values on true **T**, and *very false*, *more or less false*, *not false*, *very (not false)*, *essentially false*, etc. which are all produced from the primary term *false* and belong to the set of linguistic truth values on false **F**.

Under truth I-I, since there is no definition of false, we often prefer to refer to *not true* as *false*. In this case, we can say that all fuzzy linguistic truth values produced from *true* belong to the **T** set and all fuzzy linguistic truth values produced from *not true* belong to the **'F'** set. Clearly, the operator *not* will change an element from one set to another set.

3. FEATURES OF FUZZY LOGIC

Comparing binary logic to fuzzy logic, we have two big differences: (1) the *unique truth value* has infinitely many truth values; (2) the *universally true rule* ceases to be universally true. The two assumptions in binary logic and binary logical reasoning are not satisfied in fuzzy logic. It causes fuzzy logic to have the symbolical-numerical duality and fuzzy rule to have rule-case duality in approximate case-based reasoning.

3.1 Symbolical-Numerical Duality of Fuzzy Logic

In general, the description of a proposition in fuzzy logic may take several types:

(1) a proposition with truth value in fuzzy valued logic:
P is 0.8 true

where P is a proposition, such as "Mary likes music".

(2) a proposition with truth value in fuzzy linguistic valued logic:
P is *very true*.

(3) a proposition described by a fuzzy set:
the house is *large*,

where "the house is *large*" is a proposition described by fuzzy set *large*. *large* may be defined as follows($x = m^2$):

large(x) = 0/0, 0/10, 0.1/20, 0.2/80, ..., 0.8/1000, ...;

(4) a fuzzy set with some truth value (or confidence, see Definition 2.4) in fuzzy valued logic:
"the house is *large*" is 0.9 true;

(5) a fuzzy set with some truth value (or confidence) in fuzzy linguistic valued logic:
"the house is *large*" is *more or less true*.

Obviously, (4) and (5) are extensions of (3) + (1) and (3) + (2), respectively. (3) is a special case of (4) and (5) when the truth value always takes 1 in fuzzy valued logic or *true* in fuzzy linguistic valued logic.

All those types of propositions can be represented by two levels: the description of a proposition itself and the truth degree (or confidence degree) of the proposition which shows how sure we consider the proposition. It means that any fuzzy proposition can be discussed from two sides: the proposition itself and the truth of the proposition.

Usually people like to discuss fuzzy inference in two ways: based on fuzzy truths or based on fuzzy sets. For the former case, the inference is set up on the relation from the truth of the premise to the truth of the consequence. For the latter case, the inference is set up on the relation from the fuzzy sets in the premise to the fuzzy set in the consequence.

Any fuzzy inference based on fuzzy truths can actually be looked as a statistical structure, where every proposition is a node, with a dynamic flow from the premise(s) to the consequence(s). The structure represents the *symbolical* (binary logical) relationships among all propositions in the inference and the dynamic flow is a *numerical* flow that consists of all individual numerical truth values in those propositions. This is the *symbolical-numerical duality* of fuzzy logic.

Based on the symbolical-numerical duality, if we construct the symbolic (logic) of a fuzzy rule to a neural network *structure* and let the *weights* in the neural network deal with the numeric (semantics) of fuzzy propositions, then the neural network can implement the fuzzy inference related to the rule.

3.2 Rule-Case Duality of Fuzzy Rule

Fuzzy inference based on fuzzy sets may cause approximate reasoning. There is actually a matching law used in binary logic inference, that is the fact that P should be exactly equal to the premise P of a rule $P \rightarrow Q$. When we try to extend a knowledge-based system from binary logic to fuzzy logic, the premises and the consequences of rules may not be an exact term but a fuzzy term that is usually represented by fuzzy sets. One of the new problems which we are facing is that the matching law in binary logic cannot always be satisfied in fuzzy logic. In other words, the reasoning is often required to get an approximate conclusion by using such a fuzzy rule $P \rightarrow Q$ and a fact P', an approximation of P. The approximate conclusion is usually denoted by Q', the approximation of Q. This is so-called approximate reasoning.[8,9]

Based on the above understanding, in approximate reasoning a fuzzy rule $P \rightarrow Q$, with fuzzy premise $P \subseteq X$ and fuzzy consequence $Q \subseteq Y$, is only giving out one example or one case representing an inference relation from X (the universe of discourse of P) to Y (the universe of discourse of Q). We can do approximate reasoning from $P' \subseteq X$ (an approximation of P) to get $Q' \subseteq Y$ (an approximation of Q) by an approximate knowledge reference given in $P \rightarrow Q$. In that sense, the approximate reasoning is reasonably considered as the so-called *approximate case-based reasoning.*

A fuzzy rule $P \rightarrow Q$, where $P \subseteq X$, $Q \subseteq Y$, is an exactly correct *rule* when the matching law is satisfied (namely for any P', P' = P), otherwise it is only a given *case* related to the inference from X to Y. This is the *rule-case duality* of fuzzy rule in fuzzy inference.

It is interesting and important that the features of rule-based reasoning and case-based reasoning can be integrated in the approximate case-based reasoning on fuzzy logic sense. This also provides the possibilities to use existing methods and approaches of the rule-based reasoning and the case-based reasoning for the approximate case-based reasoning.

When we get a Q' from $P \rightarrow Q$ and P', we also get another case $P' \rightarrow Q'$. By some methods of approximate reasoning such as the *revision principle*[3,6,10], there should be some information (knowledge) kept from $R_{P \rightarrow Q}$ (the relation of $P \rightarrow Q$) to $R_{P' \rightarrow Q'}$ (the relation of $P' \rightarrow Q'$). Using $R_{P' \rightarrow Q'}$ to modify the previous relation $R_{P \rightarrow Q}$ may realize another important issue in the approximate case-based reasoning-*learning*. Such a relation $R_{P \rightarrow Q}$ can also be described as a *weight* of the rule, denoted by $W_{P \rightarrow Q}$ (see section 7). The Q' is calculated by formula:

$$Q' = f(P', W_{P \rightarrow Q}) \tag{3.1}$$

where f is a transfer function for the rule. Let $W_{P\rightarrow Q}$ be the weight of a corresponding neural network unit of the rule P→Q; the learning procedure can be implemented by using $W_{P'\rightarrow Q'}$ to modify $W_{P\rightarrow Q}$.

4. APPROXIMATE CASE BASED REASONING (ACBR) IN FUZZY LOGIC

For a rule P → Q in approximate case-based reasoning, there are two approximation families $\mathcal{P} = \{P^i \mid i = 1, ..., m\}$ and $\mathcal{Q} = \{Q^j \mid j = 1, ..., n\}$, where $P^i \subseteq X$, $Q^j \subseteq Y$, for the premise P and the consequence Q, respectively. $P \in \mathcal{P}$ and $Q \in \mathcal{Q}$, are the representatives of $\mathcal{P}$ and $\mathcal{Q}$ respectively. In that sense, the Modus Ponens in approximate reasoning:

$$\begin{array}{l} P \rightarrow Q \\ P' \\ \text{--------} \\ Q' \end{array}$$

should be interpreted as

$$\begin{array}{rl} P \rightarrow Q \approx & P' \rightarrow Q' \\ & P' \\ & \text{-----------} \\ & Q' \end{array} \tag{4.1}$$

where $P, P' \in \mathcal{P}$, $Q, Q' \in \mathcal{Q}$ and "≈" means approximately equal. The rule P' → Q' is said to be an *approximate rule* of P → Q and vice versa. Since the rule P' → Q' is not given, we use P → Q instead of P' → Q' to deduce Q' approximately. Here, the tacit agreement is that the *implication relation* $R_{P\rightarrow Q}$ (R for short) of the rule P → Q is equal to or approximately equal to the implication relation $R_{P'\rightarrow Q'}$ (R' for short) of rule P' → Q'.

In the above interpretation, R' is an approximation of R. For all approximations of R from $\mathcal{P}$ to $\mathcal{Q}$ there is a set of *approximate implication relations* $\{R_s \mid P_s \rightarrow Q_s;\ P_s \in \mathcal{P},\ Q_s \in \mathcal{Q}\}$. The *common implication relation* $\mathcal{R} = \bigcap_s (R_s)$ identifies the common part of all approximate rules. In other words, a given rule $P \rightarrow Q \in \{P_s \rightarrow Q_s \mid P_s \in \mathcal{P},\ Q_s \in \mathcal{Q}$ is only a *case rule* that does not guarantee the *universal truth* (in binary logic, because $\mathcal{P} = \{P\}$ and $\mathcal{Q} = \{Q\}$ and the common implication relation $\mathcal{R} = R$, the rule P → Q does guarantee the universal truth). Since it is almost impossible to find the set of approximate rules, the approximate reasoning is actually an *approximate case-based reasoning*. Therefore, a fuzzy rule P → Q in approximate case-based reasoning is considered as an example (or a case) for approximately describing the common implication relation $\mathcal{R}$ between two approximation families $\mathcal{P} = \{P^i \mid i = 1, ...m\}$, and $\mathcal{Q} = \{Q^j \mid j = 1, ...n\}$, where $P^i \subseteq X$, $Q^j \subseteq Y$.

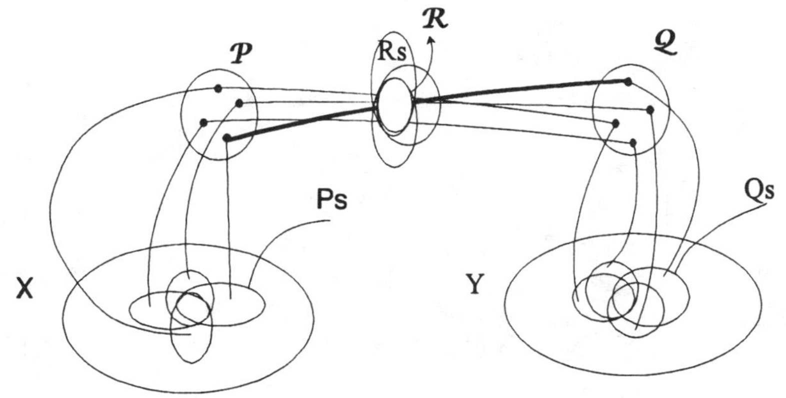

Fig. 4.1 Approximate implication relation

Three main tasks in approximate case-based reasoning are:
(1) to represent an implication relation R_1 for a rule $P_1 \rightarrow Q_1$;

(2) to use the given fact P_{t+1} and the current known approximate implication relation

$$\mathcal{R}^t = \bigcap_{k=1}^{t} (R_k) \tag{4.2}$$

for deducing Q_{t+1}, where k = t identifies the current state and k = t + 1 identifies the next state. Obviously, it is reasonable to let some kind of inference relationship in $\mathcal{R}^t$ be kept in $P' \rightarrow Q'$. The revision principle[3,6,10] has been proposed for this purpose;

(3) to find a new implication relation R_{t+1} from rule $P_{t+1} \rightarrow Q^*_{t+1}$ and then modify the approximate implication relation $\mathcal{R}^t$, where Q^*_{t+1} is a real conclusion or an expected conclusion obtained by the evaluation of the deduced Q_{t+1}.

Though $\mathcal{P}$ and $\mathcal{Q}$ are not given and it is difficult to define $\mathcal{R}$ directly, by the approximate implication relation based on the definition:

$$\mathcal{R}^{t+1} = \cap(\mathcal{R}^t, R_{t+1}) \tag{4.3}$$

the common implication relation $\mathcal{R}$ can be approached.

5. ACBR WITH TRIANGULAR MEMBERSHIP BASED RULES(TMB RULES)

Fuzzy sets represented by triangular membership are popularly used for recent fuzzy theory research and applications. For a simplified discussion, we deal with approximate case-based reasoning by the so-called *triangular memberships based rule* (TMB rule) and give out a basic neural network model based on TMB rules.

5.1 Definition of TMB Rule

Definition 5.1(TMB rule): Suppose a rule $P \rightarrow Q$ be given, where $P \subseteq X = [x_1, x_r]$ and $Q \subseteq Y = [y_1, y_r]$ are fuzzy sets defined by:

$$\mu_P(x)=\begin{cases}(x-x_1)\div(x^*-x_1) & x\in[x_1,x^*]\\ 1-(x-x^*)\div(x_r-x^*) & x\in(x^*,x_r]\end{cases} \tag{5.1}$$

$$\mu_Q(y)=\begin{cases}(y-y_1)\div(y^*-y_1) & y\in[y_1,y^*]\\ 1-(y-y^*)\div(y_r-y^*) & y\in(y^*,y_r]\end{cases} \tag{5.2}$$

then we say that

$$\mu_P(x) \rightarrow \mu_Q(y) \tag{5.3}$$

is a *triangular membership based rule* (or TMB rule for short) for approximate case-based reasoning, where $x^* \in X$, $y^* \in Y$ are called *representative points* for P and Q respectively, and $\mu_P(x^*) = 1$ as well as $\mu_Q(y^*) = 1$. (Fig. 5.1)

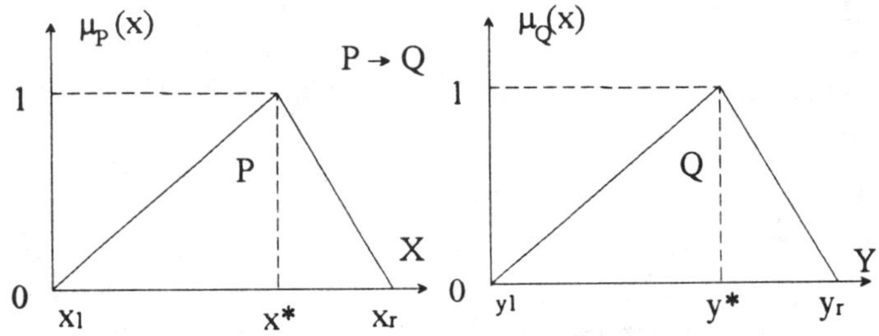

Fig. 5.1 TMB Rule P →Q

Definition 5.2 (An approximation of TMB rule): Suppose the TMB rules $P \rightarrow Q$ and $P' \rightarrow Q'$ are given, where $P \subseteq X=[x_1, x_r]$, $Q \subseteq Y = [y_1, y_r]$, $P' \subseteq X' = [x'_1, x'_r]$, $Q' \subseteq Y' = [y'_1, y'_r]$. The two rules are said to be *approximations* of each other if and only if

$X = X'$ and $Y = Y'$

are satisfied.

In this paper, we first give ACBR on the TMB rules which are approximations of each other. It means that for any $P \rightarrow Q$ and $P' \rightarrow Q'$ ($P, P' \in \mathcal{P}$, and $Q, Q' \in \mathcal{Q}$) we will always let $[x_1, x_r] = [x'_1, x'_r]$ and $[y_1, y_r] = [y'_1, y'_r]$. $X_{\mathcal{P}} = [x_1, x_r] = [x'_1, x'_r]$ and $Y_{\mathcal{Q}} = [y_1, y_r] = [y'_1, y'_r]$ are called the *valuable interval* of $\mathcal{P}$ and $\mathcal{Q}$, respectively.

5.2 TMB Rule Extended from Fuzzy Valued Logic

As a special case of the TMB rule, a fuzzy rule in fuzzy valued logic can be extended to a TMB rule by the following definition.

Definition 5.3 (Extension TMB rule): Suppose a rule $P \rightarrow Q$ and $T(P) = p_0$ as well as $T(Q) = q_0$ is given in fuzzy valued logic, then a corresponding TMB rule $\mu_P(x) \rightarrow \mu_Q(y)$ can be generated by:

$$\mu_P(x) = \begin{cases} x \div p_0 & x \in [0, p_0] \\ 1-(x-p_0) \div (1-p_0) & x \in (p_0, 1] \end{cases} \quad (5.4)$$

$$\mu_Q(x) = \begin{cases} y \div q_0 & x \in [0, q_0] \\ 1-(y-q_0) \div (1-q_0) & x \in (q_0, 1] \end{cases} \quad (5.5)$$

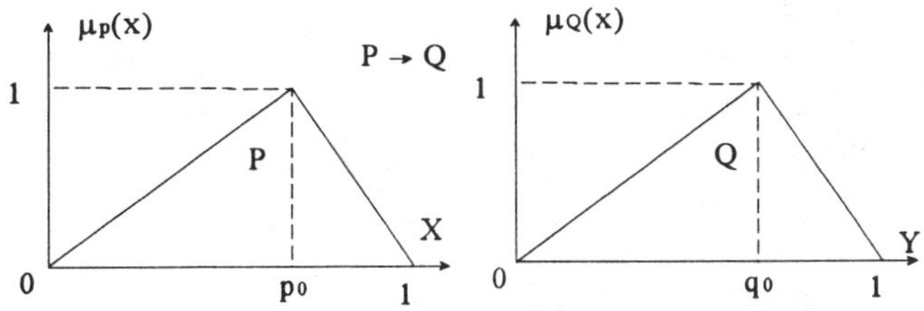

Fig. 5.2 TMB rule extended from fuzzy valued logic

where $p_0 \in X = [0, 1]$, $q_0 \in Y = [0, 1]$ are called *representative points* for P and Q respectively, and $\mu_p(p_0) = 1$ and $\mu_Q(q_0) = 1$. (Fig. 5.2)

5.3 Semantics of TMB Rules for ACBR (ACBR Semantics)

Based on the semantic revising method (SRM) of the revision principle[3], we have the following definitions.

Definition 5.4(ACBR semantics): The *ACBR semantics* of a TMB rule P → Q for the approximate case-based reasoning is defined as:

$SR_{P,Q} = R(\mu_p(x), \mu_Q(y)) = \{(s,t) \mid s=\mu_p(x), t=\mu_Q(y), s=t, x \in X, y \in Y\}$, (5.6)

where $\mu_p(x)$ and $\mu_Q(y)$ are the membership functions of P and Q respectively.

Definition 5.5(ACBR interrelation): The *ACBR interrelation* of a TMB rule for the approximate case-based reasoning is defined as:

$$IR_{P,Q} = \left\{ (x,y) \middle| \, s = \mu_P(x), t = \mu_Q(y), (s,t) \in SR_{P,Q}, \frac{ds}{dx} \times \frac{dt}{dy} > 0 \; or \left(\frac{ds}{dx} = 0, \frac{dt}{dy} = 0 \right) \right\} \quad (5.7)$$

where the point (x^*, y^*) (which lets $\mu_P(x^*) = 1$ and $\mu_Q(y^*) = 1$) is called the *semantic core* of $IR_{P,Q}$.

From Definitions 5.4 and 5.5, $W_{P \to Q}$ (*W* for short), the weight of a TMB rule P → Q for approximate case-based reasoning (*ACBR weight* for short), can be defined as:

$$W = \frac{\partial y}{\partial x} = \left(\frac{\partial y}{\partial SR_{P,Q}} \right) \div \left(\frac{\partial x}{\partial SR_{P,Q}} \right) \tag{5.8}$$

$$= \begin{cases} (y^* - y_1) \div (x^* - x_1) & x \in [x_1, x^*], y \in [y_1, y^*] \\ (y^* - y_r) \div (x_r - x^*) & x \in (x^*, x_r], y \in (y^*, y_r] \end{cases}$$

$$= \begin{cases} W_\alpha & x \in [x_1, x^*], y \in [y_1, y^*] \\ W_\beta & x \in (x^*, x_r], y \in (y^*, y_r] \end{cases}$$

Fig. 5.3 illustrates the ACBR interrelation, ACBR semantics and ACBR weight of a TMB rule.

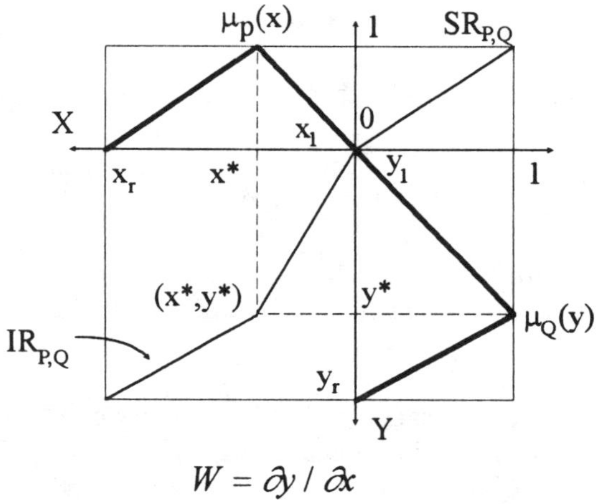

Fig. 5.3 Interrelation, semantics and weight of a TMB rule

5.4. ACBR Inference Mechanism

Formula 5.1(ACBR Inference Approach): Suppose a TMB rule P → Q is defined by:

$$\mu_P(x) = \begin{cases} (x - x_1) \div (x^* - x_1) & x \in [x_1, x^*] \\ 1 - (x - x^*) \div (x_r - x^*) & x \in (x^*, x_r] \end{cases}$$

$$\mu_Q(y) = \begin{cases} (y - y_1) \div (y^* - y_1) & y \in [y_1, y^*] \\ 1 - (y - y^*) \div (y_r - y^*) & y \in (y^*, y_r] \end{cases}$$

as well as P' be given by:

$$\mu_{P'}(x)=\begin{cases}(x-x_1)\div(x^{**}-x_1) & x\in[x_1,x^{**}]\\ 1-(x-x^{**})\div(x_r-x^{**}) & x\in(x^{**},x_r]\end{cases}\tag{5.9}$$

where the x* and y* are the representative points for P and Q respectively, and x** is the representative point for P'. Then y**, the representative point for Q' is calculated by the following transfer function:

$$y^{**} = f(x^{**}, W)\tag{5.10}$$

$$=\begin{cases}y_1+(x^{**}-x_1)\times W_\alpha & x^{**}\in[x_1,x^*]\\ y_r+(x_r-x^{**})\times W_\beta & x^{**}\in[x^*,x_r]\end{cases}$$

Then by **Definition 5.1**, we can have another new TMB rule $\mu_{P'}(x) \rightarrow \mu_{Q'}(y)$ where

$$\mu_{Q'}(y)=\begin{cases}(y-y_1)\div(y^{**}-y_1) & y\in[y_1,y^{**}]\\ 1-(y-y^{**})\div(y_r-y^{**}) & y\in(y^{**},y_r]\end{cases}\tag{5.11}$$

Fig. 5.4 illustrates an example of ACBR inference. It is important that when the above definitions and approach are used, if P' = P is given, then Q' = Q is guaranteed.

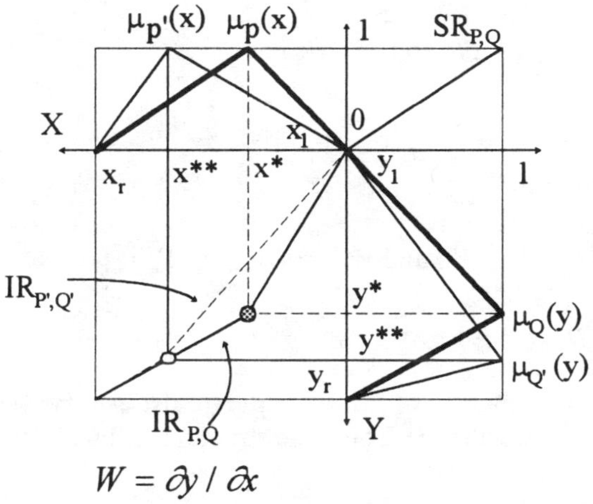

Fig. 5.4 ACBR Inference

If we construct the P' and the Q' deduced by P → Q and P' to a new TMB rule P' → Q', then we get $IR_{P',Q'}$. By some learning method, it is possible to modify the $IR_{P,Q}$ based on the $IR_{P',Q'}$. It is discussed in section 7 with the neural network implementation.

6. FUZZY INFERENCE ON NEURAL NETWORKS

There have been several models proposed for fuzzy inference on neural networks. We will give a brief review in this section.

Model 1. The unit based on truth value flow inference

The so-called *truth value flow inference* was put forward by Wang et al..[11] It provided one kind of mathematical framework for fuzzy inference. Based on truth value flow inference, a basic neural network unit has been proposed.[2] The basic idea is that a fuzzy inference by fuzzy rule P → Q on fuzzy valued logic is considered as a confidence flow from the premise P to the consequence Q. By this understanding, a neural network unit can be built as in Fig. 6.1.

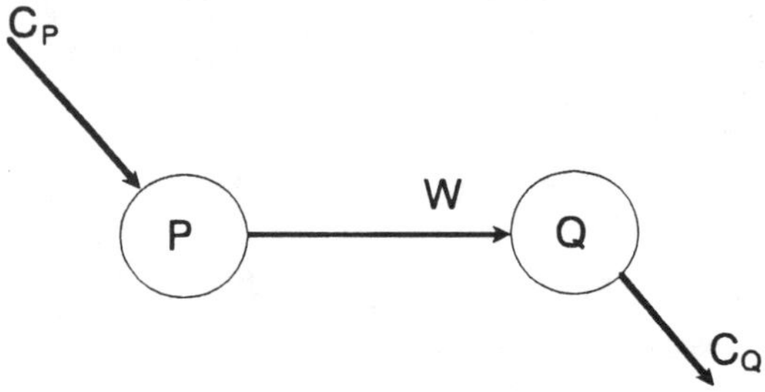

Fig. 6.1 The unit based on truth value flow inference

For truth I, we have

$$C_Q = f(C_P, W), \tag{6.1}$$

where C_P, $C_Q \in [0, 1]$, $W \in [0, \infty)$. For truth II, we have

$$C_Q = f(C_P, W) = \begin{cases} f(C_P, W_+) \\ f(C_P, W_-) \end{cases} \tag{6.2}$$

where C_P, $C_Q \in [-1, 1]$, $W = (W_+, W_-)$ and $W_+ \in (-\infty, \infty)$, $W_- \in (-\infty, \infty)$. f is defined as:

$$f(a,b) = a \times b \tag{6.3}$$

Model 2. Neural network unit of fuzzy connectionist expert system

This was proposed by Sanchez.[12] The basic structure is illustrated as Fig. 6.2,

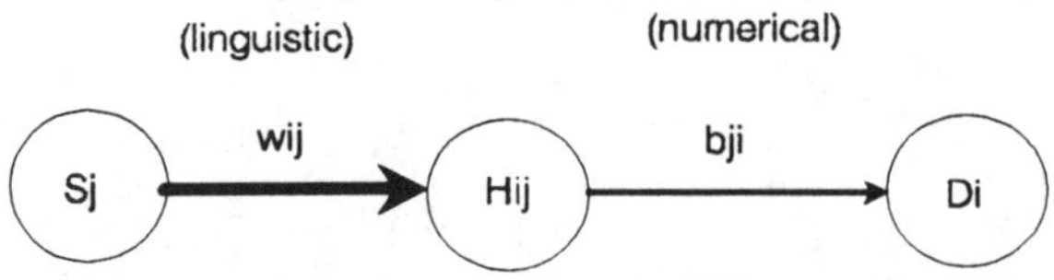

Fig. 6.2 Neural network unit of fuzzy connectionist expert system

where

$$D_i = \min_j[b_{ji} \vee \pi(w_{ji}, S_j)] \tag{6.4}$$

$\pi(w_{ji}, S_j) = \text{Sup}(w_{ji} \cap S_j)$ is the possibility measure associated with the fuzzy sets w_{ji} and S_j.[12]

In Model 2, a hidden layer is introduced between the input (premise) layer and the output (consequence) layer. The linguistic weight w_{ji} between the input layer and the hidden layer is given by original knowledge input (for example, the experience of experts). The numerical weight b_{ji} between the hidden layer and the output layer is not given and may be gotten by learning. It proposed a basic model of fuzzy connectionist network for constructing the knowledge base of an expert system from training examples.

Model 3. Unit of neural network implementation of fuzzy logic

This model was proposed by Keller et al..[13] It shows that not only the semantic relations in fuzzy rules but also the knowledge related to fuzzy concepts can be represented by the weightage in neural network. For a given fuzzy rule:

$$\text{if } A_1, A_2, \ldots, A_n \text{ then } B \tag{6.5}$$

where,

$$A_i = (a_{i1}, \ldots, a_{im}\} \qquad i = 1, 2, \ldots, n$$
$$B = \{b_1, \ldots, b_k)$$

the corresponding neural network unit can be built as in Fig. 6.3.

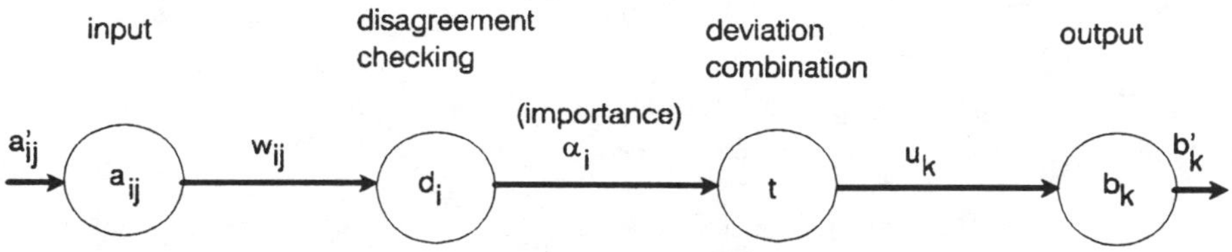

Fig. 6.3 Unit of a neural network implementation of fuzzy logic

The unit has three stages with four layers. The second stage can be seen as similar to the stage between the hidden and the output layer in the neural network unit of fuzzy a connectionist expert system. The weight α_i represents the importance of the premise A_i to the consequence B. The first stage is for constructing the input and the last stage is for summarizing the output. The weights in the first stage and the last stage are calculated by linear functions of the membership values of the premise(s) and the consequence in the rule:

$$w_{ij} = l_1(a_{ij}) \tag{6.6}$$
$$u_i = l_2(b_i) \tag{6.7}$$

The output b_i' is calculated by:

$$d_k = \max_j \{ w_{kj} * a'_{kj} \} \tag{6.8}$$

$$t = \max_i \{ \alpha_i * d_i \} \tag{6.9}$$

$$b'_i = 1 - u_i \times (1 - t) \tag{6.10}$$

There t is the deviation between real input and the premise in the rule, the operator * may correspond multiplication or to minimum. The t is used for revising the membership values of the consequence and then to get the inference result. An important property of this model is that if the input is the same as the premise of the rule (namely, t = 0), then the output should be the same as the consequence of the rule.

Model 4. Unit of fuzzy neural inference

This model was proposed by Shen et al.[1] with the concept of the so-called approximate case-based reasoning. The basic idea is to map P and Q of a fuzzy rule P → Q to two vectors $\mathbf{P} = [p_1, \ldots, p_x, \ldots, p_X]$ and $\mathbf{Q} = [q_1, \ldots, q_y, \ldots, q_Y]$, where $\mathbf{P} \in \mathbf{X}$, $\mathbf{Q} \in \mathbf{Y}$ and **X, Y** are the universes of discourse, and then identify the implication relation **R** of the rule P → Q as a X × Y relation matrix:

$$\mathbf{R} = [r_{x,y}] \tag{6.11}$$

Let the weightage matrix $\mathbf{W} = [w_{x,y}] = \mathbf{R}$ (namely, $w_{x,y} = r_{x,y}$), a two layer neural network (Fig. 6.4) can be built to represent the initial relation **R** of the given rule P → Q. The input layer has X nodes corresponding to the input vector **P** and the output layer has Y nodes corresponding to the output vector **Q**. The inference is defined as:

$$\mathbf{Q}_{t+1} = \mathbf{f}(\mathbf{P}_{t+1}, \mathcal{W}^t) \quad t \geq 1 \qquad \mathcal{W}^1 = \mathbf{W} \tag{6.12}$$

where, $\mathcal{W}^t$ is the current weight matrix. The transfer function **f** is:

$$\mathbf{f}\left(\mathbf{P}_{t+1}, \mathcal{W}^t\right) = \left[\bigoplus_{x=1}^{X}\left(w_{x,y} \times p_x\right)\right] \tag{6.13}$$

where $\mathbf{P}_{t+1} = [p_x]$, $\mathcal{W}^t = [w_{x,y}]$ for x = 1, ..., X and y = 1, ..., Y; and ⊕ is defined as:

$$\oplus(A,B) = A + B - A \times B \tag{6.14}$$

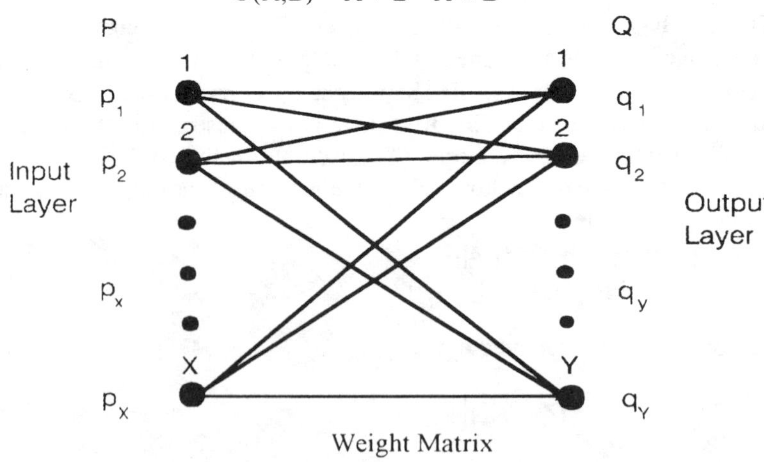

$$W = \begin{bmatrix} w_{1,1} & \bullet & \bullet & \bullet & w_{1,Y} \\ \bullet & & & & \bullet \\ \bullet & & w_{x,y} & & \bullet \\ \bullet & & & & \bullet \\ w_{X,1} & \bullet & \bullet & \bullet & w_{X,Y} \end{bmatrix}$$

Fig. 6.4 Unit of fuzzy neural inference

When an input vector $\mathbf{P}_{t+1}$ is given, based on the transfer function **f,** the output $\mathbf{Q}_{t+1}$ can be deduced. When a real or expected conclusion $\mathbf{Q}^*_{t+1}$ is obtained and mapped to the target vector $\mathbf{Q}^*_{t+1} = [q_1^*, \ldots, q_y^*, \ldots, q_Y^*]$, by the deviation of $\mathbf{Q}^*_{t+1}$ from $\mathbf{Q}_{t+1}$ and the approximate conclusion deduced by above formula, the learning algorithm for building the new weightage matrix $\mathcal{W}^{t+1}$ is defined as:

$$\mathcal{W}^{t+1} = \mathcal{W}^{t} + \Delta\mathbf{W} \tag{6.15}$$

where $\mathcal{W}^t$ is the current weight matrix after t-th learning and the delta matrix

$$\Delta\mathbf{W} = [\Delta w_{x,y}] \tag{6.16}$$

is obtained by the *delta rule:*[14]

$$\Delta w_{x,y} = \alpha\,(q_y^* - q_y) \times p_x \tag{6.17}$$

where α is a small gain constant.

This model gives an idea of representing the semantic relation in a fuzzy rule to the weightage of the corresponding neural network. The learning will be realized by modifying the semantic relation.

In Model 1, by using a neural network node to construct symbolic of a fuzzy proposition and the value passing the node to present the confidence of the proposition, the symbolical-numerical duality is considered. As a simplest version, the learning problem is not discussed.

In Model 2, by applying the learning ability of neural networks, case-based reasoning can be implemented. As the symbolical-numerical duality was not considered, the logical properties of inference are difficult to discuss.

In Model 3, by representing the information related to membership values as a corresponding weightage of a neural network, approximate reasoning can be executed and its logical properties can be discussed. As the model does not pay enough attention to the rule-case duality, the learning problem is not discussed.

Model 4 tries to consider both symbolical-numerical duality and rule-case duality and proposes a basic idea, but cannot yet discuss real implementation. Our discussion in sections 7 and 8 may be looked as a continuance of this work.

7. BASIC MODEL OF NEURAL NETWORK FOR ACBR

In this section, we propose a basic model of a neural network for implementing ACBR. It will be extended to more general cases in the next section by several examples. In this model and its extensions, we try to treat fuzzy propositions and fuzzy rules under both the symbolical-numerical duality and the rule-case duality. We have also absorbed some ideas from other models introduced in section 6.

7.1 Implementation of TMB Rule

In general, we may have a rule $P_1, P_2, \ldots, P_n \rightarrow Q$. As a TMB rule, $P_1, P_2, \ldots, P_n$ as well as Q are fuzzy sets defined as:

$$\mu_{Pi}(x_i) = \begin{cases} (x_i - x_{1i}) \div (x_i{*} - x_{1i}) & x_i \in [x_{1i}, x_i{*}] \\ 1 - (x_i - x_i{*}) \div (x_{ri} - x_i{*}) & x_i \in (x_i{*}, x_{ri}] \end{cases} \qquad 1 \le i \le n \tag{7.1}$$

$$\mu_Q(y) = \begin{cases} (y - y_1) \div (y{*} - y_1) & y \in [y_1, y{*}] \\ 1 - (y - y{*}) \div (y_r - y{*}) & y \in (y{*}, y_r] \end{cases} \tag{7.2}$$

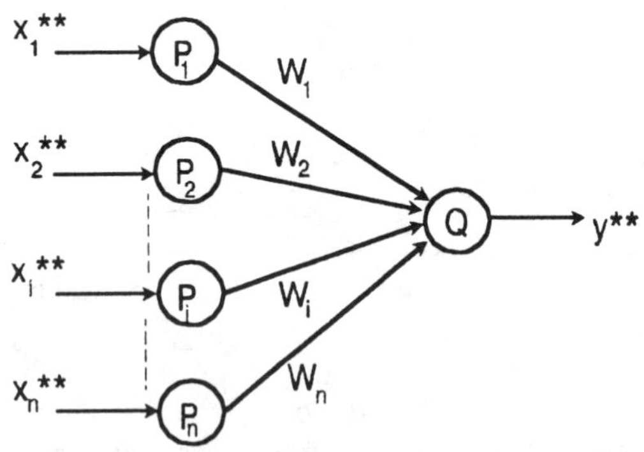

Fig. 7.1 Unit for a TMB rule in ACBR

In this case, we can build a corresponding neural network unit illustrated in Fig. 7.1, where each $\mathbf{W_i}$ (i=1, 2, ..., n) is defined as:

$$W_i = \frac{\partial y}{\partial x_i} = \left(\frac{\partial y}{\partial SR_{P_i,Q}} \right) \div \left(\frac{\partial x_i}{\partial SR_{P_i,Q}} \right) \tag{7.3}$$

$$=\begin{cases}(y^*-y_1)\div(x_i^*-x_{1i}) & x_i\in[x_{1i},x_i^*],y\in[y_1,y^*]\\(y^*-y_r)\div(x_{ri}-x_i^*) & x_i\in(x_i^*,x_{ri}],y\in(y^*,y_r]\end{cases}$$

When $P'_1, P'_2, \ldots, P'_n$ and

$$\mu_{P'_i}(x_i)=\begin{cases}(x_i-x_{1i})\div(x_i^{**}-x_{1i}) & x_i\in[x_{1i},x_i^{**}]\\1-(x_i-x_i^{**})\div(x_{ri}-x_i^{**}) & x_i\in(x_i^{**},x_{ri}]\end{cases}\quad 1\le i\le n \tag{7.4}$$

are given, the y** is calculated by:

$$y^{**}=\left(\prod_{i=1}^{n}f(x_i^{**},W_i)\right)^{\frac{1}{n}} \tag{7.5}$$

where f is the transfer function given in (5.11). The approximate conclusion $\mu_{Q'}(y)$ is deduced by formula:

$$\mu_{Q'}(y)=\begin{cases}(y-y_1)\div(y^{**}-y_1) & y\in[y_1,y^{**}]\\1-(y-y^{**})\div(y_r-y^{**}) & y\in(y^{**},y_r]\end{cases} \tag{7.6}$$

Example 7.1: Suppose a TMB rule $P_1(x_1), P_2(x_2), P_3(x_3) \to Q(y)$ be given, where $x_1 \in X_1$, $x_2 \in X_2$, $x_3 \in X_3$, $y \in Y$, and $X_1 = X_2 = X_3 = [0,1]$, $Y = [0,1]$. The representative points are $x_1^* = 0.8$, $x_2^* = 0.4$, $x_3^* = 0.6$ as well as $y^* = 0.6$. By (7.3) we can get $W_1 = (3/4, 2)$, $W_2 = (3/2, 2/3)$, $W_3 = (1,1)$.

(1) When $P_1'(x_1')$, $P_2'(x_2')$, $P_3'(x_3')$ have been input, and the representative points of P_1', P_2', P_3' are $x_1^{**} = 0.4$, $x_2^{**} = 0.2$, $x_3^{**} = 0.3$, respectively,

$$y^{**} = (0.3 \times 0.3 \times 0.3)^{1/3}=0.3$$

is calculated by (7.5). Therefore,

$$\mu_{Q'}(y)=\begin{cases}y\div 0.3 & y\in[0,0.3]\\1-(y-0.3)\div(1-0.3) & y\in(0.3,1]\end{cases} \tag{7.7}$$

is approximately deduced by (7.6).

7.2 Learning

Learning in ACBR is realized by the modification of the ACBR interrelation. The t+1 - th modification is defined as:

$$\mathbf{IR}^{t+1} = \mathcal{M}(\mathbf{IR}^{t}, IR_{t+1}) \qquad t \ge 1 \tag{7.8}$$
$$\mathbf{IR}^{1} = IR_1$$

where IR_i is the ACBR interrelation of i-th version of the rule $P_i \rightarrow Q_i$, $\mathbf{IR}^i$ is the *approximate interrelation after i-th learning*. $\mathcal{M}$ is a *modifying procedure*. $(x^{*(t+1)}, y^{*(t+1)})$, the core of $\mathbf{IR}^{t+1}$ is calculated by:

$$x^{*(t+1)} = \boldsymbol{m}(x^{*(t)}, x_{t+1}{}^{*}) \tag{7.9}$$
$$y^{*(t+1)} = \boldsymbol{m}(y^{*(t)}, y_{t+1}{}^{*}) \tag{7.10}$$

The *modification function* $\boldsymbol{m}$ is defined as:

$$\boldsymbol{m}(z_1,z_2) = (1 - \alpha) \times z_1 + \alpha \times z_2 \tag{7.11}$$

where α is a small gain constant.

Fig. 7.2 illustrates an example of the modification of ACBR interrelation based on (7.9)~(7.11).

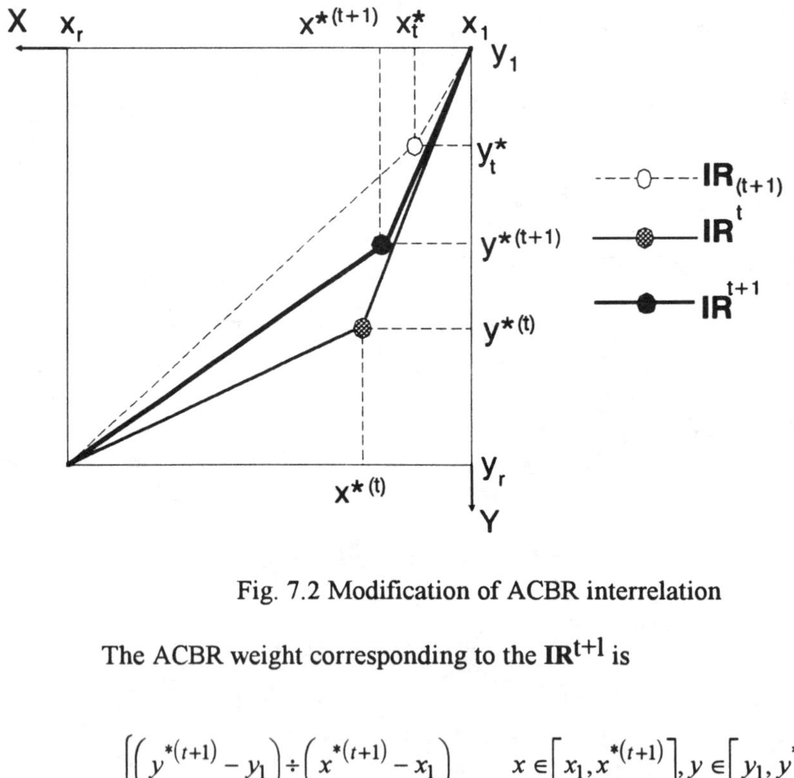

Fig. 7.2 Modification of ACBR interrelation

The ACBR weight corresponding to the $\mathbf{IR}^{t+1}$ is

$$W^{t+1} = \begin{cases} \left(y^{*(t+1)} - y_1\right) \div \left(x^{*(t+1)} - x_1\right) & x \in \left[x_1, x^{*(t+1)}\right], y \in \left[y_1, y^{*(t+1)}\right] \\ \left(y^{*(t+1)} - y_r\right) \div \left(x_r - x^{*(t+1)}\right) & x \in \left(x^{*(t+1)}, x_r\right], y \in \left(y^{*(t+1)}, y_r\right] \end{cases} \tag{7.12}$$

When the learning is considered, the representative point of Q_{t+1} is calculated by:

$$y_{t+1}{}^{*} = f(x_{t+1}{}^{*}, W^t) \tag{7.13}$$

and the TMB rule $P_{t+1} \rightarrow Q_{t+1}$ is deduced as:

$$\mu_{P_{t+1}}(x)=\begin{cases}(x-x_1)\div(x_{t+1}*-x_1) & x\in[x_1,x_{t+1}*]\\ 1-(x-x_{t+1}*)\div(x_r-x_{t+1}*) & x\in(x_{t+1}*,x_r]\end{cases} \quad (7.14)$$

$$\mu_{Q_{t+1}}(y)=\begin{cases}(y-y_1)\div(y_{t+1}*-y_1) & y\in[y_1,y_{t+1}*]\\ 1-(y-y_{t+1}*)\div(y_r-y_{t+1}*) & y\in(y_{t+1}*,y_r]\end{cases} \quad (7.15)$$

The ACBR interrelation IR_{t+1} with the semantic core (x_{t+1}*, y_{t+1}*) can be similarly gotten by Definition 5.5.

8. APPLICATIONS

Section 7 discussed the basic model for ACBR inference on TMB rules which are the approximations of each other. In that case, we only need the information of representative points to describe a TMB rule. However, to describe a premise or a consequence of a TMB rule in more general case, we have to keep three points of information: the *left terminal point*, the *representative point* and the *right terminal point*. (Fig. 8.1)

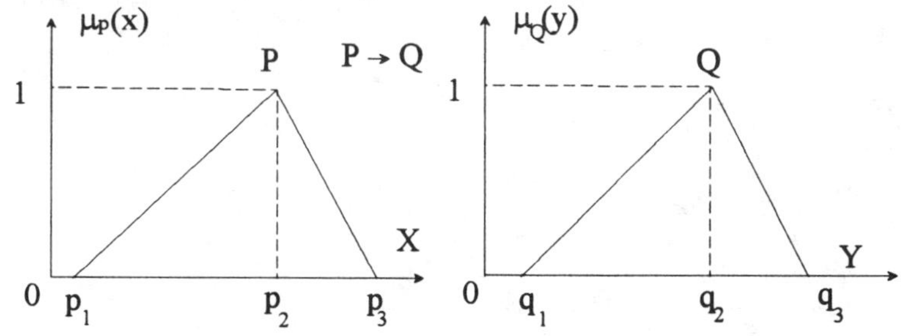

p_1 - left terminal point　　q_1 - left terminal point
p_2 - representative point　　q_2 - representative point
p_3 - right terminal point　　q_3 - right terminal point

Fig. 8.1 A general TMB rule

In this section, we first give Example 8.1 for applying basic model to general TMB rules. Example 8.2 illustrates a three layer neural network unit representing an ACBR inference. It can be also looked as an extension of the basic model introduced in section 7, where the ACBR inference is divided as two stages: the first stage is for retrieving deviation between real input and the premise of a rule, the second stage is for revising the consequence of the rule based on the deviation to get an approximate consequence. By using the idea shown in Example 8.1, it can be used for any TMB rule. For the application based on case knowledge, not only the fuzzy relation in a rule but also the uncertainty of a given case itself should be considered. Example 8.3 is given for this kind of applications, where we add one more stage between deviation retrieving and consequence revision to deal with the uncertainty of case knowledge. The weights of the stage are decided by learning.

Example 8.1 (Extending basic model to general TMB rules):

Suppose a rule $P \to Q$ and a fact P' be given, where $P, P' \subseteq X = [x_1, x_r]$ and $Q \subseteq Y = [y_1, y_r]$ are fuzzy sets defined as:

$$\mu_P(x) = \begin{cases} (x-a_1) \div (a_2 - a_1) & x \in [a_1, a_2] \\ 1-(x-a_2) \div (a_3 - a_2) & x \in (a_2, a_3] \end{cases} \tag{8.1}$$

$$\mu_Q(y) = \begin{cases} (y-b_1) \div (b_2 - b_1) & y \in [b_1, b_2] \\ 1-(y-b_2) \div (b_3 - b_2) & y \in (b_2, b_3] \end{cases} \tag{8.2}$$

$$\mu_{P'}(x) = \begin{cases} (x-c_1) \div (c_2 - c_1) & x \in [c_1, c_2] \\ 1-(x-c_2) \div (c_3 - c_2) & x \in (c_2, c_3] \end{cases} \tag{8.3}$$

$a_1, a_2, a_3, c_1, c_2, c_3, \in X$ and $b_1, b_2, b_3 \in Y$ (Fig. 8.2). To get an approximate consequence Q', we first generate three TMB rules:

$A_1 \to B_1$
$A_2 \to B_2$
$A_3 \to B_3$

and define them as:

$$\mu_{Ak}(x) = \begin{cases} (x-x_1) \div (a_k - x_1) & x \in [x_1, a_k] \\ 1-(x-a_k) \div (x_r - a_k) & x \in (a_k, x_r] \end{cases} \quad k = 1,2,3 \tag{8.4}$$

$$\mu_{Bk}(y) = \begin{cases} (y-y_1) \div (b_k - y_1) & y \in [y_1, b_k] \\ 1-(y-b_k) \div (y_r - b_k) & y \in (b_k, y_r] \end{cases} \tag{8.5}$$

then we generate three fuzzy sets C_1, C_2, C_3 as:

$$\mu_{Ck}(x) = \begin{cases} (x-x_1) \div (c_k - x_1) & x \in [x_1, c_k] \\ 1-(x-c_k) \div (x_r - c_k) & x \in (c_k, x_r] \end{cases} \quad k = 1,2,3 \tag{8.6}$$

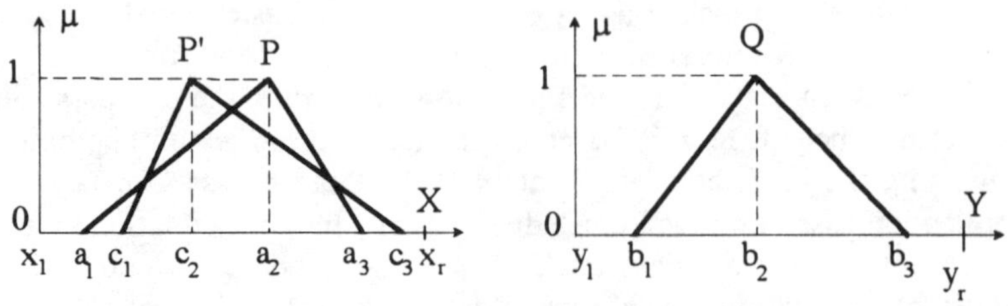

Fig. 8.2 General TMB rule

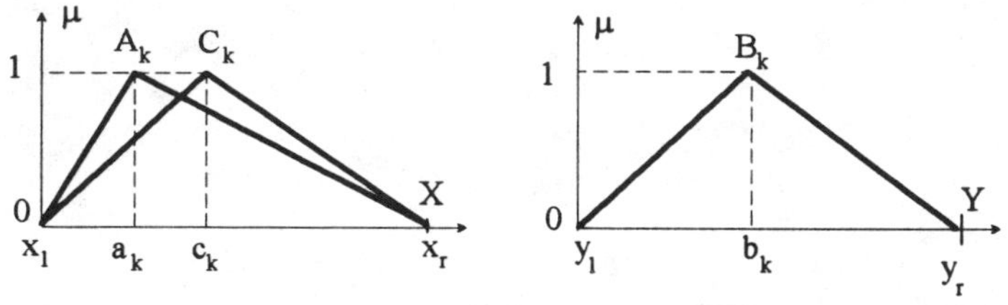

Fig. 8.3 Generated TMB rule and fact

Obviously, for each of rule $A_k \rightarrow B_k$ and fact C_k (k = 1, 2, 3), we can use the ACBR mechanism defined in section 5 to reduce an approximate consequence D_k (Fig. 8.4).

Combining the representative points d_1, d_2, d_3 of D_1, D_2, D_3, the approximate consequence Q' from P → Q and P' can be gotten as:

$$\mu_{Q'}(y) = \begin{cases} (y-d_1) \div (d_2 - d_1) & y \in [d_1, d_2] \\ 1-(y-d_2) \div (d_3 - d_2) & y \in (d_2, d_3] \end{cases} \tag{8.7}$$

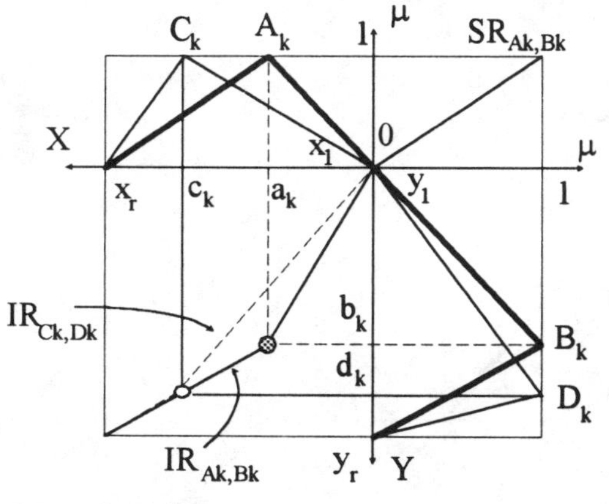

$W = \partial y / \partial x$

Fig. 8.4 Inference on generated rule and fact

Example 8.2 (Dividing ACBR inference into two stages):

Suppose a TMB rule $P_1, P_2, \ldots, P_n \rightarrow Q$ be given, where $P_i \subseteq X = [x_1, x_r]$ and $Q \subseteq Y = [y_1, y_r]$. We represent P_i and Q as vectors:

$$\mathbf{P_i} = [p_{i1}, p_{i2}, p_{i3}] \qquad i=1,2,\ldots,n \tag{8.8}$$

$$\mathbf{Q} = [q_1, q_2, q_3] \tag{8.9}$$

where p_{i1}, p_{i2}, p_{i3} are the left terminal point, representative point and the right terminal point respectively of P_i, and q_1, q_2, q_3 are the left terminal point, representative point and the right terminal point of Q_i, respectively. The basic unit is illustrated in Fig. 8.4. The weights of the unit may be generated by formulas:

$$w_{ik} = -p_{ik} \div |X| \quad i = 1,2,\ldots, n \qquad k=1,2,3 \tag{8.10}$$

$$v_k = q_k \div |Y| \qquad k=1,2,3 \tag{8.11}$$

The output of the node h_k in hidden layer is:

$$h_k = \frac{1}{n}\sum_i \left(p'_{ik} \div |X| + w_{ik}\right) \quad (8.12)$$

and the output of q_k is:

$$q'_k = (h_k + v_k) \times |Y| \quad (8.13)$$

The vector of the approximate consequence is:

$$Q' = [q'_1, q'_2, q'_3] \quad (8.14)$$

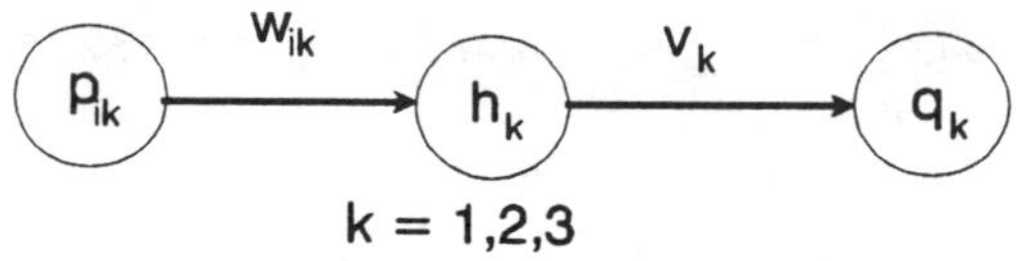

Fig. 8.5 Two stages of ACBR inference

When a rule $P_1, P_2 \rightarrow Q$ and facts P'_1, P'_2 are given(Fig. 8.6), by the above formulas we can build a neural network as Fig. 8.7 and get the inference result as:

$h_1 = (p_{11} \div 10 + w_{11} + p_{21} \div 10 + w_{21}) \div 2 = (0.4 + 0.3) \div 2 = 0.35$

$h_2 = (p_{12} \div 10 + w_{12} + p_{22} \div 10 + w_{22}) \div 2 = (0.4 + 0.5) \div 2 = 0.45$

$h_3 = (p_{13} \div 10 + w_{13} + p_{23} \div 10 + w_{23}) \div 2 = (0.2 + 0.5) \div 2 = 0.35$

$q'_1 = (h_1 + v_1) \times 20 = (0.35 + 0.1) \times 20 = 9$

$q'_2 = (h_2 + v_2) \times 20 = (0.45 + 0.3) \times 20 = 15$

$q'_3 = (h_3 + v_3) \times 20 = (0.35 + 0.55) \times 20 = 18$

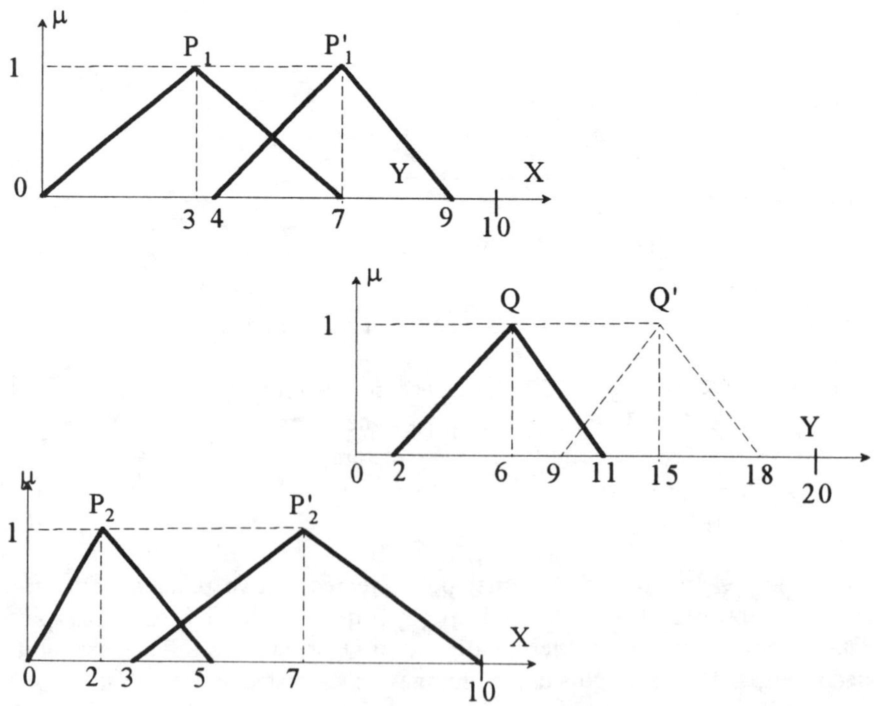

Fig. 8.6 Rule, facts and the inference result of Example 8.2

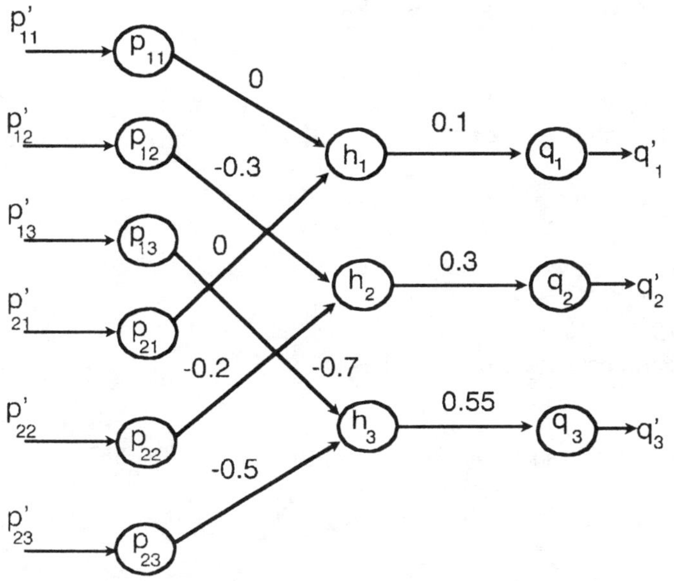

Fig. 8.7 Neural network unit of Example 8.2

Example 8.3(ACBR based on uncertain training cases):

The basic neural network unit is illustrated as Fig. 8.8, where w_{ik} and v_{jk} are decided by:

$$w_{ik} = -p_{ik} \div |X| \quad i = 1,2 ,..., n \quad k=1,2,3 \tag{8.15}$$

$$v_{jk} = q_{jk} \div |Y| \quad j = 1,2 ,..., m \quad k=1,2,3 \tag{8.16}$$

and c_{ijk} is gotten by the learning based on clustering algorithm,[4,5] it describes some kind of probability distribution of importance of each P_i to Q_j. The index number k is related to the left terminal point (when k=1), the representative point (when k=2) or the right terminal point (when k=3) of corresponding P_i or Q_j, respectively. The inference is defined by:

$$m_{ik} = p'_{ik} + |X| + w_{ik} \tag{8.17}$$

$$h_{jk} = \frac{1}{\xi} \sum_{i} m_{ik} \times c_{ijk} \tag{8.18}$$

$$q'_{jk} = (h_{jk} + v_{jk}) \times |Y| \tag{8.19}$$

where ξ is the number of c_{ijk}'s that are non-zero.

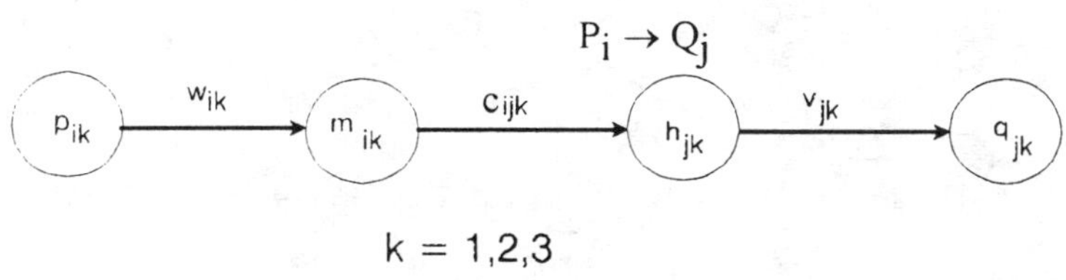

Fig. 8.8 Three stages inference of ACBR

Suppose a set of TMB rules be given as follows:

$P_1, P_2 \rightarrow Q_1$

$P_2, P_3 \rightarrow Q_2$

$P_2, P_3 \rightarrow Q_1$

where $P_i \subseteq X = [0,10]$ and $Q_j \subseteq Y = [0,10]$ are all linguistic values as shown in Fig. 8.9 and

$P_1 = P2 = P_3 = \textit{sure}$

$Q_1 = Q_2 = \textit{sure}$

Representing them as vectors, we have

$\mathbf{P}_1 = \mathbf{P}_2 = \mathbf{P}_3 = [5, 7.5, 10]$

$\mathbf{Q}_1 = \mathbf{Q}_2 = [5, 7.5, 10].$

The corresponding neural network is illustrated as in Fig. 8.10.

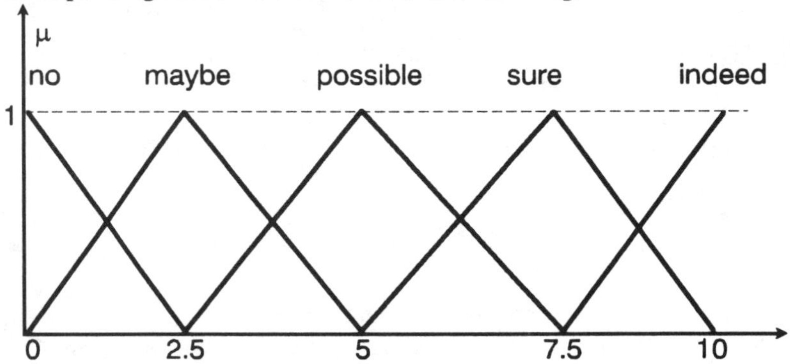

Fig. 8.9 Linguistic values of Example 8.3

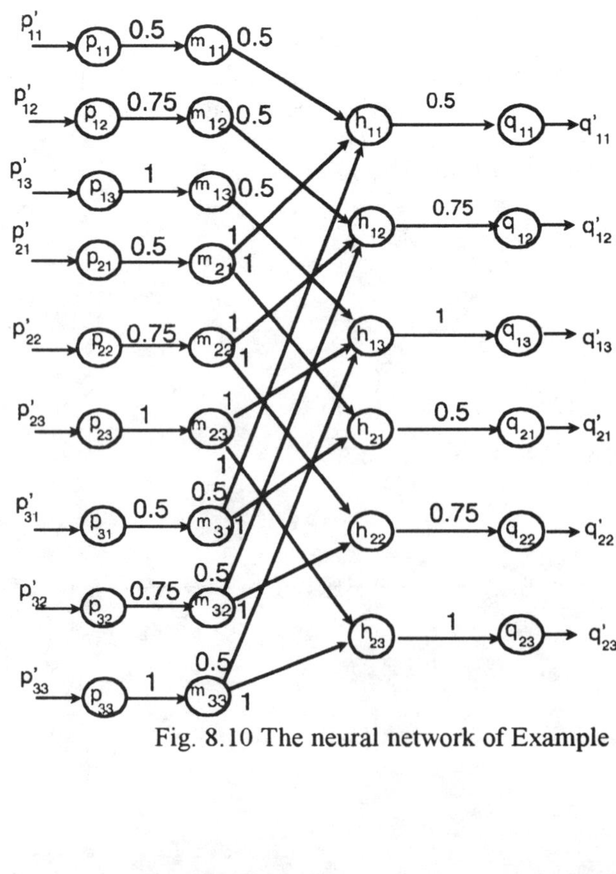

Fig. 8.10 The neural network of Example 8.3

When P'_1 = *maybe*, P'_2 = *sure* and P'_3 = *possible* are give, we have vectors as:

$\mathbf{P_1}' = [0, 2.5, 5]$
$\mathbf{P_2}' = [5, 7.5, 10]$
$\mathbf{P_3}' = [2.5, 5, 7.5]$

By formulas (8.15) - (8.17), we can get the results as:

$\mathbf{Q_1}' = [3.33, 5.83, 8.33]$
$\mathbf{Q_2}' = [3.75, 6.25, 8.75]$

The $\mathbf{Q_1}'$ and $\mathbf{Q_2}'$ can be approximately said as *very possible* and *almost sure*, respectively.

9. CONCLUSION

The main contributions in this work are: (1) distinguishing the different truth types in fuzzy valued logic and fuzzy linguistic valued logic; (2) pointing out the symbolical-numerical duality of fuzzy logic and the rule-case duality of fuzzy rules; (3) proposing approximate case-based reasoning for dealing with the two kinds of duality; (4) designing the approximate case-based reasoning mechanism based on TMB rules; and (5) implementing the approximate case-based reasoning on neural networks.

REFERENCES

1. Z. Shen, H. C. Lui & L. Ding, "A Proposal for Approximate Case-based Reasoning in Neural Networks," *Proceedings of the 2nd International Conference on Fuzzy Logic and Neural Networks*, pp. 221-224(1992).
2. L. Ding, Z. Shen & H. C. Lui, "Weight of Fuzzy Rule in Approximate Case-based Reasoning," *Proceedings of International Symposia on Information Science*, pp. 75-78(1992).
3. Z. Shen, L. Ding, H.C. Lui, P. Z. Wang & M. Mukaidono, "Revision Principle for Approximate Reasoning--Based on Semantic Revising Method," *IEEE Proceedings of ISMVL - the 22nd International Symposium on Multiple-valued Logic*, pp. 467-473(1992).
4. H. C. Lui, A. H. Tan, J. H. Lim & H. H. Teh, "Practical Application of A Connectionist Expert System - The INSIDE Story," *Operational Expert System Applications in the Far East*, Edit by J. K. Lee, et al., PERGAMON PRESS, pp. 162-175(1991).
5. J. H. Lim, H. C. Lui, A. H. Tan & H. H. Teh, "INSIDE: A Connectionist Case-based Diagnostic Expert System That Learns Incrementally," *Proceedings of International Joint Conference on Neural Networks,* pp. 1693-1698(1991).
6. L. Ding, Z. Shen & M. Mukaidono, "A New Method for Approximate Reasoning," *IEEE Proceedings of ISMVL - the 19th International Symposium on Multiple-valued Logic,* pp. 179- 185(1989).
7. Y. Tsukamoto, "An Approach to Fuzzy Reasoning Method," *Advances in Fuzzy Set Theory and Applications*, Edit by M. M. Gupta, et al., North-Holland, pp. 137-149(1979).
8. L. A. Zadeh, "The Concept of a Linguistic Variable and Its Application to Approximate Reasoning (I); (II); (II)," *Information Science,* 8, pp. 199-249; 8, pp. 301-357; 9, pp. 43-80(1975).

9. L. A. Zadeh, "A Theory of Approximate Reasoning," *Fuzzy Sets and Systems*, 1, pp. 3-28(1977).
10. L. Ding, Z. Shen & M. Mukaidono, "Revision Principle for Approximate Reasoning-Based on Linear Revising Method," *Proceedings of the 2nd International Conference on Fuzzy Logic and Neural Networks*, pp. 305-308(1992).
11. P. Z. Wang, H. M. Zhang, X. T. Peng & W. Xu, "Truth-Valued-Flow Inference," *BUSEFAL* No. 38, pp. 130-139(1989).
12. E. Sanchez, "Fuzzy Connectionist Expert Systems," *Proceedings of the International Conference on Fuzzy Logic & Neural Networks*, pp. 31-35(1990).
13. J. M. Keller, R. R. Yager & H. Tahani, "Neural Network Implementation of Fuzzy Logic," *Fuzzy Sets and Systems*, 45, pp. 1-12(1992).
14. R. Hecht-Nielsen, *Neurocomputing*, Addison-Wesley, 1990.
15. M. Mizumoto, "Comparison of Various Fuzzy Reasoning Methods," *Preprints of the 2nd IFSA Congress*, pp. 2-7(1987).

SYNTHESIS OF FUZZY, ARTIFICIAL INTELLIGENCE AND NEURAL NETWORKS FOR HIERARCHICAL INTELLIGENT CONTROL

Toshio FUKUDA* and Takanori SHIBATA**

* Dept. of Mechano-Informatics and Systems, Nagoya University,
1 Furo-cho, Chikusa-ku, Nagoya, 464-01, JAPAN
Phone +81-52-781-5111 ext. 4478
Fax +81-52-781-9243
E-mail: fukuda@mein.cc.nagoya-u.ac.jp

** Bio-Robotics, Robotics Department
Mechanical Engineering Laboratory
Ministry of International Trade and Industry
1-2 Namiki, Tsukuba 305, Japan
Phone: +81-298-58-7299
Fax: +81-298-58-7201
E-mail: shibata@mel.go.jp

This chapter presents a new structure of intelligent control for robotic motion. This structure is analogous to the human cerebral control structure for intelligent control. Therefore, the system has a hierarchical structure as an integrated approach of Neuromorphic and Symbolic control, including an applied neural network for servo control, a knowledge based approximation, and a fuzzy set theory for a human interface. The neural network in the servo control level is numerical manipulation, while the knowledge based part is symbolic manipulation. In the neuromorphic control, the neural network compensates for the nonlinearities of the system and uncertainty in its environment. The knowledge base part develops control strategies symbolically for the servo level with a-priori knowledge. The fuzzy logic combined with the neural network is used between the servo control level and the knowledge based part to link numerals to symbols and express human skills through learning.

1. INTRODUCTION

Autonomous robots, which perform tasks without human operators, are required in many fields. The autonomous robots have to carry out tasks in various environments by themselves like human beings. They have to be intelligent to determine their own actions in unknown environments by themselves based on sensory information. This chapter introduces a control scheme for autonomous robots, which this chapter refers to as an *intelligent control* scheme.

The autonomous robots should have hierarchical control structures while mimicking human's cerebral control structure in order to achieve some tasks without human operators [1, 2]. Therefore, it is necessary for the autonomous robots to have reasoning mechanisms, sensor system and adaptive servo controllers (Fig. 1). The reasoning mechanism produces control strategies symbolically for complex and composite tasks by using knowledge and data base systems with sensory information at a high level. Moreover, at a servo level, the controllers have to adapt to the environment having uncertainty. While the reasoning mechanism deals with symbols, the servo controller deals with numerals (Fig. 2).

The Artificial Intelligence (AI) techniques have been used to synthesize knowledge-based systems as expert systems. For intelligent control, there were some examples of *symbolic control* which uses symbolic reasoning mechanisms for higher-level control [2 - 4]. However, it is difficult to classify sensed data in order to map numerical data set into symbolic data set for understanding of process state. On the other hand, at the servo level, adaptive control like the Model Reference Adaptive Control (MRAC) and the Self Tuning (ST) Regulator has become available to complicated and composite systems having much uncertainty [5]. However, the adaptive control has drawbacks such as exponentially complicated calculation for many unknown parameters and limitation on applicability to nonlinear systems [1]. In order to solve these problems, many groups have used a neural network for adaptive control [6 - 10]. We have shown applicability of the neural network-based controllers to force control, stabbing control, and hybrid position/force control of robotic manipulators [7, 8]. Then, these research have shown that a neural network in a servo controller can compensate for nonlinearities and uncertainties of robotic system and its environment. The neural network-based control is referred to as *neuromorphic control*. The neuromorphic control is effective for a specific learned task [11]. However, characteristics of the environment changes largely, the neuromorphic controller has to learn again so as to adapt to the environment. Therefore, it is desirable for the neuromorphic controller to store experiences of various tasks in various environments. That is, it is necessary for the neuromorphic control to expand task range so as to be intelligent control.

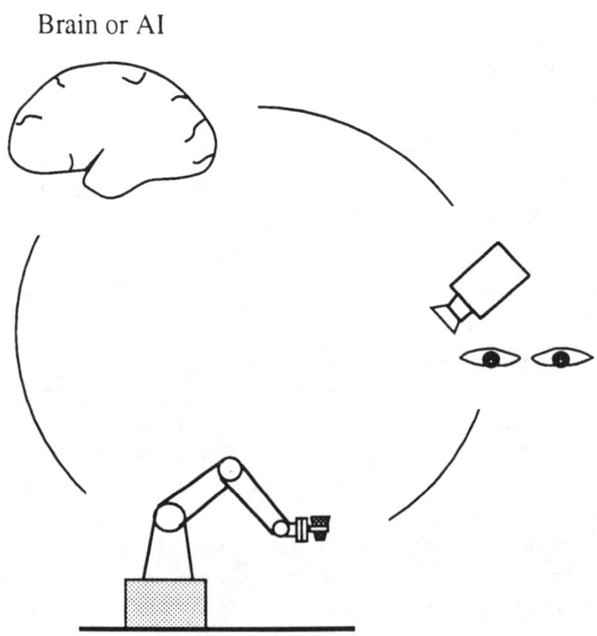

Fig. 1 Concept of intelligent control

For an intelligent autonomous robotic system, we have proposed Hierarchical Intelligent Control, which is a hybrid control of the neuromorphic control and the symbolic control [11, 12] (Fig. 2). The control strategy is developed by symbolic reasoning with knowledge-base system, and a neuromorphic controller is used at the servo level in order to adapt to the environment having uncertainties. The system used neural networks to link numerical real world and symbolical AI world. The system comprised two levels of feedback loops: a learning level of control process by the symbolic control and an adaptation level of dynamic process by the neuromorphic control. At the high level, the system planed a control strategy approximately and symbolically. At the servo level, the neural network compensated for vagueness of the symbolic control strategy and for uncertainty of the environment. However, the system was not always effective with respect to a cost of time for planning or scheduling since the reasoning mechanism took much time.

When there are changes in the characteristic of the environment at the same task, human beings change the way of task by using *skills* without symbolical reasoning. For autonomous robots, it is not always necessary to reason symbolically in the same case so as to save the cost of time. In this case, it is enough only to change the way of task without the planning. That is, it is enough only to tune the control references for the servo controller by using skills when the environment changes. In the past few years, several methods have developed for modeling and acquiring human skills and transferring them to robots and telerobots [13, 14]. For this purpose, Asada showed that the neural network can acquire human skills by using teaching data [15]. However, the neural network can not cluster or classify the input space for the higher level to understand extracted human skills.

This chapter presents an architecture of the Hierarchical Intelligent Control for intelligent robot motion. The system uses the AI technique, the neural network, and the fuzzy logic [16]. This chapter describes how to combine them for intelligent control of robot motion.

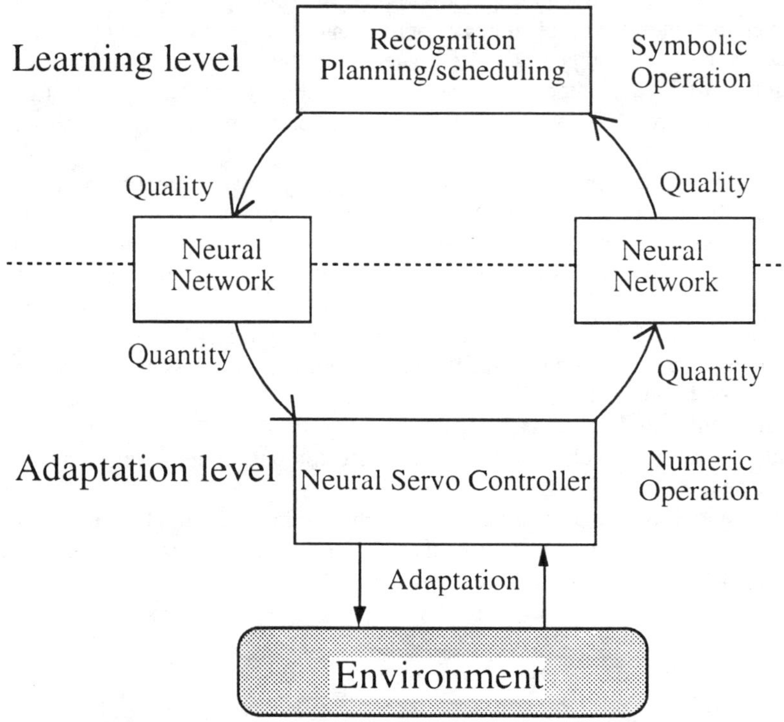

Fig. 2 Concept of Hierarchical Intelligent Control

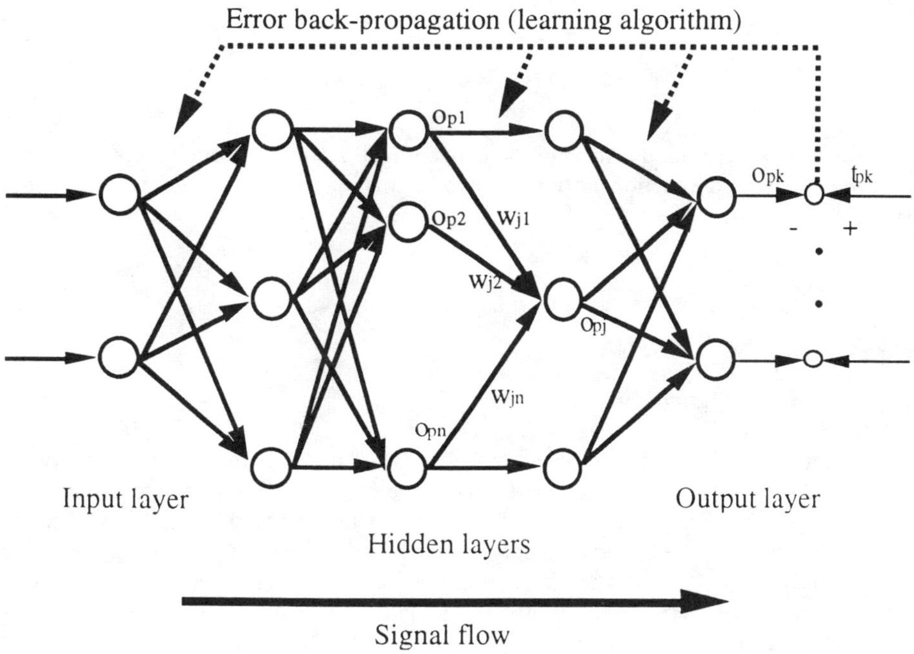

Fig. 3 Feed-forward neural network.
The solid line indicates the physical signal flow,
while the dotted line indicates the back propagation.

2. Neural Network and Fuzzy Logic

2.1. Neuron Model

Each neuron model building up a network simulates a biological neuron. The neuron unit consists of multiple inputs, x, and one output, y, and its internal state is given as the weighted sum of input signals. The output of the neuron unit is as follows:

$$y_{(t)}=f\left(\sum_{i=1}^{n} w_i \cdot x_{i(t)}-\theta\right) \quad , \qquad (1)$$

where w_i is a weight of connection, θ is a bias of the neuron unit, t is time, and n is the number of inputs. The weight coefficient w_i that represents the strength of connection indicates the synapse load. It takes a positive value for excitation and a negative one for inhibition.

The neuron output function f(x) often uses the two-valued function of 1 and 0 using threshold fend-limiter, or the sigmoid function that is a continuous and nonlinear function [25]. A conventional sigmoid function is represented by the following expressions:

$$f_{(x)}=\frac{1}{1+e^{-ax}} \qquad (0 < f_{(x)} <1). \qquad (2)$$

A tanh function is often used instead of the conventional sigmoid function since the output value of the sigmoid function is positive while that of the tanh function being both positive and negative. The tanh function is as follows:

$$f_{(x)}=\frac{1-e^{-ax}}{1+e^{-ax}} \qquad (-1 < f_{(x)} < 1). \qquad (3)$$

2.2. Learning Algorithm

Figure 3 illustrates an example of a feed-forward neural network with only forward connections. For the input/output of neurons in the hidden and output layers, nonlinear function is used. Since a form of the feed-forward net indicated a multi-layering of perceptrons, it is also called a perceptron type. In the case of the original perceptron, limit of its capability was that only the connection weights were learned between the hidden layer and output layer [19]. In the 1980s, Rumelhart et al. proposed a learning algorithm with the teacher of the feed-forward net to solve this problem by multi-layering using a back-propagation algorithm [20]. In the 1970s, Werbos originally developed a back-propagation algorithm [21], and its first practical application was for estimating a dynamic model to predict nationalism and social communications. However, Werbos' work remained almost unknown in the scientific community. The back-propagation is a generalization of the Least Mean Squares algorithm. In this algorithm, an error function is defined and equal to the mean square difference between the desired output and the actual output of the feed-forward net. In order to minimize this error function, the back-propagation algorithm uses a gradient search technique. It is called the generalized delta rule [20].

2.3. Standard Delta Rule

The learning procedure involves the presentation of a set of pairs of input and output patterns. The system first uses the input vector to produce its own output vector and then compares this with the desired output, or target vector. If there is no difference, no learning take a place. Otherwise, the weights are changed to reduce the difference. In this case, with no hidden units, this generates the standard delta rule. The rule for changing weights following presentation of an input/output pair p is given by

$$\Delta_p w_{ji} = \eta(t_{pj}-o_{pj})i_{pi} = \eta\delta_{pj}i_{pi} \qquad (4)$$

where t_{pj} is the target input for jth component of the output pattern for pattern p, o_{pj} is the jth element of the actual output pattern produced by the presentation of input pattern p, i_{pi} is the value of the ith element of the input pattern, $\delta_{pj} = t_{pj} - o_{pj}$, and $\Delta_p w_{ji}$ is the change to be made to the weight from the ith to the jth unit following presentation of pattern p. There are many ways of derivation the delta rule. The standard delta rule essentially implements gradient descent in sum-squared error for linear functions. In this case, without hidden units, the error surface is shaped like a bowl with only one minimum, so gradient descent is guaranteed to find out the best set of weights. With hidden units, however, it is not so obvious how to compute the derivatives, and the error surface is not concave upwards, so there is the danger of getting stuck in local minima. The next section describes an efficient way of computing the derivatives.

2.4. Generalized Delta Rule

Here, we consider the layered feed-forward networks. In these networks, the input units are the bottom layer and the output units are the top layer. There can be many layers of hidden units in between, but every unit must send its output to higher layers than its own and must receive its input from lower layers than its own. Given an input vector, the output vector is computed by a forward pass which computes the activity levels in the earlier layers.

The back-propagation algorithm performs steepest descent on a surface in a weight space whose height at any point in weight space is equal to the error measure. In order to show the algorithm, let

$$E_p = \frac{1}{2}\sum_j (t_{pj}-o_{pj})^2 \tag{5}$$

be the measure of the error on input/output pattern p and let $E = \sum E_p$ be the overall measure of the error.

Here, we define the weighted sum of the output of the previous layer,

$$s_{pj} = \sum_i w_{ji} o_{pi} \tag{6}$$

as the state of the unit. The output,

$$o_{pj} = f_j(s_{pj}) \tag{7}$$

uses the sigmoid function which is non-decreasing and differentiable function described before. To get the correct generalization of the delta rule, w_{ji} is set as

$$\Delta_p w_{ji} \propto -\frac{\partial E_p}{\partial w_{ji}}. \tag{8}$$

It is useful to see this derivative as resulting from the product of two parts: one part reflecting the change in error as a function of the change in the net input to the unit and one part representing the effect of changing a particular weight on the net input. Thus we can write

$$\frac{\partial E_p}{\partial w_{ji}} = \frac{\partial E_p}{\partial s_{pj}} \frac{\partial s_{pj}}{\partial w_{ji}} \tag{9}$$

By equation (6) we see that the second factor is

$$\frac{\partial s_{pj}}{\partial w_{ji}} = \frac{\partial}{\partial w_{ji}} \sum_k w_{jk} o_{pk} = o_{pi} \quad . \tag{10}$$

Now let us define

$$\delta_{pj} = -\frac{\partial E_p}{\partial s_{pj}} \quad . \tag{11}$$

Equation (9) thus has the equivalent form

$$-\frac{\partial E_p}{\partial w_{ji}} = \delta_{pj} o_{pi} \ . \quad . \tag{12}$$

This says that to implement gradient descent in E, we should make the weight changes according to

$$\Delta_p w_{ji} = \eta \delta_{pj} o_{pi} \tag{13}$$

just as in the standard delta rule. The interesting result is that there is a simple recursive computation of these δ's which can be implemented by propagating an error signal backward through the network.

To compute (11), the chain rule is applied to write this partial derivative as the product of two factors, one factor reflecting the change in error as a function of the output of the unit, and one reflecting the change in the output as a function of changes in the input,

$$\delta_{pj} = -\frac{\partial E_p}{\partial s_{pj}} = -\frac{\partial E_p}{\partial o_{pj}} \frac{\partial o_{pj}}{\partial s_{pj}} \quad . \tag{14}$$

By equation (7) we see that

$$\frac{\partial o_{pj}}{\partial s_{pj}} = f'_j(s_{pj}) \quad , \tag{15}$$

which is simply the derivative of the compressing function f_j for the jth unit, evaluated at the net input s_{pj} to that unit. To compute the first factor, there are two cases. First, assume that unit u_j is an output unit of the network. In this case, it follows from the definition of E_p that

$$\frac{\partial E_p}{\partial o_{pj}} = -(t_{pj} - o_{pj}) \quad . \tag{16}$$

Substituting for the two factors in Equation (14), we can get

$$\delta_{pj} = (t_{pj} - o_{pj}) f'_j(s_{pj}) \tag{17}$$

for any output unit u_j. If u_j is not an output unit, the chain rule is used to write

$$\sum_k \frac{\partial E_p}{\partial s_{pk}} \frac{\partial s_{pk}}{\partial o_{pj}} = \sum_k \frac{\partial E_p}{\partial s_{pk}} \frac{\partial}{\partial o_{pj}} \sum_i w_{ki} o_{pi}$$
$$= \sum_k \frac{\partial E_p}{\partial s_{pk}} w_{kj} = - \sum_k \delta_{pk} w_{kj} \qquad (18)$$

In this case, substituting for the two factors in Equation (14) yields

$$\delta_{pj} = f'_j(s_{pj}) \sum_k \delta_{pk} w_{kj} \quad , \qquad (19)$$

whenever u_j is not an output unit. Equations (17) and (19) give a recursive procedure for computing the δ's for all units in the network, which are then used to compute the weight changes in the network according to Equation (13). This procedure constitutes the generalized delta rule for a feed-forward network.

2.5. Back-propagation Algorithm

This section describes the back-propagation algorithm on training. The Back propagation training algorithm is an iterative gradient algorithm designed to minimize the mean square error between the actual output of a feed-forward net and the desired output. It requires continuous differentiable non-linearities. Figure 4 shows a flow chart of the back-propagation training algorithm. The following assumes that a sigmoid logistic non-linearity is used .

Step 1. Initialize Weights and Offsets

Set all weights and units offsets to small random values.

Step 2. Present Input and Desired Output

Present a continuous valued input vector and specify the desired outputs. If the net is used as a classifier then all desired outputs are typically set to zero except for the output corresponding to the class the input is from. That desired output is 1. The input could be new on each trial or samples from a training set could be presented cyclically until weights stabilize.

Step 3. Calculate Actual Outputs

Use the sigmoid nonlinearity and formulas as in Fig. 3 to calculate outputs.

Step 4. Adapt Weights

Use a recursive algorithm starting at the output units and working back to the first hidden layer. Adjust weights by

$$w_{ji}(t+1) = w_{ji}(t) + \Delta w_{ji}(t) \quad , \qquad (20)$$

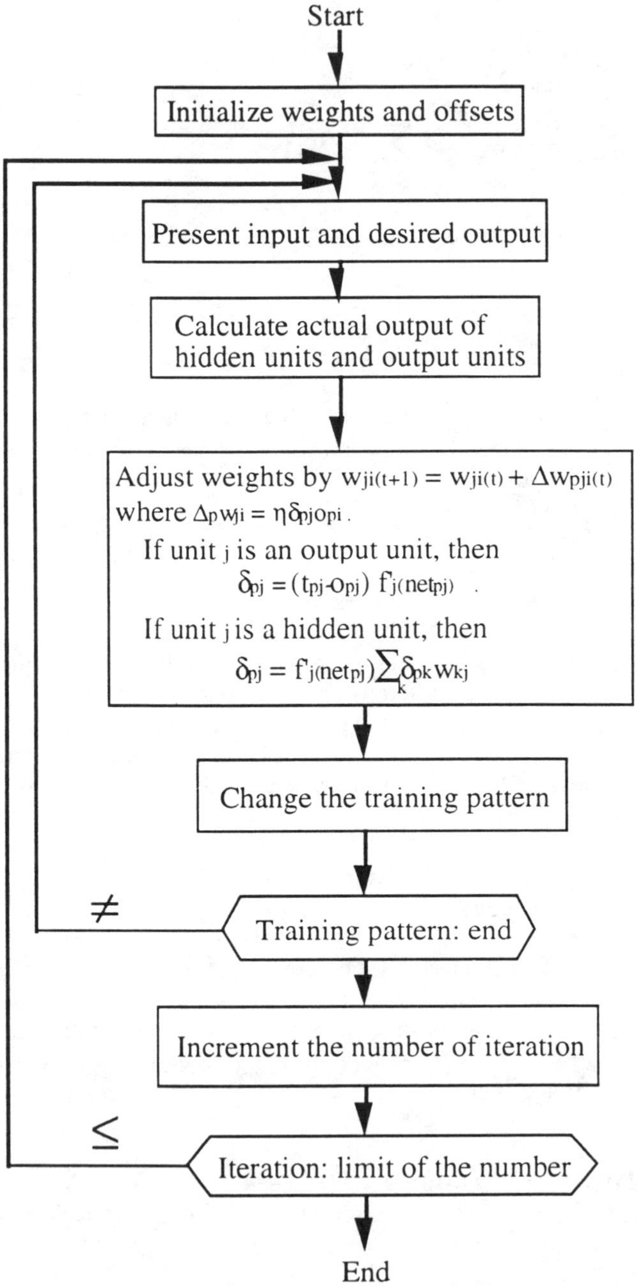

Fig. 4 Flow chart of the back-propagation training algorithm

where $\Delta_p w_{ji} = \eta\delta_{pj}o_{pi}$ as shown in Equation (4).
In this equation $w_{ji(t)}$ is the weight, η is a gain term which we call a learning rate, and δ_{pj} is an error term for unit j. If the unit j is an output unit, then δ_{pj} can be computed by Equation (17),

$$\delta_{pj} = (t_{pj}-o_{pj})f'_j(s_{pj})$$

If the unit j is an internal hidden unit, then δ_{pj} can be computed by Equation (19),

$$\delta_{pj} = f'_j(s_{pj})\sum_k \delta_{pk}w_{kj}$$

Internal unit thresholds are adapted in a similar manner by assuming they are connection weights on links from auxiliary constant-valued inputs. Convergence is sometimes faster if a momentum term is added and weight changes are smoothed.

These expressions are used for a single teacher signal. In fact, a pattern of several input signals and teacher signals is given for iterative learning. If the error becomes small and learning succeeds, correct output can be obtained for the learned pattern. In addition, if the unlearned pattern is close to the learned pattern, the output value close to the correct output can be obtained by the generalization capability. However, successful learning is not always guaranteed and convergence may take time, or learning may be trapped as a local minimum on the way to the most rapid drop. In order to improve convergence and find the global minimum, some research has been done (A. R. Jacobs, 1988 [22]), (N. Baba, 1989 [23]), and (T. Fukuda, T. Shibata et al., 1990 [8]). Problems include the coding of variables and how to determine the number of neurons (particularly for the hidden layer) in the network.

Nevertheless, it is very meaningful that the hidden layer was introduced and multi-layered to allow problems such as EX-OR to be solved. This cannot be implemented by perceptrons.

2.6. Fuzzy Sets

The fuzzy set is characterized as extension of binary crisp set [16, 24]. Crisp sets allow only full membership or no membership at all, whereas fuzzy sets allow partial membership. In other words, an element may partially belong to a set. In a crisp set, the membership or nonmembership of an element x in set A is described by a characteristic function $\mu_A(x)$, where

$$\mu_A(x) = \begin{cases} 1 & \text{if } x \in A \\ 0 & \text{if } x \notin A. \end{cases} \quad (21)$$

Fuzzy set theory extends this concept by defining partial memberships, which can take values ranging from 0 to 1:

$$\mu_A : X \rightarrow [0, 1], \quad (22)$$

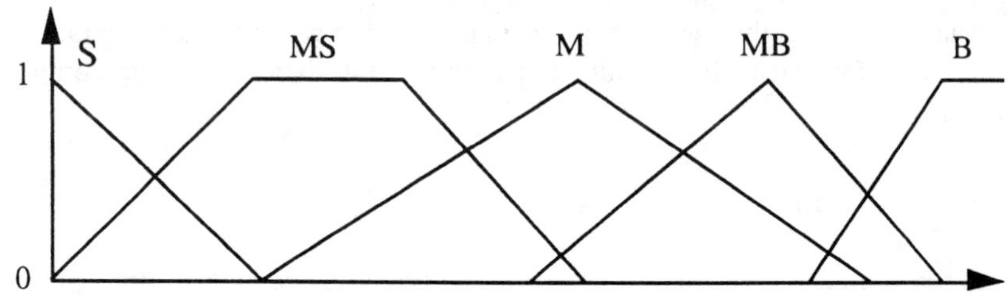

Fig. 5 Membership functions for fuzzy inference

where X refers to the universal set defined in a specific problem. Assuming that A and B are two fuzzy sets with membership functions of μ_A and μ_B, then the following operations can be defined on these sets. The complement of a fuzzy set A is a fuzzy set $\overline{A}$ with a membership function

$$\mu_{\overline{A}} = 1 - \mu_A(x) \tag{23}$$

The union of A and B is a fuzzy det with the following membership function:

$$\mu_{A \cup B} = \max\{ \mu_A, \mu_B \}. \tag{24}$$

The intersection of A and B is a fuzzy set:

$$\mu_{A \cap B} = \min\{ \mu_A, \mu_B \}. \tag{25}$$

One identifies the main parameters for evaluation and determines a term set which is the right level of granularity for describing the values of each linguistic variables. For example, a term set including linguistic values such as {Small, Medium, Large} may not be satisfactory in certain domains, and may instead require the use of a five-term such as {Small, Medium Small, Medium, Medium Large, Large}.

2.7. Inference Rules

Because of the partial matching attribute of fuzzy rules and the fact that the preconditions of rules do overlap, more than one fuzzy rule can fire at a time. The methodology which is used in determining which value should be taken as the result of firing of several rules can be referred to as conflict resolution. The following example, using *n* rules, illustrates this process. Assume that we have the following rules:

Rule 1: IF X is A_1 and Y is B_1 THEN Z is C_1
Rule 2: IF X is A_2 and Y is B_2 THEN Z is C_2

:   :

Rule n: IF X is A_n and Y is B_n THEN Z is C_n. (26)

Each rule has an antecedent, or *if*, part containing several preconditions, and a consequent, or *then*, part which prescribes the value. Now, if we have x_0 and y_0 as the inputs for fuzzy values X and Y, then their truth values are represented by $\mu_{A1}(x_0)$ and $\mu_{B1}(y_0)$ respectively for rule 1, where μ_{A1} and μ_{B1} represent the membership function for A_1 and B_1, respectively. Similarly for rule n, we have $\mu_{An}(x_0)$ and $\mu_{Bn}(y_0)$ as the truth values of the preconditions:

$$h_1 = \mu_{A1}(x_0) \bullet \mu_{B1}(y_0). \tag{27}$$

Similarly for rule n,

$$h_n = \mu_{An}(x_0) \bullet \mu_{Bn}(y_0), \tag{28}$$

Traditionally, fuzzy logic uses a minimum operator. In the simplified fuzzy logic, however, we use multiplier instead of the minimum operator. Defuzzification procedure to compute $\mu^{-1}(h)$ is also simple. The above equations show that, as a result of input values x_0 and y_0, rule 1 is recommending a evaluation value C_1 and rule 2 is recommending a evaluation value C_2. The combination of the above rules produces a nonfuzzy evaluation value z*, which is calculated using a weighted averaging approach:

$$z^* = \frac{\sum_{i=1}^{n} h_i C_i}{\sum_{i=1}^{n} h_i} \tag{29}$$

where n is the number of rules, and C_i is the amount of evaluation value recommended by rule i.

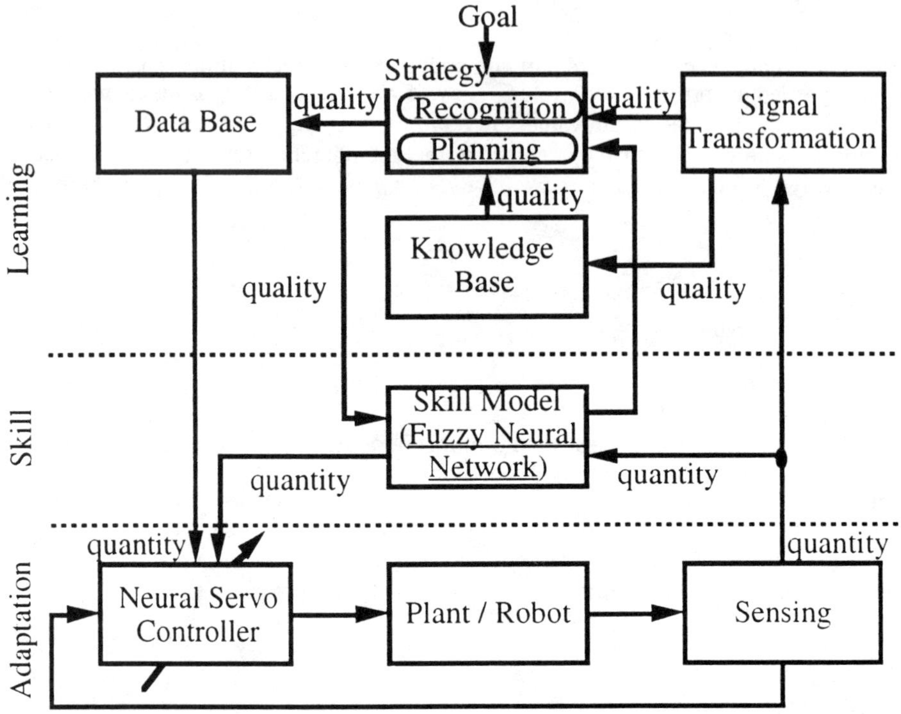

Fig. 6 Hierarchical Intelligent Control with skill based control by using Fuzzy Neural Network

3. Concept of Hierarchical Intelligent Control

The Hierarchical Intelligent Control system is a hybrid system of the neural network, the fuzzy logic, and the AI. The system comprises three levels: a *learning* level, a *skill* level, and an *adaptation* level (Fig. 6). Therefore, there are three feed-back loops. The learning level is based on the expert system for a reasoning mechanism and has a hierarchical structure: recognition and planning to develop a control strategy. The recognition level uses neural networks and fuzzy logic combined with the neural network as nodes of decision tree [12, 17, 18]. In the case of the neural network, input is numeric quantity sensed by some sensors and output is symbolic quality which indicates a process state (Fig. 7). In the case of the fuzzy neural network, inputs and outputs are numeric quantities and the fuzzy neural network clusters input signals by using membership functions. That is, the fuzzy neural network can transform numerical quantity into symbolic quality by using membership functions. Both the neural network and the fuzzy neural network are trained with the training data of a-priori knowledge obtained from human experts. As the result, the neural network and the fuzzy neural network can transform various sensed data from numerical quantities to symbolic qualities, and perform *sensor fusion* and production of *meta-knowledge* at the learning level. The important information is sensed actively on using the knowledge base. The sensors of vision, weight, force, touch, acoustic, and others can be used as nodes of decision tree for recognition of the environment. Then, the planning level reasons symbolically for a

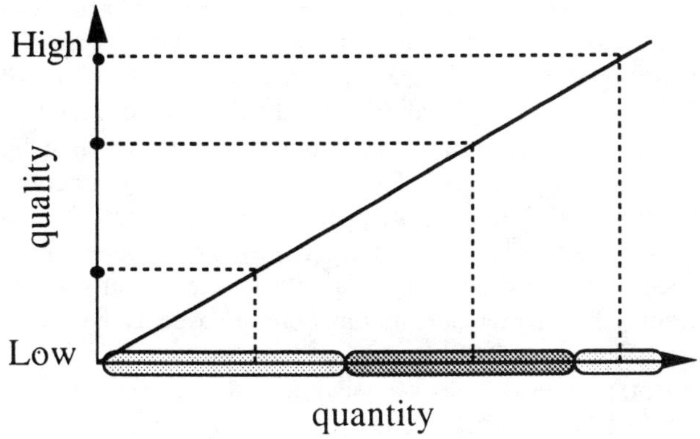

(a) 'If .. , then ..' rules: straight line function

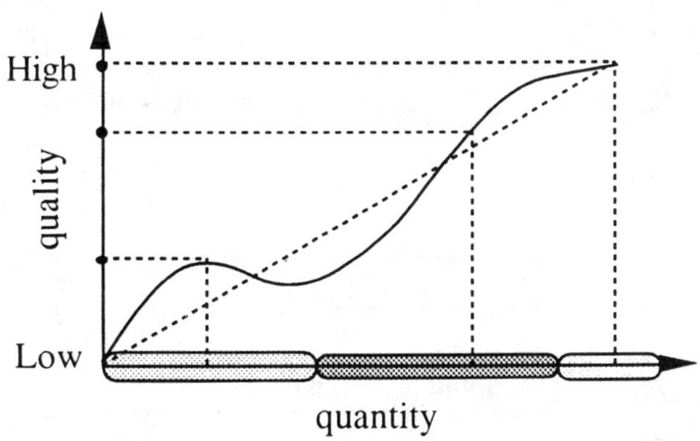

(b) neural network: sigmoid-like curve function acquired through learning

Fig. 7 Signal transformation from quantity to quality

strategic plan or schedule of robotic manipulation, such as task, trajectory, force, and other planning in conjunction with the knowledge base. The system can include another *common sense* for robotic manipulation. Thus, the learning level reasons an unknown fact from a-priori knowledge and sensed information. Then, the learning level produces a control strategy for the adaptation level. Following the control strategy, the learning level selects initial data set for a servo controller at the adaptation level from a data base which maintains some gains and initial values of interconnection weights of the neural network in the servo controller. Moreover, the recent sensed information from the skill level and the adaptation level updates the learning level through long-term learning process with human instruction.

In the same task and different environments, it is necessary to change control references depending on the environment for the servo controller at the adaptation level. At the skill level as a middle level, the fuzzy neural network is used for specific task following the control strategy produced at the learning level in order to generate appropriate control references. Input signals into the fuzzy neural network are numerical values sensed by some specific sensors and some symbols which indicate the control strategy produced at the learning level. Output of the fuzzy neural network is the control reference for the servo controller at the adaptation level. This output is based on the skill extracted from a human expert through learning training sets obtained from human experts. At the same moment, the fuzzy neural network clusters the input signals in the shape of membership functions. These membership functions are used as the symbolic information for the learning level.

At the adaptation level, a neural network in the servo controller adjusts control law to current status of dynamic process [1]. Particularly, compensation for nonlinearity of the system and uncertainties included in the environment must be dealt with by the neural network. Thus, the neural network in the adaptation process has to work more rapidly than that in the learning process. We have shown that a neural network-based controller, the Neural Servo Controller, is useful to the nonlinear dynamic control with uncertainty such as the force control of a robotic manipulator [8]. Eventually, the neural networks and the fuzzy neural networks connect the neuromorphic control with the symbolic control for Hierarchical Intelligent Control while combining human skills.

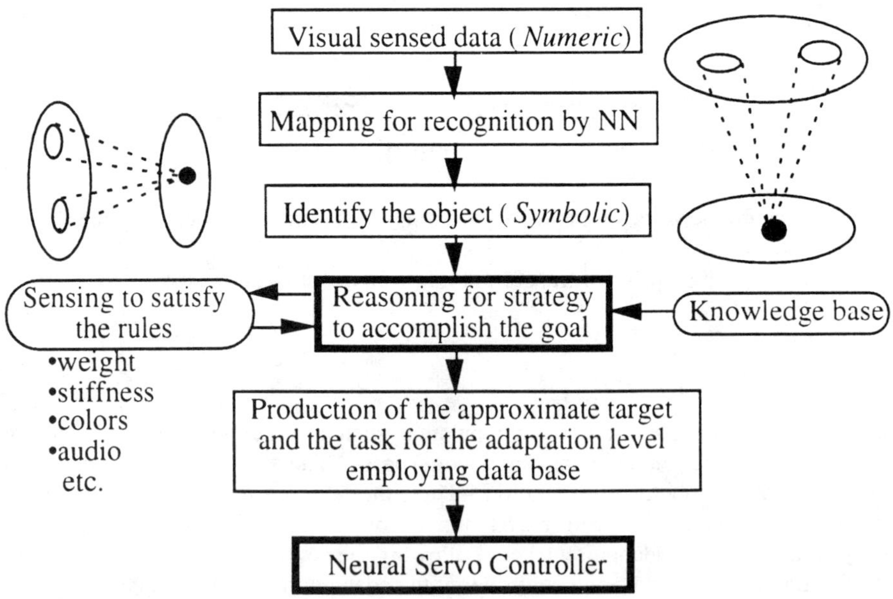

Fig. 8 Flow chart of reasoning: At first, the system identifies the object roughly by visual sensed data. Then, in order to investigate the states and characteristics, other sensing are made actively. Thus, the system recognizes the object and then reasons symbolically with knowledge base.

4. Symbolic Control for Strategy

Symbolic control uses reasoning mechanism to produce a control strategy while using knowledge. The knowledge is acquired from a human expert and results of control at the adaptation level through long-term learning process. Here, it is assumed that the manipulator grasps and carries objects whose characteristics are unknown. At first, the recognition level identifies the object approximately by visual sensing. There, the trained neural network is used for an associative memory. The recognition level reasons symbolically from the knowledge concerned with the object, such as what the states and characteristics of the object are like, and then important information should be obtained by sensing. The recognition level produces the strategy to achieve the goal according to the states and the characteristics of the object based on the knowledge and the skill (Fig. 8). The control strategies select data of initial weights of the neural network in the controller at the adaptation level. The control strategy also selects the fuzzy neural network at the skill level, if it is necessary to determine control references. Therefore, the control strategy includes vagueness in itself because of its symbolic expression. The neuromorphic control compensates for the vagueness by its adaptability.

4.1. Symbolic Control

Figure 8 shows the flow chart of reasoning that produces the control strategies. In this section, the complicated task is a robot manipulator grasping and carrying objects whose characteristics are unknown. At first, the recognition level identifies an object roughly by visual sensing while using the knowledge base. There, the neural networks are used as associative memories. The recognition level reasons what the states and characteristics of the object are by using the knowledge about the object and important information is obtained by active sensing. The recognition level produces the control strategy to achieve the goal according to the states and characteristics of the object based on knowledge and skill. The complicated task of this section considers the case of a robot carrying a container which has some liquid in it.

1) By vision, the recognition level recognizes the object as a container at first.
2) Data about the materials are also included in the knowledge base. It is necessary to examine what the container is made of, for example, paper, plastic, or glass.
3) Therefore, the robot examines the stiffness of the container.
4) After knowing that the object is a container, it is also necessary to judge how much liquid the container has in it.
5) Therefore, the robot examines the weight of the container.

In this way, the recognition level recognizes the object more precisely so that the planning level can perform planning for the control strategies successfully using knowledge and skill.

6) To grasp the container without breaking it or to carry it without spilling the liquid out of the container. Thus, the learning level produces the control strategies for the adaptation level by reasoning with sensory information, knowledge, and skill. Following sections discuss the sensing methods, the recognition algorithm, the knowledge base of the planning level, and the reasoning method.

4.2. Sensing Methods and Recognition Algorithm in Learning Level

Conventional AI method employs "if..then.." rules to infer a fact from a prior knowledge. However, when numerical sensory information is mapped into a state described in symbols, there are many difficulties in deciding rules and the number of assortments. On the other hand, there is a possibility to infer the unknown fact from the knowledge and multiple sensor-fusion in some period.

The neural networks are used as nodes of a decision tree for recognition. These neural networks are trained with input-target training data by a prior knowledge in order to transform various sensed data from numerical quantities to symbolic qualities. This is because the neural networks have the mapping ability. The neural networks store the knowledge in weights of interconnections through learning and also develop inference rules. In general, human beings identify objects from their shapes at first. Therefore, processed image data about a head and a side from CCD cameras are used as numerically sensed visual information. Data comprise the following preprocessed parameters:
A) area,
B) perimeter,
C) roundness,
D) normalized variance of the radius,
E) its maximum value,
F) ruggedness,
G) ratio between length and width,
and lengths in the directions of
H) x axis and
I) y axis.
In this section, since objects such as balls, cans, and containers are dealt with, those preprocessed parameters are enough to express features of the objects. The prior knowledge as training data comprises sets of numerical sensed data and symbols corresponding with the states. In order to acquire mapping rules in the neural networks, inputs and outputs for the neural networks as training data of visual information are expressed as followings:
Input data:

$$<in\_1\_P>::=<A_h, B_h, \ldots I_h, A_s, B_s, \ldots I_s>, \quad (30)$$

where
P: name of the object,
in_1_P: name of input data,
A, B, ... I: processed numerical image data,
h: image data of the object from the head,
s: image data of the object from the side.

The output of the neural network is used as a flag for the knowledge base, and belongs to the output sub-space $\mathbf{X_{ij}}$. *i* means the sort of sensor, and *j* means the order in the neural networks. In the case of visual information, since the visual sense is the first level for recognition, i is 1 and output sub-space is defined as follows:

$$\begin{aligned}
\mathbf{X_{1j}} &= \{ x_{1j1}, x_{1j2}, \ldots , x_{1jn}\}, \\
x_{1j1} &= [1\ 0\ \ldots\ 0]^T, \\
x_{1j2} &= [0\ 1\ \ldots\ 0]^T, \\
&\vdots \\
x_{1jn} &= [0\ 0\ \ldots\ 1]^T,
\end{aligned} \quad (31)$$

where x_{1jk} is the symbolic output pattern for training data or knowledge, and **n** is the number of units in the output layer of **j**th neural network in first level. Plural neural networks are used in the same level to memorize many things . The training data set of the numerical input and the symbolical output are described as follows:

<object P>::=<in_1_P, x_{1jk}> (32)

After the identification of the object by visual sensing, it is necessary to investigate the object minutely. Therefore, the system senses actively states and characteristics, such as stiffness, weight, colors, and so forth by using knowledge. In this section, the system senses the stiffness and the weight by using the knowledge base. For the stiffness sensing, the end-effector grasps the object with a little force, and then senses the stiffness by force sensor at the tips using the data base. For weight sensing, the manipulator lifts up the object a little, and then senses the weight by a force sensor at the wrist. The numerically sensed data is transformed symbolically by the trained neural network or the "if ... then ..." rules.

4.3. Reasoning for Control Strategies

After the recognition of states and characteristics, the learning level reasons symbolically for the control strategy, using the knowledge base of the robotic manipulation. In the case of complicated tasks, it is necessary to use the knowledge base and the data base. The domains of the knowledge base and data base are specific and limited depending on the tasks. Since this paper deals with grasping and carrying tasks, knowledge consists of characteristics, such as fragility and slipperiness, and manipulation, such as *move fast* or *move slowly*. Therefore, the learning level can produce appropriate control strategies by using the data base. The data base maintains tasks, targets, and initial states for the NSC in the adaptation level. The adaptable ranges of the NSC are not so wide that a finite number of data has to be prepared to accomplish the control strategies. The control strategy expresses the process when the manipulator grasps the object slowly at first so as to avoid impact, and then grasp it firmly if it is a container made of glass. Moreover, if the container have some liquid, then it must carry the container slowly to avoid spilling the liquid. This paper uses "*Prolog*" for reasoning. Some of knowledge is written in Prolog for recognition, as follows:

```
object(container, paper).
characteristic(paper, [very_soft, very_light, slippery, fragile]).
attribute(container, much_water, heavy).
```

Some of programs are written in Prolog for planning, as follows:

```
carry(Y, F, M) :- stiffness(Y, F1),
                  weight(Y, F2, M1),
                  fragility(Y, F3),
                  slipperiness(Y, F4, M2),
                  contain(Y, M3),
                  F = [F1, F2, F3, F4],
                  M = [M1, M2, M3].
```

Y is the name of object, F is the approximate target force for grasping, and M is the manner of manipulation decided in carrying the object.

Consequently, by sensing and reasoning, the learning level produces the approximate target force for the end-effector, the objective trajectory and the control law for the NSC depending on states and characteristics of the object. In the case that the approximate target is not desirable, such as a strong pushing force for a fragile object which may cause excess displacement, the learning level adjusts the target force to the process state. In this way, for complicated tasks, the learning level recognizes the characteristics of the environment by active sensing while using the knowledge base. Then, it produces the control strategies by symbolic reasoning. Thus, the symbolic control is carried out.

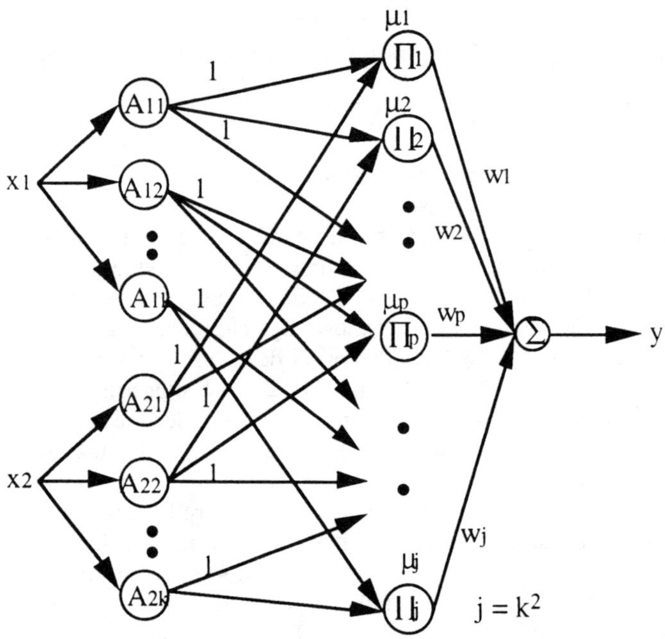

Fig. 9 Model of Fuzzy Neural Network

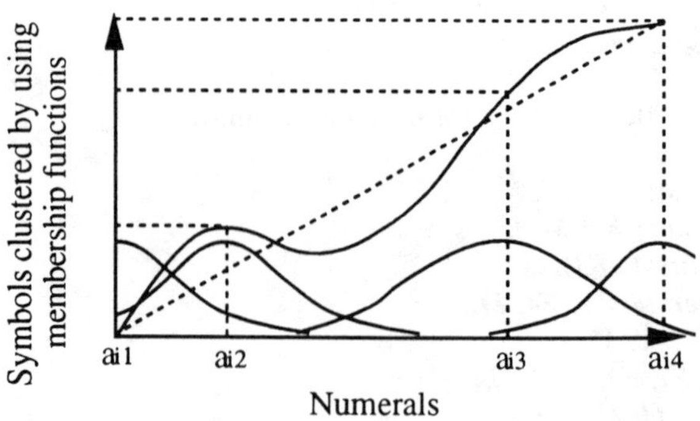

Fig. 10 Signal transformation from quantity to quality by using the membership functions in the fuzzy neural network at the skill level

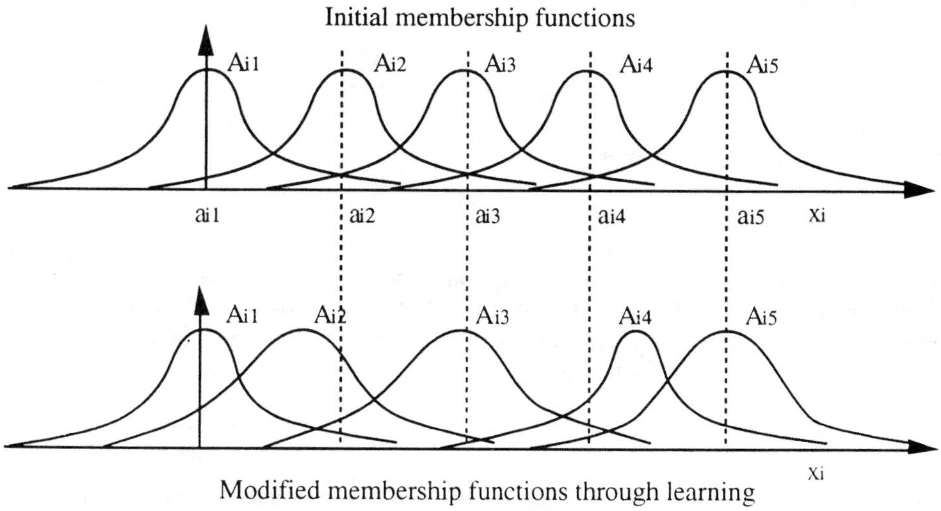

Fig. 11 Modified membership functions through learning of FNN

5. Skill Based Control by Fuzzy Neural Network

As shown in section 2, the neural network has abilities of nonlinear mapping, parallel processing, and learning. On the other hand, the fuzzy logic is characterized as extension of binary Boolean logic [16]. The fuzzy logic is a class in which transition from membership to non-membership is gradual rather than abrupt. Both the neural network and the fuzzy logic have some drawbacks. The neural network can produce mapping rules from empirical training-data sets through learning, but the mapping rules in the network are not visible and are difficult to understand. On the other hand, since the fuzzy logic does not have learning capability, it is difficult to tune the rules. In order to solve these difficulties, recently, much research has been trying to synthesize the fuzzy logic and the neural network [17, 18, 24], and it is often called fuzzy neural network. The fuzzy neural network is used at the skill level in order to express human skills through learning. The fuzzy neural network clusters the input parameters into fuzzy sub-space and identifies input-output relationships by using a weighted network. The fuzzy neural network in Fig. 9 was proposed by Ichihashi [18] The membership functions and weights are modified through learning. After learning, the system can acquire mapping rules as human skills from the fuzzy neural network.

5.1. Learning Scheme for Fuzzy Neural Network

The fuzzy neural network clusters the input parameters into fuzzy sub-space and identifies input-output relationships by using a weighted network. The fuzzy neural network is based on the simplified fuzzy inference. A_{ip} is the membership function for *i*th input variable x_i ($i = 1 \dots n$) in *p*th rule. w_p is the consequence of the *p*th rule. Individual rules' results are given by μ_p as follows:

$$\mu_p = \Pi\, A_{ip}(x_i) = A_{1p}(x_1) \times A_{2p}(x_2) \ldots \times A_{np}(x_n) \quad . \tag{33}$$

Then, the output of the FNN is as follows:

$$y = \Sigma\ \mu_p \bullet w_p \quad . \tag{34}$$

In order to obtain the inference rules in the fuzzy neural network, the back-propagation algorithm is used for learning to minimize difference δ between the desired output y_d obtained from human experts and the actual output y of the fuzzy neural network. A energy function to be minimized is defined as follows:

$$E = (y_d - y)^2 / 2 \quad . \tag{35}$$

Then, change of weights in the network based on the back-propagation is as follows:

$$\Delta w_p = \tau \bullet \partial E/\partial w_p = \tau \bullet \mu_p \bullet \delta \quad , \tag{36}$$

where τ is a learning rate.

This fuzzy neural network uses a Gaussian base function (Fig. 11) as a membership function as follows:

$$A_{ip}(x_i) = \exp(- (x_i - a_{ip})^2 / b_{ip}) \quad . \tag{37}$$

The parameters of a_{ip} and b_{ip} are defined depending on the rules. The Gaussian base is both continuous and differentiable (i.e., possess derivatives) function. In the back-propagation algorithm, this characteristics are required. In order to modify or tune membership functions at the input layer through learning (Fig. 9), following equations are used based on the back-propagation:

$$\Delta a_{ip} = \tau \bullet \partial E/\partial a_{ip} = \tau \bullet \mu_p \bullet w_p \bullet 2(x_i - a_{ip}) / b_{ip} \bullet \delta \quad , \tag{38}$$

and

$$\Delta b_{ip} = \tau \bullet \partial E/\partial b_{ip} = \tau \bullet \mu_p \bullet w_p \bullet (x_i - a_{ip})^2 / b_{ip}^2 \bullet \delta \quad . \tag{39}$$

Figure 11 shows an example of modified membership functions through learning. The experience of the neuron is stored in a fuzzy relation and its output is composed of the current inputs and the past experiences. The neuron can be either fuzzy or non-fuzzy, and crisp values are the special cases of the fuzzy ones.

5.2. Skill Based Control in Hierarchical Intelligent Control

In complex tasks, symbolic control develops a control strategy by reasoning after recognition of the control process. Then, following the control strategy, the fuzzy neural network produces control references for the servo controller at the adaptation level. In this section, we deals with grasping task (Fig. 12) in order to show the effectiveness of the skill based control by using the fuzzy neural network for the Hierarchical Intelligent Control of robotic manipulators. This section examine whether the fuzzy neural network can model and acquire human skills and cluster teaching data by symbolic expression. This section defines that in the skill based control, the system senses the environment and decides the control reference for the servo controller at the skill level without symbolic reasoning, like reflex action of human. We simulate skill based control and train the fuzzy neural network by using teaching data acquired from human experts. We input two kinds of numerical image data into the fuzzy neural network specifically and we use the output from the fuzzy neural network as the control references of grasping force.

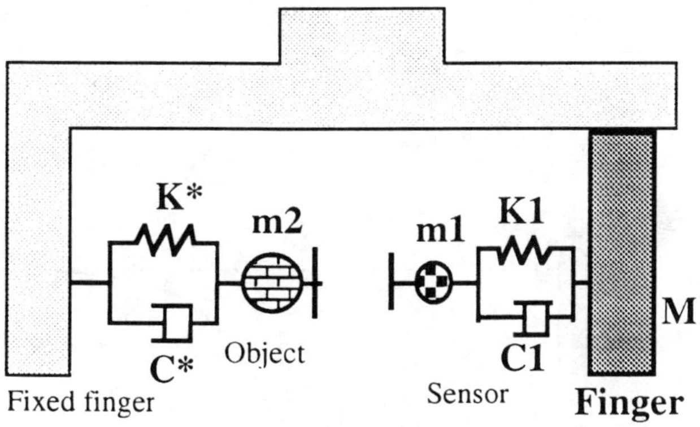

Fig. 12 Mathematical model of a end-effector and a object for simulation of grasping task

5.3. Example

In order to investigate effectiveness of the skill based level, this section deals with grasping task. We have shown effectiveness of the reasoning mechanism at the learning level for the control strategy [12, 27] and of the neural network-based controller at the adaptation level. Processed image data from a CCD camera is used as numerical sensed visual information [11]. This section assumes the objects are foods, specifically tomato, and image data as input into the fuzzy neural network comprised Red and Green about object. As human experts, we determined teaching data for output of the fuzzy neural network which was used as control reference of servo controller. We trained the fuzzy neural network by using 15 input-output patterns, and 3 rules for each input space ($a_{ik} \in \{0, 0.5, 1\}$ and $b_{ik} = 0.2$ as initial values). After training, the fuzzy neural network succeeded in modeling of human skills while tuning membership functions, and in clustering the input space by using a_{ik}.

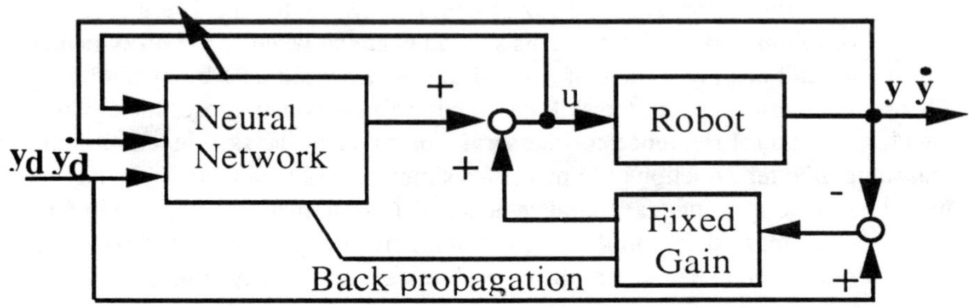

Fig. 13 Indirect inverse control: nonlinear regulator type

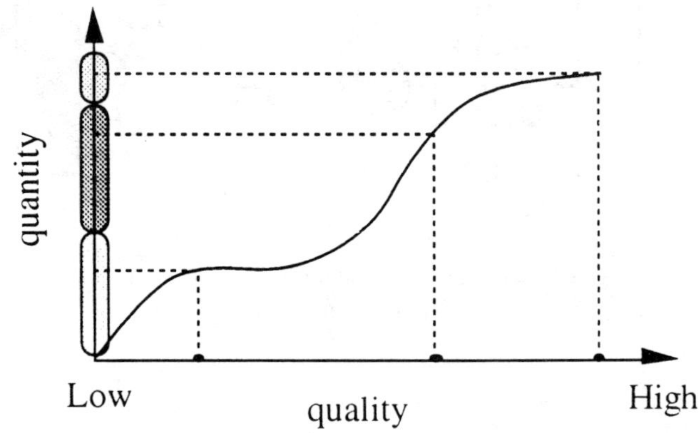

Fig. 14 Signal transformation from quality to quantity:
the neural network-based controller compensate for vagueness
in the control strategies by its adaptability.

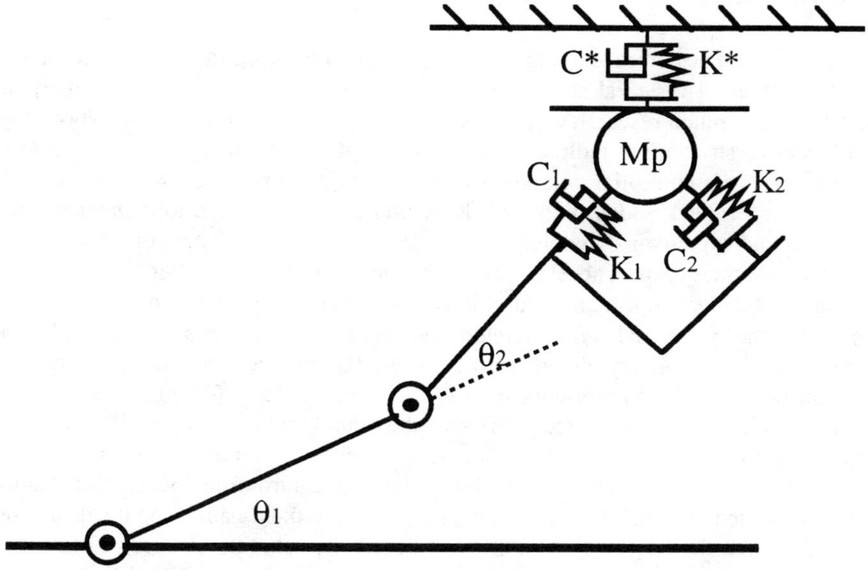

Fig. 15 Equivalent model of a robotic manipulator and its environment

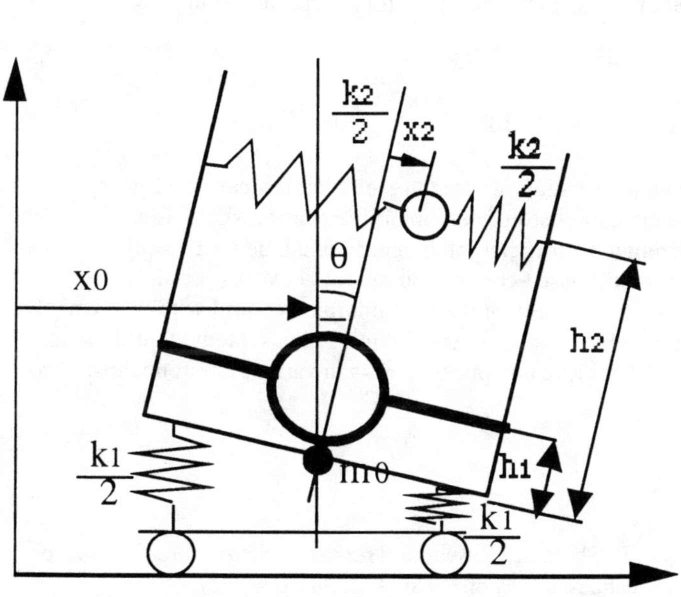

Fig. 16 Equivalent model of the container and liquid based on the Housner model [29].

6. NEUROMORPHIC CONTROL

This section describes learning method of a neural network in a servo controller at the adaptation level. The neural network-based control is referred to as the neuromorphic control. There are much research of adaptive control by using the neural network [25]. Figure 13 shows a structure of indirect inverse control of robotic manipulator, which uses a neural network as a controller combined with local regulators with fixed gain [8, 26]. The linear state feedback added output of the neural network is used to compensate for nonlinearities and unknown dynamics. Input data for the neural network consists of positions and velocities (expect acceleration). The outputs of the neural network consist of input torque to each actuator. The neural network is trained by using an error between actual output of the system and control reference value like feedback error learning [6]. The neural network has time delay elements so as to be dynamic configuration. In indirect inverse control, the neural network is used to compensate for nonlinearity and uncertainties. We have shown effectiveness of the neural servo controller [8, 26]. The controller adapts to the environment in short-term when the desirable initial state of the neural network is given [11]. The learning level gives the control references, initial values of the interconnected weights of the neural network, and the fixed gain using the data base. Therefore, the controller adapts to its environment easily by using the control strategy.

Effectiveness of Hierarchical Intelligent Control is shown in [11, 12, 28], while applied to hybrid position and force control of robotic manipulators for grasping and suppression control of sloshing. Figure 15 shows a equivalent model of a robotic manipulator and its environment for hybrid position and force control. Figure 16 shows a equivalent model of a container with liquid for suppression control.

7. CONCLUDING REMARKS

This chapter presents a structure of hierarchical intelligent control for robot motion. The system comprises of neuromorphic control, skill based control and symbolic control. The system uses artificial intelligence technique for reasoning mechanism, fuzzy neural network for skill based control and neural networks for adaptive control. Several methods are shown for connecting the neuromorphic control with the symbolic control by using the neural network and fuzzy logic. The system would be useful for the autonomous robot to achieve complicated tasks in various environments.

REFERENCES

[1] T. Fukuda and T. Shibata, "Research Trends in Neuromorphic Control," Journal of Robotics and Mechatronics, Vol. 2 No. 4, pp. 4-18 (1991)

[2] J. Rasmussen, "Skills, Rules, and Knowledges; Signals, Signs, and Symbols, and Other Distinctions in Human Performance Models," IEEE Trans. on Systems, Man, and Cybernetics, Vol. SMC-13, pp. 257-266, (1983)

[3] G. Saridis, "Intelligent Robotic Control," IEEE Trans. on Automatic Control, Vol. 28, No. 5, pp. 311-321 (1983)

[4] H. Asada and S. Hirai, "Towards A Symbolic-Level Force Feedback: Recognition of Assembly Process States," Proc. of Int'l. Symp. on Robotic Research, pp. 341-346

(1990)

[5] J. E. Slotine and W. Li, "On the adaptive control of robot manipulators,' Proc. of Robotics and Automation (1986)

[6] M. Kawato, Y. Uno, M. Isobe, and R. Suzuki, "A Hierarchical Neural -Network Model for Voluntary Movement with Application to Robots," IEEE Control Systems Magazine, 8(2), pp. 8-16 (1988)

[7] M. Tokita, T. Mitsuoka, T. Fukuda, T. Kurihara, "Force Control for Robotic Manipulator by the Neural Network," JRSJ, Vol. 7 No. 2, pp. 47-51 (1989)

[8] T. Fukuda, T. Shibata, M. Tokita and T. Mitsuoka, "Neural Network Applications for Robotic Motion Control; Adaptation and Learning," Proc. of Int'l Joint Conf. on Neural Networks '90 - San Diego, pp. 447-451 (1990)

[9] K. S. Narendra and K. Parthasarathy, "Identification and Control of Dynamical Systems Using Neural Networks," IEEE Trans. on Neural Networks, vol. 1, pp. 4-27 (1990)

[10] P. J. Werbos, "Neurocontrol and Related techniques," Hand Book of Neural Computing Applications, Academic Press, Inc., pp. 345-380 (1990)

[11] T. Fukuda, T. Shibata et al., "Adaptation and Learning for Hierarchical Intelligent Control," Proc. of Int'l Joint Conf. on Neural Networks '91-Singapore, Vol. 2, pp. 1033-1038 (1991)

[12] T. Shibata, T. Fukuda et al., "New Strategy for Hierarchical Intelligent Control of Robotic manipulator: - Hybrid Neuromorphic and Symbolic Control -," Proc. of Int'l Joint Conf. on Neural Networks '91-Singapore, Vol. 1, PP. 107-112 (1991)

[13] G. Hirzinger and K. Landzettel, "Sensory Feedback Structures for Robots with Supervised Learning," Proc. of Int'l Conf. on Robotics and Automation, pp. 627-635, (1987)

[14] H. Asada and H. Izumi, "Automatic Program Generation from Teaching Data for the Hybrid Control of Robots," IEEE Trans. on Robotics and Automation, Vol. 5, No. 2, pp. 163-173 (1989)

[15] H. Asada and S. Liu, "Transfer of Human Skills to Neural Net Robot Controllers," Proc. of Robotics and Automation, pp. 2442-2448 (1991)

[16] L. A. Zadeh, "Fuzzy Sets," Information and Control, Vol. 8, pp. 228, (1965)

[17] M. M. Gupta and J. Qi, "On Fuzzy Neuron Models," Proc. of Int'l Joint Conf. on Neural Networks '91-Seattle, Vol. 2, pp. 431-436, (1991)

[18] H. Ichihashi, "Learning in Hierarchical Fuzzy Models by Conjugate Gradient Method using Backpropagation Errors," Proc. of Intelligent System Symposium, pp. 235-240 (1991)

[19] M. Minsky and S. Papert, "Perceptrons, Cambridge," MA: MIT Press, Introduction, pp. 1-20 (1969)

[20] D. E. Rumelhart, G. E. Hinton, and R. J. Williams, "Parallel Distributed Processing," MIT Press (1986)

[21] P. Werbos, "Beyond Regression: New Tools for Prediction and Analysis in the Behavioral Sciences," Ph. D. thesis, Harvard University, Cambridge, MA, Aug. (1974)

[22] A. R. Jacobs, "Increased Rates of Convergence Through Learning Rate Adaptation," Neural Networks, Vol. 1, pp. 295-307 (1988)

[23] N. Baba, "A New Approach for Finding the Global minimum of Error Function of Neural Networks," Neural Networks, Vol. 2, pp. 367-373 (1989)

[24] H. R. Berenji and P. Khedkar, "Learning and Tuning Fuzzy Logic Controllers Through Reinforcements," Trans. on Neural Networks, Vol. 3, No. 5, pp. 724-740 (1992)

[25] T. Fukuda and T. Shibata, "Theory and Applications for Neural Networks for Industrial Control Systems," Trans. on Industrial Electronics, Vol. 39, No. 6, pp. 472-489 (1992)

[26] T. Fukuda, T. Shibata, T. Mitsuoka and M. Tokita, "Neuromorphic Control -Adaptation and Learning, Trans. on Industrial Electronics," Vol. 39, No. 6, pp. 497-503 (1992)

[27] T. Shibata and T. Fukuda, "Hierarchical Intelligent Control for Robotic Motion." Trans. on Neural Networks, (to be appear)

[28] T. Shibata, T Fukuda, et al., "Skill Based Control by using Fuzzy Neural Network for Hierarchical Intelligent Control," Proc. of Int'l Joint Conf. on Neural Networks '92-Baltimore, Vol. 2, pp. 81-86 (1992)

[29] G. W. Housner, Dynamic Pressures on Accelerated Fluid Containers, Bull. Seism. Soc. Am., 15 (1971)

Neural Learning of Robot Inverse Kinematics Transformations

M.M. Gupta and D.H. Rao

Intelligent Systems Research Laboratory, College of Engineering
University of Saskatchewan, Saskatoon, Canada, S7N 0W0

ABSTRACT

The accuracy of a robot arm is determined by its ability to move in a given particular task space to specific Cartesian positions that are not necessarily pretaught. As a consequence, the inverse kinematics is an important problem as it must be solved in real-time in order to position the end-effector at an appropriate Cartesian location. However, it is a difficult and challenging problem for it involves the determination whether or not at least one mathematical set of robot joint angle values exists that will produce a desired coordinate configuration. The mathematical solutions should be checked against the physical constraints associated with the manipulator. Many times, a solution many not be physically realizable in a constrained environment. The advent of artificial neural networks has made it possible to obtain general learning schemes which can be used to arrive at feasible solutions to inverse kinematics problem in a constrained environment independent of a robotic structure. In this paper, we present such a learning scheme using a *dynamic neural processor* (DNP). This neural model functionally mimics the subpopulation of biological neurons. For analytical simplicity, only two subpopulations of neurons, namely excitatory and inhibitory, are assumed to coexist. The DNP is a neural network structure consisting of two *dynamic neural units* coupled as excitatory and inhibitory neurons. It is demonstrated in this study that the DNP would avoid time consuming numerical calculations and provide, more or less, instant recall of the learned associations. The learning and adaptive nature of this neural approach is demonstrated for two- and three-linked robots.

Keywords: Inverse kinematics, robotic manipulator, learning and adaptation, dynamic neural units.

1. INTRODUCTION

The successful operation of an autonomous intelligent robot depends upon its ability to cope with a variety of unexpected events in a constrained task environment. Such an autonomous robot would need only to be presented with a goal, it would achieve its objective through continuous interaction with the work-space, and with continuous feedback about its operation. An autonomous robot with certain level of intelligence would be able to learn tasks not easily handled by existing machines. Also, such an autonomous robot would continue to adapt and perform these tasks with increasing efficiency even under changing and unpredictable environmental conditions. Such independent or autonomous robots would prove useful where direct human interaction would be hazardous, tedious, or impossible.

In order to design such an intelligent autonomous robot, it is necessary to base the design of artificial adaptive systems on their biological counterparts. By doing so, it is hoped that some of the robust and adaptive capabilities of the biological system can be realized in the artificial system. The lack of these robust and adaptive abilities in robots is due to the fact that the biological method of processing information is fundamentally different from that used in the conventional robot controllers. Also, the design of conventional automatic control systems involves the construction of a mathematical model of the dynamic system to be controlled and the application of off-line analytical techniques for an optimal control.

Biological control systems are quite successful at dealing with uncertainty, complexity, and strong nonlinearities. They smoothly coordinate many degrees of freedom during the execution of complex tasks within unstructured environments, solving complex planning and redundancy management problems with ease.[1] They are usually very complex and defy exact mathematical formulation of their operation. The incredible flexibility and adaptability of biological control mechanisms is the necessary proof that such a robust control system is possible as well as it provides a framework from which a comparable autonomous robotic system can be developed. From the perspective of a control system engineer, it is not necessary to emulate the precise neuro-physiological behavior of biological control processes. Rather, it is desirable to replicate some of the computational operations involved in biological learning and adaptation. In this way, the scientific principles derived from the understanding of biological processes may be employed for the design of more effective autonomous intelligent robots.

If robots are to emulate the capabilities found in biological controls, then a great deal can be learned from the structural, functional, and behavioral aspects of biological neural systems. Biological neuronal morphology (structure), indeed, provides a clue and a challenge in the design of intelligent robots which can emulate capabilities for dealing with uncertainty, thereby capable of executing complex tasks within unstructured environments.

In this paper we describe a biologically motivated neural system called the *dynamic neural processor* (DNP) which emphasizes the dynamic properties of subpopulation of neurons. For simplicity only two subpopulations of neurons, namely excitatory and inhibitory, are assumed to coexist. The basic functioning nodes of this neural structure are the *dynamic neural units* (DNUs). The DNU is a dynamic model of biological neuron comprising of delay operators (memory elements), feedforward and feedback weights followed by a time-varying nonlinear activation function. Two such DNUs are configured to function as excitatory and inhibitory neurons.

In the next section, we briefly describe the problem of inverse kinematics in robotics. The architectural details and the learning and adaptive algorithm of the DNU are described in Section 3. The implementation scheme of the above algorithm is also presented in this section. The dynamic neural processor (DNP) is developed in Section 4. Solutions to the inverse kinematics problem are developed for two- and three-linked robotic structures in Section 5. Extensive computer simulation results are also presented in this section, followed by conclusions in the last section.

2. THE PROBLEM OF INVERSE KINEMATICS IN ROBOTICS

A robot manipulator (or arm) comprises of 'links' connected by 'joints' in the form of an open chain with one end fixed while the other end is free. The free end is also referred to as the end-effector. The joints are rotary (revolute) and/or prismatic (linear). A revolute joint allows relative rotation, while a prismatic joint allows a linear motion between the two links. The number of joints determines the degrees-of-freedom (DOF) for the manipulator. The work (task)-space of a robot arm is the total volume swept out by the end-effector as the arm executes all possible motions.[2] Of course, not all the points in the work plane are reachable by the arm because of the constraints imposed by its geometry.

The *forward kinematics* problem is concerned with the determination of the configuration (position and orientation) of the end-effector for a given set of the joint variables such as angles and lengths. The joint variables are the angles in the case of rotational joints, and the link extension in the case of prismatic or sliding joints. Thus, the forward kinematics problem uniquely relates the given set of joint variable (the angles between the links **θ**, and the link lengths **L**) vector, **η**, to a set of task-space coordinate vector, **X**, by a relation

$$\mathbf{X} = \varphi[\boldsymbol{\eta}] \tag{1}$$

where φ is a continuously differentiable nonlinear function and $\boldsymbol{\eta} = [\boldsymbol{\theta}, \mathbf{L}]^T$. Equation (1) has a unique solution in the sense that for a given set of joint angles **θ** and link lengths **L**, the configuration of the end-effector within the task-space can be uniquely determined. The more difficult problem, which is of primary practical importance in robot manipulation and the subject of this paper, is the *inverse kinematics transformation*

$$\boldsymbol{\eta} = \varphi^{-1}[\mathbf{X}]. \tag{2}$$

In other words, the inverse kinematics problem may be stated as follows:[2] Given a desired position and orientation of the end-effector, determine a set of joint variables that achieve the desired position and orientation. The accuracy of a robot manipulator pertains to its ability to move in a given particular task space to specific Cartesian positions that are not necessarily pretaught. As a consequence, the inverse kinematics problem can become quite important as it must be solved in order to position the end-effector at an appropriate Cartesian location. In our further discussion, we assume rigid robot structures which means that the link lengths are constant.

To elaborate the problem of inverse kinematics, let us consider a two-linked robot shown in Fig. 1.

Now consider a two-linked planar robot with two revolute joints shown in Fig. 1a. The relationship between the two angles θ_1, θ_2 and the end point coordinates, x and y form the kinematic equations of the two-linked robot. Specifically, the coordinates x and y are defined as:[3]

$$\begin{aligned} x &= L_1 \cos(\theta_1) + L_2 \cos(\theta_1 + \theta_2), \text{ and} \\ y &= L_1 \sin(\theta_1) + L_2 \cos(\theta_1 + \theta_2). \end{aligned} \tag{3}$$

The above two equations are the 'forward' kinematic equations of the robot in which the lengths (L_1, L_2) are fixed while the point coordinates (x, y) of the end-effector are uniquely determined given the two variable joint angles (θ_1, θ_2). The inverse relationship, namely, the definition of the joint angles with respect to the coordinates are:

$$\theta_2 = \tan^{-1}\left[\frac{S2}{C2}\right]; \tag{4:a}$$

$$\text{where } S2 = \sin(\theta_2) = \pm\sqrt{1\text{-}C2},$$

$$C2 = \cos(\theta_2) = \left[\frac{x^2 + y^2 - L_1^2 - L_2^2}{2L_1 L_2}\right],$$

and,

$$\theta_1 = \tan^{-1}\left[\frac{y}{x}\right] - \tan^{-1}\left[\frac{L_2 S_2}{L_1 + L_2 C2}\right]. \qquad \text{(b)}$$

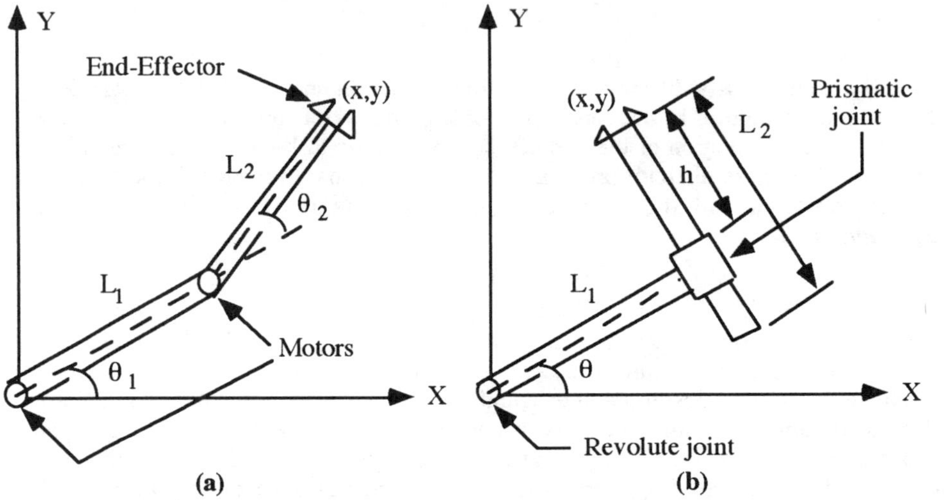

Fig. 1: (a) A two-linked planar robot with two revolute joints and fixed link lengths L_1 and L_2. The angles θ_1 and θ_2 are the joint variables (angles) which determine the position and orientation of the end-effector; (b) A two-linked planar robot with one revolute and one prismatic joints. The angle θ and length h are the joint variables.

The structure shown in Fig. 1b has its first link capable of rotating about the origin of the frame at an angle θ, while a second link of total length L_2, positioned at the end of the first link at a distance of L_1, is able to slide (translate) a distance h at right angles to the first. The Cartesian positions in terms of the joint variables (the forward kinematic equations) are given by:[3]

$$x = L_1 \cos(\theta) - h\cos(\theta), \text{ and}$$
$$y = L_1 \sin(\theta) + h\cos(\theta). \qquad (5)$$

The corresponding inverse kinematic equations are:

$$h = \sqrt{x^2 + y^2 - L_1^2}, \text{ and} \qquad (6:a)$$

$$\theta = \tan^{-1}\left[\frac{L_1 y - h x}{L_1 x + h y}\right] \qquad (b)$$

In this paper, we confine our discussion to the robot structure with two revolute joints shown in Fig. 1a and it will be referred to as Fig. 1 unless otherwise specified.

The equations in θ_1 and θ_2, Eqns. (4a) and (4b), are called the 'inverse' kinematic equations. These equations are nonlinear in nature because they not only contain the trigonometric functions, but also squared and cross product terms. The periodicity and symmetry of the tangent function and the multiple roots of the squared terms result in an ambiguous determination of the end point location. Let us assume that we want to use the robot shown in Fig. 1 to reach a certain object in its work-space. In order to do so, we have to determine the values of joint angles θ_1 and θ_2 to make the robot move its end-effector to coincide with the desired location. For the robot of Fig. 1, there are at least two configurations which means that two sets of joint angles θ_1 and θ_2 can lead to the same end-effector position as shown in Fig. 2.

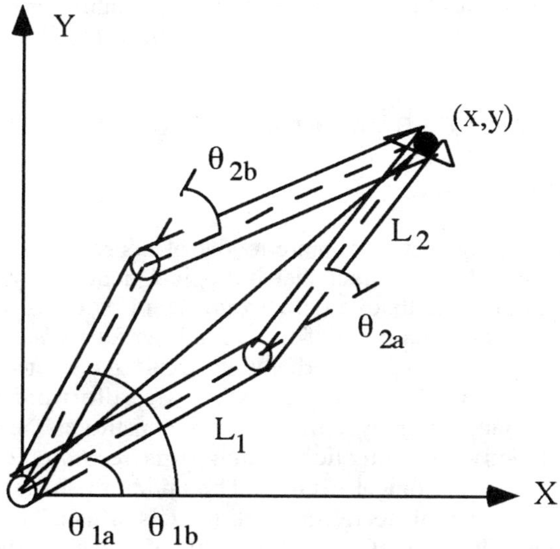

Fig. 2: The desired location (x, y) can be reached by two configurations indicating the existence of two inverse kinematic solutions. The joint angles θ_{1a} and θ_{2a} represent the first configuration, while θ_{1b} and θ_{2b} represent the second configuration for given position (x, y) of the end-effector.

The problem becomes more severe as the number of links in the robot increases. Therefore, the inverse kinematics problem involves the questions of existence, uniqueness and the method of solution. More specifically, the existence question involves the ability to determine whether or not at least one mathematical set of robot joint angle values exists that will produce a desired configuration. The mathematical solutions that we may arrive at should be checked against the constraints associated with the manipulator in order to determine whether or not they represent physical solutions as well[3] One way of obtaining the solution to this problem is to constrain the movement of the two links to certain physically convenient angular ranges which will usually avoid the occurrence of multiple solutions.

Neural networks have been used to obtain solution to the inverse kinematics problem in robotics[4,5] The type of architecture most commonly used in robotics and control applications is the multi-layered feedforward or non-recurrent neural networks. This neuronal model has no feedback connections, that is, connections through weights extending from the outputs of a layer to the inputs of the same or the previous layers. Also, these networks have no memory and therefore their output is solely determined by the current inputs and the values of the weights. Although this neural network structure has been used successfully for solving inverse kinematics problem in robotics, system identification and control of dynamic systems, the major disadvantage of feedforward neural networks, besides having no feedback and memory, is the very slow learning rate. Attempts to accelerate the learning process by increasing the values of the gains in back-propagation algorithms may result into unstable computation.

The neural structure described in this chapter, called the *dynamic neural processor* (DNP), has been proposed by the authors[6,7] based on the fact that all nervous processes of any complexity are dependent upon the interaction of excitatory and inhibitory neurons.[8] The basic functioning elements in the dynamic neural processor are the *dynamic neural units* (DNUs) also proposed by the authors.[9,10] Two such neural units are coupled to function as excitatory and inhibitory neurons. The architectural details and the learning and adaptive algorithm of the DNU are described in the following section.

3. THE DNU: THE BASIC COMPUTING ELEMENT

3.1 Architectural Details

The authors have proposed a dynamic model of neuron called the *dynamic neural unit* (DNU) motivated by the fact that biological neuronal systems always function with a continuous feedback. One example of such a system is the *reverberating circuit* in the neuronal pool of the central nervous system (CNS). The topology of the DNU is analogous to the reverberating circuit in the sense that the DNU comprises of memory elements (delay operators), and feedforward and feedback synaptic weights. The output of this dynamic structure constitutes the argument to a time-varying nonlinear activation function which signifies intercellular current conduction mechanism in the biological neuron. The optimum shape of a nonlinear activation function in conventional neural network models is usually determined by trial and error techniques. This heuristic selection of the shape of sigmoidal function may limit the application of neural networks to complex systems involving nonlinear dynamics. Because, an improper selection of sigmoidal gain may result in an undesirable or even an unstable response. Recently, Yamada and Yabuta have considered determining the optimum shape of the sigmoidal function and have applied to some linear and simple nonlinear problems. [11] Independently, it was proposed in[10,12] that the parameter which controls the shape of the nonlinear function can be considered as one of the adjustable parameters of the neural structure in addition to the synaptic weights. This component contributes to what is called, by the authors, as *somatic adaptation*. The effect of somatic adaptation on neural network's performance as applied to the control of unknown nonlinear systems is discussed in.[13]

The structure of DNU as proposed in[10] comprises of two distinct operations: (i) the synaptic operation, and (ii) the somatic operation. The synaptic operation involves the determination of the optimum feedforward and feedback weights, while the somatic operation determines the optimum gain (shape) of the nonlinear activation function for a given task. The DNU, shown in Fig. 3, consists of two delay elements and two feedforward and feedback paths weighted by the synaptic weights $\mathbf{a}_{ff}$ and $\mathbf{b}_{fb}$ respectively representing a second-order structure with two poles and two zeros (complex or real), followed by a time-varying nonlinear function.

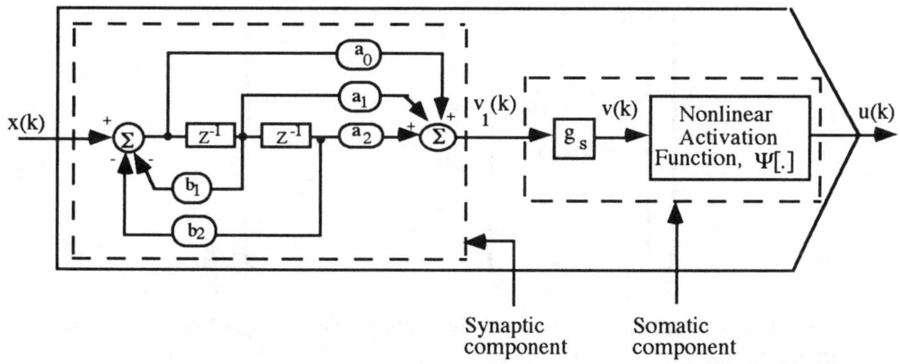

Fig. 3: The basic structure of DNU which consists of synaptic and somatic components. z^{-1} is the unit delay operator, k is the discrete-time index, and $\mathbf{a_{ff}} = [a_0, a_1, a_2]$ and $\mathbf{b_{fb}} = [b_1, b_2]$ are respectively the adjustable feedforward and feedback weights contributing to the forward and feedback signals. The output of this dynamic structure, $v_1(k)$, is passed through the somatic gain (slope), g_s, and a nonlinear activation function Ψ[.] which produces a neural output $u(k) \in \mathbf{R}^1$ in response to a neural input signal vector $x(k) \in \mathbf{R}^n$.

As shown in Fig. 3, the synaptic component of the DNU is represented by a second-order dynamic structure which can be expressed in transfer relation form as

$$w(k, \mathbf{a_{ff}}, \mathbf{b_{fb}}) = \frac{v_1(k)}{x(k)} = \frac{[a_0 + a_1 z^{-1} + a_2 z^{-2}]}{[1 + b_1 z^{-1} + b_2 z^{-2}]}. \tag{7}$$

where $x(k) \in R^n$ is the neural input, $v_1(k) \in R^1$ is the synaptic output of the dynamic structure, $u(k) \in R^1$ is the neural output, k is the discrete-time index, z^{-1} is the unit delay operator, and $\mathbf{a_{ff}} = [a_0, a_1, a_2]$ and $\mathbf{b_{fb}} = [b_1, b_2]$ are the vectors of adaptable feedforward and feedback weights respectively. Alternatively, from Eqn. (7), the synaptic output may be described by the following difference equation

$$v_1(k) = -b_1 v_1(k-1) - b_2 v_1(k-2) + a_0 x(k) + a_1 x(k-1) + a_2 x(k-2) \tag{8}$$

The vectors of signals and adaptable weights of the dynamic neuron are defined now as follows:

$$\gamma^T(k, v_1, x) \triangleq \left[v_1(k-1) \;\; v_1(k-2) \;\;\; x(k) \;\; x(k-1) \;\; x(k-2) \right], \quad \text{and} \tag{9}$$

$$\zeta^T_{(\mathbf{a_{ff}}, \mathbf{b_{fb}})} \triangleq \left[-b_1 \;\; -b_2 \;\; a_0 \;\; a_1 \;\; a_2 \right]^T \quad \text{(T denotes transpose).} \tag{10}$$

Using (9) and (10), the synaptic output, Eqn. (8), is rewritten as

$$v_1(k) = \gamma(k, v_1, x)\, \zeta^T_{(\mathbf{a_{ff}}, \mathbf{b_{fb}})}$$

$$= \left\{ \begin{bmatrix} (v_1(k-1)) \\ (v_1(k-2)) \\ (x(k)) \\ (x(k-1)) \\ (x(k-2)) \end{bmatrix} [-b_1 \ -b_2 \ a_0 \ a_1 \ a_2] \right\} \tag{11}$$

The nonlinear mapping operation on the synaptic signal $v_1(k)$ yields a neural output u(k) which is given by

$$u(k) = \Psi[g_s \, v_1(k)] \tag{12}$$

where Ψ[.] is a nonlinear activation function, usually a sigmoidal function. Many different forms of mathematical functions can be used to model the nonlinear behavior of the biological neuron. Therefore, many forms of nonlinear input functions can be derived. But it is the geometrical shape that is more important than the concrete mathematical function. The sigmoidal form is taken to be the characteristic of an acceptable nonlinear input function. Any function Ψ(z) is said to belong to the class of sigmoidal functions if

(i) Ψ(z) is a monotonically increasing continuous function on the interval (-∞, ∞) $\forall\, z_1, z_2 \in (-\infty, \infty)$, if $z_1 < z_2$ then $\Psi(z_1) < \Psi(z_2)$;

(ii) Ψ(z) approaches or attains two asymptotes

$$\lim_{z(k)\to -\infty} \Psi[z(k)] = 0, \text{ and } \lim_{z(k)\to \infty} \Psi[z(k)] = 1.$$

(iii) Ψ(z) has one and only one inflection poinT.[8]

There are several points to be made concerning the sigmoidal shape of the nonlinear input function. First, the phenomenological significance of the sigmoidal shape is intuitively clear. A low level of excitation in a population of threshold elements will fail to excite any elements, while very strong excitation can excite all of the elements in the neural population. Second, a number of experimental studies have shown that single cell response curves are sigmoidal functions of excitation. Finally, the sigmoidal function compresses the range of neural response so that the output lies between zero and one.

To extend the neural activity for both the excitatory and inhibitory inputs, a bipolar sigmoidal function is used which is defined as a hyperbolic tangent function:

$$\Psi[v(k)] = \frac{\left[e^{(g_s v_1)} - e^{-(g_s v_1)} \right]}{\left[e^{(g_s v_1)} + e^{-(g_s v_1)} \right]} = \tanh\, [g_s \, v_1(k)] = \tanh\, [v(k)] \tag{13}$$

where $v(k) = g_s \, v_1(k)$, and g_s is the somatic gain, the parameter which controls the slope of the activation function. In the limit as $g_s \to \infty$, the sigmoidal function tends to become the sign (binary) function with an infinite slope at $v = 0$, and a zero slope for $v \neq 0$. Figure 4 shows $\Psi[v]$ and its derivative $\Psi'[v]$ for an increasing value of g_s.

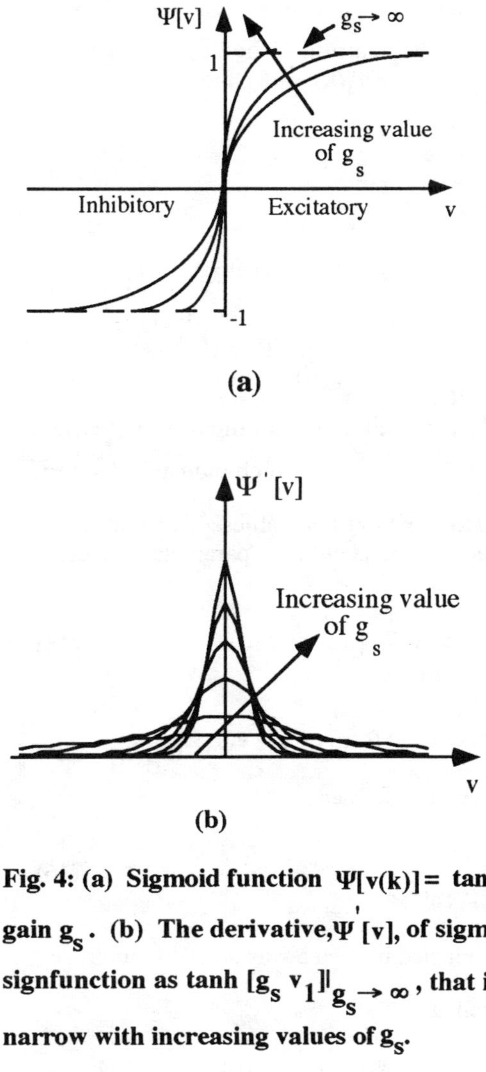

Fig. 4: (a) Sigmoid function $\Psi[v(k)] = \tanh[g_s \, v_1]$, for different values of somatic gain g_s. (b) The derivative, $\Psi'[v]$, of sigmoidal function which tends to become a signfunction as $\tanh[g_s \, v_1]|_{g_s \to \infty}$, that is, the slope $\Psi'[v]$ tends to become very narrow with increasing values of g_s.

In the next subsection, we will derive the learning and adaptive algorithm for the adjustable parameters of DNU, namely the feedforward and feedback weights, $(\mathbf{a}_{ff}, \mathbf{b}_{fb})$, which contribute towards synaptic learning and the sigmoidal gain, g_s, which contributes towards somatic learning.

3.2 The Learning and Adaptive Algorithm

The learning process in a neural network paradigm involves the determination of feedforward and feedback weights, and somatic gain which minimize the error function in some optimal fashion. In an iterative learning scheme, the control sequence is modified in each learning iteration to cause the neural output u(k) to approach the desired state $u_d(k)$. If the error, e(k), can be reduced to an infinitesimally small value as the number of learning iterations increase, the learning scheme is said to be convergent; that is $u(k) \to u_d(k)$ as $k \to \infty$ or, let $e(k) = u_d(k) - u(k)$, then

$$\lim_{k \to \infty} [e(k)] \to 0 \tag{14}$$

for arbitrary initial conditions. The components of the parameter vector $\Omega_{(\mathbf{a}_{ff},\mathbf{b}_{fb},\, g_s)}$ and error e(k) vary with every learning trial k. To obtain $\Omega_{(\mathbf{a}_{ff},\mathbf{b}_{fb},\, g_s)}(k+1)$ requires only the information set $\{e(k-m), e(k), \Omega_{(\mathbf{a}_{ff},\mathbf{b}_{fb},\, g_s)}(k)\}$, where m = 1,2, ... which determines the size of the learning window. As the number of learning trials increases, the information set reduces to only $\{\Omega^*_{(\mathbf{a}_{ff},\mathbf{b}_{fb},\, g_s)}(k), e^*(k)\}$ which indicates that the DNU parameters and the error have converged to some optimal values. To achieve this a performance index, which has to be optimized with respect to the parameter vector, is defined as

$$J = E\left\{F\left[e(k;\Omega_{(\mathbf{a}_{ff},\mathbf{b}_{fb},\, g_s)})\right]\right\} \tag{15}$$

where E is the expectation operator. A commonly used form of $F\left[e(k;\Omega_{(\mathbf{a}_{ff},\mathbf{b}_{fb},\, g_s)})\right]$ in Eqn. (15) is a quadratic (even) function of the error defined as

$$J = \frac{1}{2} E\left\{e^2(k;\Omega_{(\mathbf{a}_{ff},\mathbf{b}_{fb},\, g_s)})\right\}. \tag{16}$$

Each component of the vector $\Omega_{(\mathbf{a}_{ff},\mathbf{b}_{fb},\, g_s)}$ is adapted in such a way so as to minimize J based on the steepest-descent algorithm written as

$$\Omega_{(\mathbf{a}_{ff},\mathbf{b}_{fb},\, g_s)}(k+1) = \Omega_{(\mathbf{a}_{ff},\mathbf{b}_{fb},\, g_s)}(k) + \delta\Omega_{(\mathbf{a}_{ff},\mathbf{b}_{fb},\, g_s)}(k) \tag{17}$$

where $\Omega_{(\mathbf{a}_{ff},\mathbf{b}_{fb},\, g_s)}(k+1)$ is the new parameter vector, $\Omega_{(\mathbf{a}_{ff},\mathbf{b}_{fb},\, g_s)}(k)$ is the present parameter vector, and $\delta\Omega_{(\mathbf{a}_{ff},\mathbf{b}_{fb},\, g_s)}(k)$ is an adaptive adjustment in the parameter vector. In the steepest-descent method, the adjustment of the parameter vector is made proportional to the negative of the gradient of the performance index J, that is,

$$\delta\Omega_{(\mathbf{a}_{ff},\mathbf{b}_{fb},\, g_s)}(k) \propto (-\nabla J) \tag{18:a}$$

$$\text{where, } \nabla J = \frac{\partial J}{\partial \Omega_{(\mathbf{a}_{ff},\mathbf{b}_{fb},\, g_s)}} \quad \text{and, therefore,}$$

$$\delta\Omega_{(\mathbf{a}_{ff},\mathbf{b}_{fb},\, g_s)}(k) = -\,\text{dia}[\mu]\ \frac{\partial J}{\partial \Omega_{(\mathbf{a}_{ff},\mathbf{b}_{fb},\, g_s)}} = -\,\text{dia}[\mu]\ \nabla J \tag{b}$$

where, dia[μ] is the matrix of individual adaptive gains. In the above equation, the dia[μ] and the constraint on the parameter vector are respectively

$$\text{dia}[\mu] = \begin{bmatrix} \mu_{a_i} & 0 & 0 \\ 0 & \mu_{b_j} & 0 \\ 0 & 0 & \mu_{gs} \end{bmatrix}, \text{ and} \tag{19:a}$$

$$\| \Omega_{(\mathbf{a}_{ff},\mathbf{b}_{fb},\, g_s)}(k) \| = \left| \sum_{i=1}^{n} \sqrt{\Omega^2_{i\,(\mathbf{a}_{ff},\mathbf{b}_{fb}, g_s)}(k)} \right| = 1 \tag{b}$$

where μ_{a_i}, i = 0,1,2, and μ_{b_j}, j =1,2, are the individual gains of the adaptable parameters of DNU which determine the stability and the speed of convergence to the optimal values, and ||.|| represents the norm of $\Omega_{(\mathbf{a}_{ff},\mathbf{b}_{fb},\, g_s)}(k)$.

Thus, the feedforward parameters a_{ff_i}, i = 0, 1, 2, and the feedback parameters b_{fb_j}, j = 1, 2, are adapted according to the following set of equations

$$a_{ff_i}(k+1) = a_{ff_i}(k) - \mu_{a_i} \frac{\partial J}{\partial a_{ff_i}(k)}, \ i = 0, 1,2, \text{ and} \tag{20:a}$$

$$b_{fb_j}(k+1) = b_{fb_j}(k) - \mu_{b_j} \frac{\partial J}{\partial b_{fb_j}(k)}, \ j =1,\ 2 \tag{b}$$

Similarly, the somatic gain is adapted according to the following equation

$$g_s(k+1) = g_s(k) - \mu_{g_s} \frac{\partial J}{\partial g_s(k)} . \tag{c}$$

As it was discussed earlier, the DNU provides two distinct operations, namely the synaptic and the somatic operations. The synaptic learning operation involves the determination of the optimum weights, a_{ff_i} and b_{fb_j}, based on Eqns. (20a) and (20b). The second operation adjusts the somatic gain, thereby providing the optimum slope of the sigmoidal function, Eqn. (20c). Therefore, to achieve the synaptic and somatic operations separately it is necessary to obtain the gradient of performance index with respect to the synaptic weights $\mathbf{a}_{ff}$ and $\mathbf{b}_{fb}$, and the somatic gain g_s.

Let us represent the synaptic weight vector as $\phi_{(\mathbf{a}_{ff},\mathbf{b}_{fb})}$. From the definitions of the performance index, Eqn. (16), and the error signal, Eqn. (14), the gradient of performance index with respect to $\phi_{(\mathbf{a}_{ff},\mathbf{b}_{fb})}$ is given by:

$$\frac{\partial J}{\partial \phi_{(\mathbf{a}_{ff},\mathbf{b}_{fb})}} = \frac{1}{2} E\left[\frac{\partial[u_d(k) - u(k)]^2}{\partial \phi_{(\mathbf{a}_{ff},\mathbf{b}_{fb})}}\right] = E\left\{ e(k)\left[- \frac{\partial u(k)}{\partial \phi_{(\mathbf{a}_{ff},\mathbf{b}_{fb})}}\right]\right\}$$

$$= E\left\{ e(k)\left[- \frac{\partial \Psi(v)}{\partial \phi_{(\mathbf{a}_{ff},\mathbf{b}_{fb})}}\right]\right\} = E\left\{ -e(k)\left[\frac{\partial \Psi(v)}{\partial v}\frac{\partial v}{\partial \phi_{(\mathbf{a}_{ff},\mathbf{b}_{fb})}}\right]\right\} \quad \text{(by chain rule)}$$

$$= E\left\{ -e(k)\left[g_s \frac{4}{\left[e^{(g_s v_1)} - e^{-(g_s v_1)}\right]^2}\frac{\partial v}{\partial \phi_{(\mathbf{a}_{ff},\mathbf{b}_{fb})}}\right]\right\}$$

$$= E\left\{ -e(k)\left[g_s \operatorname{sech}^2[v_1(k)]\, s\phi_{(\mathbf{a}_{ff},\mathbf{b}_{fb})}(k)\right]\right\}$$

$$= E\left\{ -e(k)\left[\operatorname{sech}^2[v(k)]\, s\phi_{(\mathbf{a}_{ff},\mathbf{b}_{fb})}(k)\right]\right\} \tag{21}$$

where $s\phi_{(\mathbf{a}_{ff},\mathbf{b}_{fb})}(k) = \dfrac{\partial v(k)}{\partial \phi_{(\mathbf{a}_{ff},\mathbf{b}_{fb})}} = g_s \dfrac{\partial v_1(k)}{\partial \phi_{(\mathbf{a}_{ff},\mathbf{b}_{fb})}}$ represents a vector of parameter-state (or sensitivity) signals. The parameter-state signals for feedforward and feedback weights are respectively given by the relations:

$$s\phi_{\mathbf{a}_{ff_i}}(k) = g_s\,[x\,(k-i)],\quad i = 0, 1, 2, \quad \text{and} \tag{22}$$

$$s\phi_{\mathbf{b}_{fb_j}}(k) = -\,g_s\,[v_1\,(k-j)],\quad j = 1,2. \tag{23}$$

Proof of Eqns. (22) and (23): From Eqn. (11) we can write

$$s\phi_{\mathbf{a}_{ff_i}}(k) = g_s \frac{\partial}{\partial a_{ff_i}(k)}\left\{\left[\begin{matrix}(v_1(k-1))\\(v_1(k-2))\\(x(k))\\(x(k-1))\\(x(k-2))\end{matrix}\right]\left[-b_1\ -b_2\ a_0\ a_1\ a_2\right]\right\}$$

$$= g_s \frac{\partial}{\partial a_{ff_i}(k)}\left\{\left[\begin{matrix}(x(k))\\(x(k-1))\\(x(k-2))\end{matrix}\right]\left[a_0\ a_1\ a_2\right]\right\},\ i = 0, 1,$$

Thus, the individual parameter-state signals for the feedforward weights are
For $i = 0$, $\quad s_{a_0}(k) = g_s\,[x\,(k)]$,

For i = 1, $\quad s_{a_1}(k) = g_s\,[x\,(k-1)]$, and

For i = 2, $\quad s_{a_2}(k) = g_s\,[x\,(k-2)]$.

Therefore, the parameter-state signals for the feedforward weights are

$$s_{\phi_{a_{ff_i}}}(k) = g_s\,[x\,(k-i)],\quad i = 0, 1, 2.$$

Similarly, to obtain the parameter-state signals for feedback weights we have from Eqn. (11) that

$$s_{\phi_{b_{fb_j}}}(k) = g_s \frac{\partial}{\partial b_{ff_j}(k)} \left\{ \begin{bmatrix} (v_1(k-1)) \\ (v_1(k-2)) \\ (x(k)) \\ (x(k-1)) \\ (x(k-2)) \end{bmatrix} \begin{bmatrix} -b_1 & -b_2 & a_0 & a_1 & a_2 \end{bmatrix} \right\}.$$

Thus, the individual parameter state signals for the feedback weights are

For j = 1, $\quad s_{b_1}(k) = -\,g_s\,[v_1(k-1)]$, and

For j = 2, $\quad s_{b_2}(k) = -\,g_s\,[v_1(k-2)]$.

Therefore, the parameter state signals for the feedback weights may be written as

$$s_{\phi_{b_{fb_j}}}(k) = -\,g_s\,[v_1(k-j)],\quad j = 1,2.$$

As seen from Eqns. (22) and (23), the parameter-state signals for the feedforward weights may be obtained by tapping the node signals from the controller structure, while the generation of the parameter-state signals for the feedback weights requires an additional sensitivity structure with only feedback weights as shown in Fig. 5.

Similarly, the gradient of performance index with respect to somatic gain (slope) g_s, $\frac{\partial J}{\partial g_s}$, is given by

$$\frac{\partial J}{\partial g_s} = \frac{1}{2} E \left[\frac{\partial [u_d(k) - u(k)]^2}{\partial g_s} \right] = E \left\{ -e(k) \left[\frac{\partial \Psi(v)}{\partial v} \frac{\partial v}{\partial g_s} \right] \right\}$$

$$= E \left\{ -e(k) \left[g_s\, \mathrm{sech}^2[v_1(k)] \frac{\partial v}{\partial g_s} \right] \right\}$$

$$= E \left\{ -e(k) \left[g_s\, \mathrm{sech}^2[v_1(k)]\, v_1(k) \right] \right\}. \qquad (24)$$

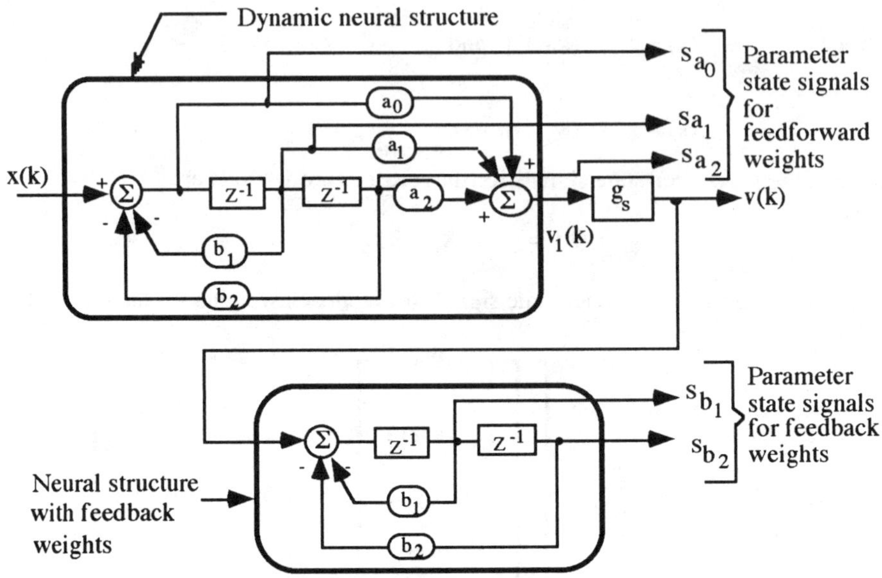

Fig. 5: Obtaining the parameter-state signals from dynamic structure of the DNU.

From Eqns. (20)-(24), the learning and adaptive algorithm to update the DNU parameters can be written as

$$a_{ff_i}(k+1) = a_{ff_i}(k) + \mu_{a_i} E\left\{e(k)\, g_s\, \mathrm{sech}^2[v(k)]\,[x(k-i)]\right\},\quad i = 0,1,2, \qquad (25{:}a)$$

$$b_{fb_j}(k+1) = b_{fb_j}(k) + \mu_{b_j} E\left\{-\,e(k)\, g_s\, \mathrm{sech}^2[v(k)]\,[v_1(k-j)]\right\},\quad j = 1,2, \qquad (b)$$

$$g_s(k+1) = g_s(k)\left[1 + \mu_{g_s} E\left\{e(k)\, \mathrm{sech}^2[v_1(k)]\, v_1(k)\right\}\right] \qquad (c)$$

In the above equations the expectation operator E[.] is defined as follows: The expectation of a random process x with probability density function p(x) is defined as $E[x] = \int_{-\infty}^{\infty} x\, p(x)\, dx$. Thus, E[x] is an averaging process and, for an ergodic process, can be approximated by the time average $E[x(t)] = \lim_{T \to \infty} \frac{1}{2T} \int_{-T}^{T} x(t)\, dt$. For a discrete case, $E[x(k)] = \frac{1}{K} \sum_{i=t+1-K}^{t} x(i)$.

Equations (25a) and (25b) determine the optimum synaptic weights, referred to as the synaptic operation, while Eqn. (25c) determines the optimum sigmoidal gain, referred to as the somatic operation. The schematic diagram for implementing the above learning and adaptive algorithm is shown in Fig. 6.

The DNU described in this section forms the basic component for the dynamic neural processor elucidated in the next section.

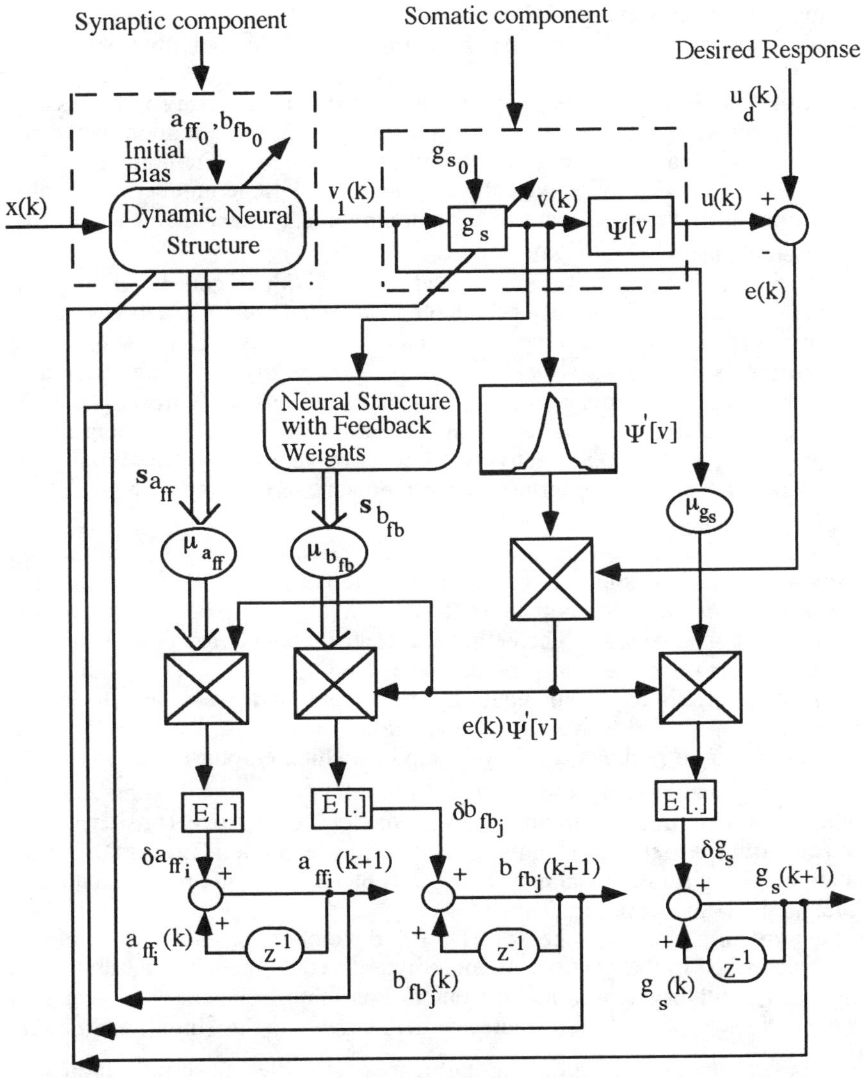

Fig. 6: The implementation scheme of the learning and adaptive algorithm. z^{-1} represents the unit delay, δa_{ff_i} and δb_{fb_j} are the adaptive terms respectively for the feedforward and feedback weights of the dynamic structure of DNU, and δg_s for the slope of the nonlinear function .

4. THE DYNAMIC NEURAL PROCESSOR (DNP)

The neural network structures described in the existing literature often consider the behavior of a single neuron as the basic computing unit for describing neural information processing operations. Each computing unit in the network is based on the concept of an *idealized* neuron. An ideal neuron is assumed to respond optimally to the applied inputs. However, experimental studies in neurophysiology show that the response of a biological neuron appear random, and only by averaging many observations is it possible to obtain predictable results.[8] This observed variability in the response of a neuron is a function of uncontrolled or extraneous electrical signals from many neurons in other parts of the nervous system. As well, this variability is also enhanced by the intrinsic fluctuations of electrical membrane potential within the neuron.

In general, a biological neuron is an unpredictable mechanism for processing information. However, mathematical analysis has shown that these random cells can transmit reliable information if they are sufficiently redundant in numbers. It is postulated, therefore, that the collective activity generated by large numbers of locally redundant neurons is more significant in a computational context than the activity generated by a single neuron.[8,14,15,16,17]

Furthermore, the study of neurons at the individual cell level may be appropriate to emulate some functions of the biological neural network. However, it is not necessarily suited for investigations of those parts which are associated with higher cortical functions, such as sensory information processing and the attendant complexities of learning, memory storage, and vision. This shift in emphasis is warranted since the sensory information is introduced into the nervous system in the form of large scale spatio-temporal activity in sheets of neural cells; the number of neural cells involved is simply too large for any approach starting at the single cell to be tractable.[8]

The total neural activity results from a collective assembly of cells called *neural population*. Each neural population may be further divided into several coexisting *subpopulations*. A subpopulation contains a large class of similar neurons that lie in close spatial proximity. The neurons in each subpopulation are assumed to receive a common set of inputs and provide corresponding outputs. The individual synaptic connections within any subpopulation are random, but dense enough to ensure at least one mutual connection between any two neurons.

In view of these remarks, the neural models which emphasize the dynamic properties of subpopulation of neurons have been developed.[6, 7, 17] It is assumed in these models that two neural subpopulations, namely excitatory (positive) and inhibitory (negative), exist. It is envisaged that these models present a better representation of biological neuronal morphology. A block schematic representation of such a neural model is shown in Fig. 7.

The dynamic neural processor (DNP), developed based on the above hypothesis, is described in this section. It comprises of two dynamic neural units; one functioning as an excitatory neuron and the other as an inhibitory neuron as shown in Fig. 8. The notations $x_\lambda(k)$ and $u_\lambda(k)$ represent the neural stimulus (input) and neural response (output) of the neural computing unit respectively, the subscript λ indicates either an excitatory **E**, or inhibitory **I**, state, $x_{t\lambda}(k)$ denote the total input to the neural units, $w_{\lambda\lambda}$ represent the strength of self-synaptic connections (w_{EE}, w_{II} in Fig. 8), and $w_{\lambda\lambda'}$ represent the strength of synaptic connections from the λ neural unit to the λ' neural unit (w_{IE}, w_{EI} in Fig. 8). The two state variables that describe the dynamic activity exhibited by the neural computing units are defined as the proportion of active excitatory (positive) neurons, $u_E(k+1)$, and active inhibitory (negative) neurons, $u_I(k+1)$.

The functional dynamics exhibited by a neural computing unit, the DNU, is defined by a second-order difference equation. The state variables $u_E(k+1)$ and $u_I(k+1)$ generated at time (k+1) by the excitatory and inhibitory neural units of the proposed neural processor are modeled as:

$$u_E(k+1) = \mathbf{E}\,[u_E(k)\,,\; v_E(k)],\ \text{and} \tag{26:a}$$
$$u_I(k+1) = \mathbf{I}\,[u_I(k)\,,\; v_I(k)] \tag{b}$$

where $v_E(k)$ and $v_I(k)$ represent the proportion of neurons in the neural unit that receive inputs greater than an intrinsic threshold, and **E** and **I** represent the excitatory and inhibitory actions of the neurons. The neurons that receive inputs greater than a threshold value is given by a nonlinear function of $v_\lambda(k)$. This nonlinear function, $\Psi\left[v_\lambda(k)\right]$, is related to the distribution of neural thresholds within the neural uni.t[17].

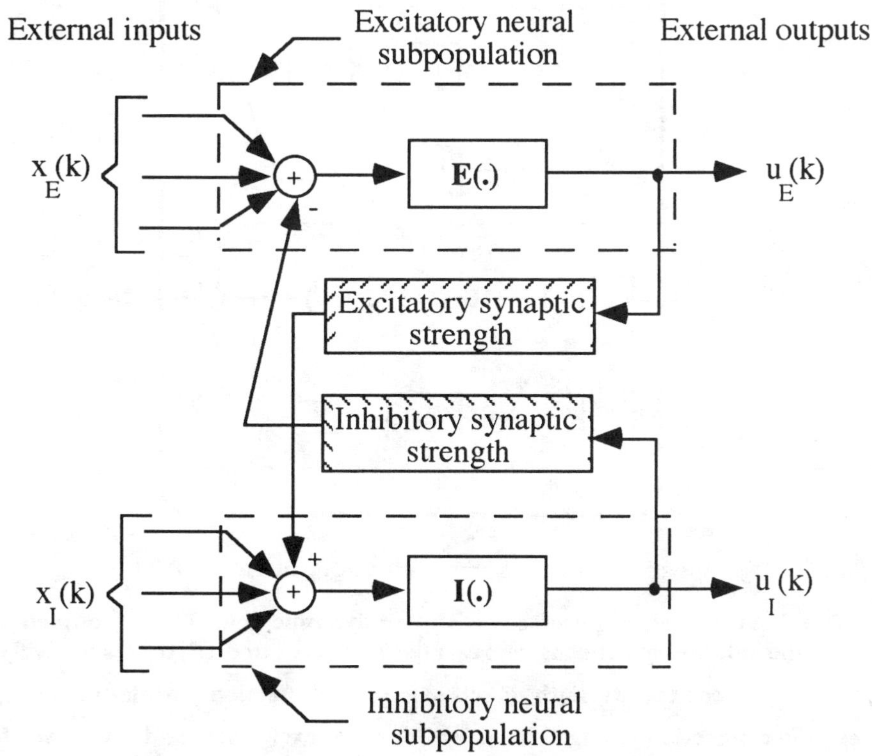

Fig. 7: A simplified schematic diagram of the dynamic neural processor comprising of excitatory (positive) and inhibitory (negative) neural units.

The total inputs incident on the excitatory and inhibitory neural units are respectively given by:

$$x_{tE}(k) = w_E\, x_E(k) + w_{EE}\, u_E(k-1) - w_{IE}\, u_I(k-1) - \theta_E,\ \text{and} \tag{27:a}$$
$$x_{tI}(k) = w_I\, x_I(k) - w_{II}\, u_I(k-1) + w_{EI}\, u_E(k-1) - \theta_I \tag{b}$$

where w_E and w_I are scaling factors of the excitatory and inhibitory neural inputs respectively, w_{EE} and w_{II} represent the self-synaptic connection strengths, and w_{IE} and w_{EI} represent the inter-neuron connection strengths as described above.

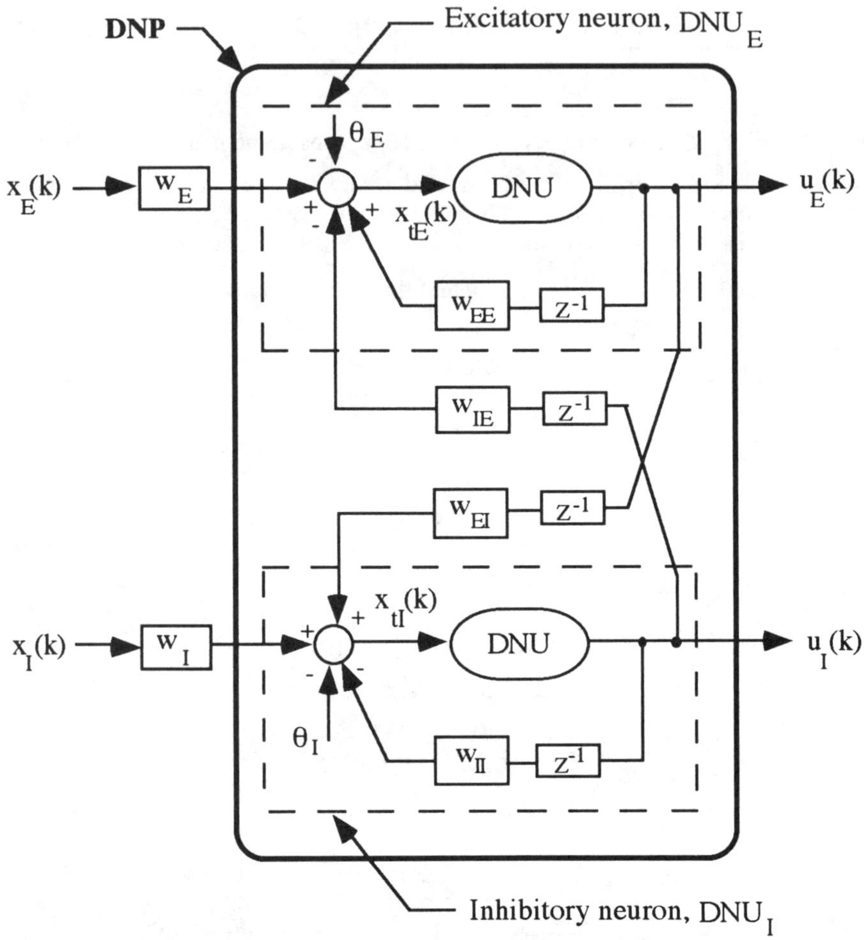

Fig. 8: The dynamic neural processor with two dynamic neural units coupled as excitatory and inhibitory neurons represented as DNU_E and DNU_I respectively. w_{EE}, w_{II} represent the strength of self-synaptic connections, while w_{IE}, w_{EI} represent the connection strength from inhibitory to excitatory, and excitatory to inhibitory neurons respectively. w_E and w_I are the scaling factors for excitatory and inhibitory neural inputs. θ_E and θ_I represent the internal thresholds of excitatory and inhibitory neurons respectively.

Alternatively, Eqns. (27a) and (27b) may be written in a matrix form as

$$\begin{bmatrix} x_{tE}(k) \\ x_{tI}(k) \end{bmatrix} = \begin{bmatrix} w_{EE} & -w_{IE} \\ w_{EI} & -w_{II} \end{bmatrix} \begin{bmatrix} u_E(k-1) \\ u_I(k-1) \end{bmatrix} + \begin{bmatrix} w_E & 0 \\ 0 & w_I \end{bmatrix} \begin{bmatrix} x_E(k) \\ x_I(k) \end{bmatrix} - \begin{bmatrix} \theta_E \\ \theta_I \end{bmatrix}. \quad (27c)$$

From Eqns. (27a) and (27b), the responses of neural units, $u_\lambda(k)$, interms of the stimulus (input) $x_\lambda(k)$, the input $x_{t\lambda}(k)$, and the strength of synaptic connections $w_{\lambda\lambda}$ and $w_{\lambda\lambda'}$ are:

Excitatory neuron:

$$u_E(k) = \frac{1}{w_{EE}} \left[x_{tE}(k) - w_E x_E(k) + w_{IE} u_I(k-1) + \theta_E \right] + \Delta u_E(k) \text{, and} \tag{28a}$$

Inhibitory neuron:

$$u_I(k) = \frac{-1}{w_{II}} \left[x_{tI}(k) - w_I x_I(k) - w_{EI} u_E(k-1) + \theta_I \right] + \Delta u_I(k) \tag{b}$$

where $\Delta u_E(k)$ and $\Delta u_I(k)$ represent the change in neural activities of excitatory and inhibitory neurons during the process of learning, and these are determined by the changes in synaptic and somatic components of the DNUs based on the algorithm derived in Section 3. The response of the dynamic neural processor, u(k), is the total summation of the individual responses, $u_\lambda(k)$, of excitatory and inhibitory neurons in the neuronal subpopulation, and is given by

$$u(k) = u_E(k) + u_I(k) . \tag{29}$$

However, as may be seen from Fig. 8 and observed from Eqn. (28), response of the neural units not only depends on the inputs and the synaptic connections, but also on the structure of the DNU. An analysis of the dynamic activity generated by a spatially distributed dynamic neural processor is extremely complex because of the immense number of inherent nonlinearities and spatial feedback interactions involved. Therefore, the investigation of such a complex process would begin by examining a spatially isolated DNU described in the preceding section.

In the next section we use the neural model developed in this section to obtain solution to the problem of inverse kinematics transformations.

5. NEURAL LEARNING OF THE ROBOT INVERSE KINEMATICS TRANSFORMATIONS

As discussed in Section 2, the artificial neural models have features which make them strong candidates to model biological systems, such as motor control. Indicating the usage of neural networks for motor control modeling, many researchers have considered robots as a 'fitting bench' for implementing the proposed theories. This is basically because: [18]

(i) biological systems provide great motivation by virtue of their striking performance;

(ii) robots face the same physical laws and environmental conditions as those faced by biological systems carrying out similar tasks. In solving the problems for the robot, a set of plausible algorithms can be generated for a biological system. In observing biological phenomena, possible behaviors for a robot can be explored.

Because of their parallel and distributed computations, neural networks have the ability to learn associations between example patterns. These patterns could represent, for example, the task space coordinates and the corresponding joint angles of the model leg. The association between these two sets of patterns basically amounts to the inverse kinematics computations in the robotic paradigm.

The motion and control of a complex mechanism such as the human skeleton, involves great dynamic complexity. Consequently, its mathematical description results in a high order system of coupled nonlinear equations. Analytical solutions of such equations are often not possible. To ease the problem, it is better to simplify the mechanical model used to represent the motor apparatus, up to the level where only the major (or principal) characteristics of interest are preserved. Indeed, to better understand locomotion and its neural control, the human leg has been approximated as a two-link structure.[18, 19]

It has been postulated that a body-centered space coordinate system exists which features motor learning and control [20] Furthermore, the target positions are located within these spatial coordinate systems with reference to 'anchor points'. In the case of the upper limbs, the shoulder joints are considered as plausible anchor points.[18] In this section, the dynamic neural processor (DNP), described in the preceding section, is employed to obtain the inverse kinematics transformations for two - and three - linked robots.

5.1 Inverse Kinematics For a Two-Linked Robot

In the context of the above observations, consider the two-link leg shown in Fig. 9a. The joints at which rotary motion occurs (within limits) are analogous to the hip and the knee joints of the human leg. The free tip of the second link, also called the 'end point', describes the trajectories based on a Cartesian coordinate system. The origin of the coordinate system is the first (hip) joint, which is assumed to be fixed in space, while the end point coordinates (x, y) are located with respect to the two perpendicular axes, X, Y. The hip joint is considered as the anchor (fixed) point. The position of the leg can also be restricted using the angles formed at the two joints with the reference axes as shown. The relationship between these two angles, defined as θ_1, θ_2 and the end point coordinates, x and y, form the kinematic equations of the two-link leg as described by Eqn. (3).

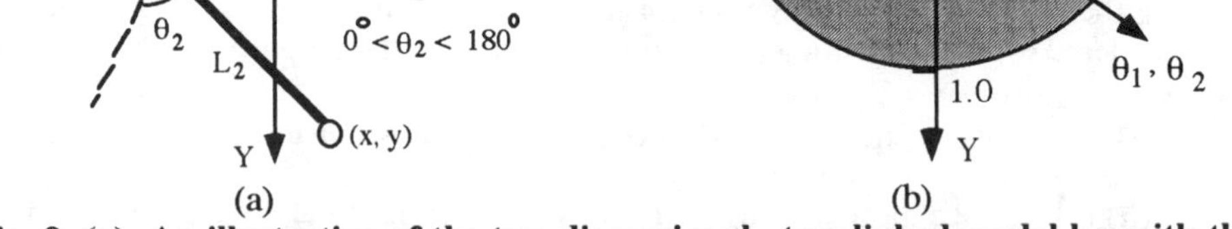

Fig. 9: (a) An illustration of the two-dimensional, two-linked model leg with the joint angles and their angles. (b) Constraints on the task space of the two-linked leg.

As was mentioned in Section 2 that one way of obtaining the solution to inverse kinematics problem is to constrain the movement of the two links to certain convenient angular ranges which will usually avoid the occurrence of multiple solutions. Incidentally, the structure of the human leg, with its hinge like joint at the knee, only permits a constrained motion of the shank. Taking into consideration the above mentioned constraints, the two joints of the two-link leg model were constrained to move within the specific angular ranges: -30° < θ_1 < 180°, and 0° < θ_2 < 180°. The introduction of these constraints results in a task space, that is, the allowed area of motion, as shown in Fig. 9b.

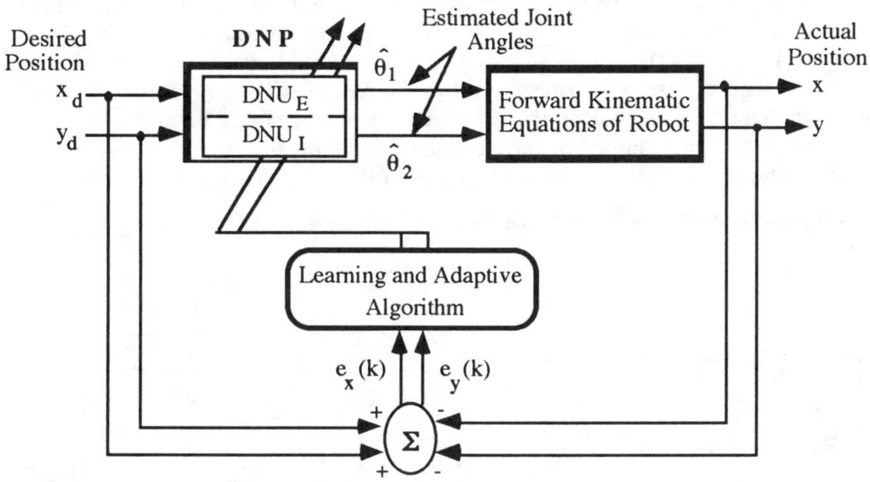

Fig. 10: A schematic diagram of the learning methodology employed for learning the inverse kinematics of model leg.

The basic learning procedure employed is shown in Fig. 10. The kinematic equations for the two-link leg serve as the model of the leg structure. A number of joint angle values, θ_1 and θ_2, serve as inputs for the model which outputs the corresponding end point (end-effector) coordinates, x and y. In this study, the joint angles were within the permitted ranges mentioned above, and were spaced 10 degrees apart. Thus, there were 21 values for θ_1 and for each such value there were 18 values for θ_2. The input units which are related to the end point coordinates had activations in the range of -1.0 and 1.0. The output units which relate to the joint angles, however, had their activations scaled to 0 and 1.0. The minimum value 0 corresponded to an angle of -30 degrees for θ_1, and 0 degrees for θ_2, while the maximum value of 1.0 corresponded to 180 degrees for both the angles θ_1 and θ_2. The learning and adaptive algorithm, derived in the earlier section, was implemented for the proposed neural processor. The external synaptic weights were arbitrarily set as follows: $w_{EE} = 0.5$, $w_{EI} = 2$, $w_{II} = 0.5$, $w_{IE} = 1.5$. The components of the scaling vector, $\mathbf{w}=[w_E \ w_I]^T$, were set to 1. During computer simulation studies random end-effector values (x-y coordinates) within the robot task space were presented to inputs of the DNP. The outputs of the neural processor, which are the estimates of the joint angles, were transformed into their corresponding coordinate values using the forward kinematic equations so as to compare the targeted and observed outputs. The DNP weights were adjusted using the learning and adaptive algorithm derived in Section 3.2. This was repeated until the observed outputs (the actual end-effector positions) were consistent with the targeted values (the desired positions).

Simulation Studies

During the learning procedure, the joint angle values were selected at random and applied to the network until the output error decreased to a pre-determined value of 0.05. These simulation experiments were repeated for more than 100 different values. The desired inputs and the network's (DNP's) outputs obtained are shown in Fig. 11a.

Figure 11b shows the actual and the learned joint angle values of a two-linked robot. From this figure, it may be observed that the neural processor could learn the desired joint angles that were within the robot work space. A large error between the desired and the actual joint angles occurred when the neural processor was presented with learning patterns having the joint angle value of θ_2 either 0° or 180°. The reason for the neural processor's inability to learn these patterns is briefly explained below.

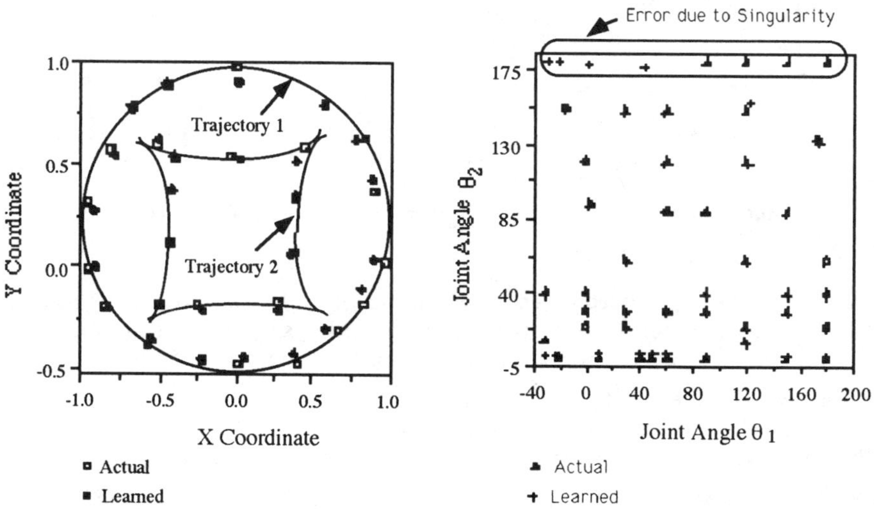

(a) (b)

Fig. 11:(a) Illustration of the actual and the learned trajectories.
(b) Illustration of the actual and the learned joint angle values for arbitrary positions. Also shown in this figure is the large error in learning due to singular configurations of the robot links which occur at θ_2 = 0° or 180°. However, the error is more significant for θ_2 =180°, that is, when the second link is fully contracted than for θ_2 = 0°, that is, when the link is fully stretched.

The forward kinematic equations of a two-linked robot are given by Eqn. (3). The manipulator Jacobian which represents the relationship between the joint displacements and the end-effector location is given by[21]

$$\mathbf{J} = \begin{bmatrix} -L_1 \sin\theta_1 - L_2 \sin(\theta_1+\theta_2) & -L_2 \sin(\theta_1+\theta_2) \\ L_1 \cos\theta_1 + L_2 \cos(\theta_1+\theta_2) & L_2 \cos(\theta_1+\theta_2) \end{bmatrix} \tag{30}$$

Singularity occurs when the determinant of the manipulator Jacobian **J** is zero. From Eqn. (30) we have det (**J**) = $\sin\theta_2$. Therefore, singular configurations occur for θ_2 = 0° or 180°, that is, when the robot arm is fully extended or fully contracted. Form Fig. 11b, the learning error is quite significant when the arm is fully contracted compared to when the arm is fully extended.

Figures 12 (a) through (d) show the error responses during the learning process and the joint angle trajectories for different targeted values of the space coordinates. In Figs. 12 (e) and (f) are shown the results for a case where the desired angles (θ_1, θ_2) were changed during the learning process from (30°, 120°) to (120°, 30°) at k = 150. It is seen from this example that the adaptation in the processor's output was almost instantaneous.

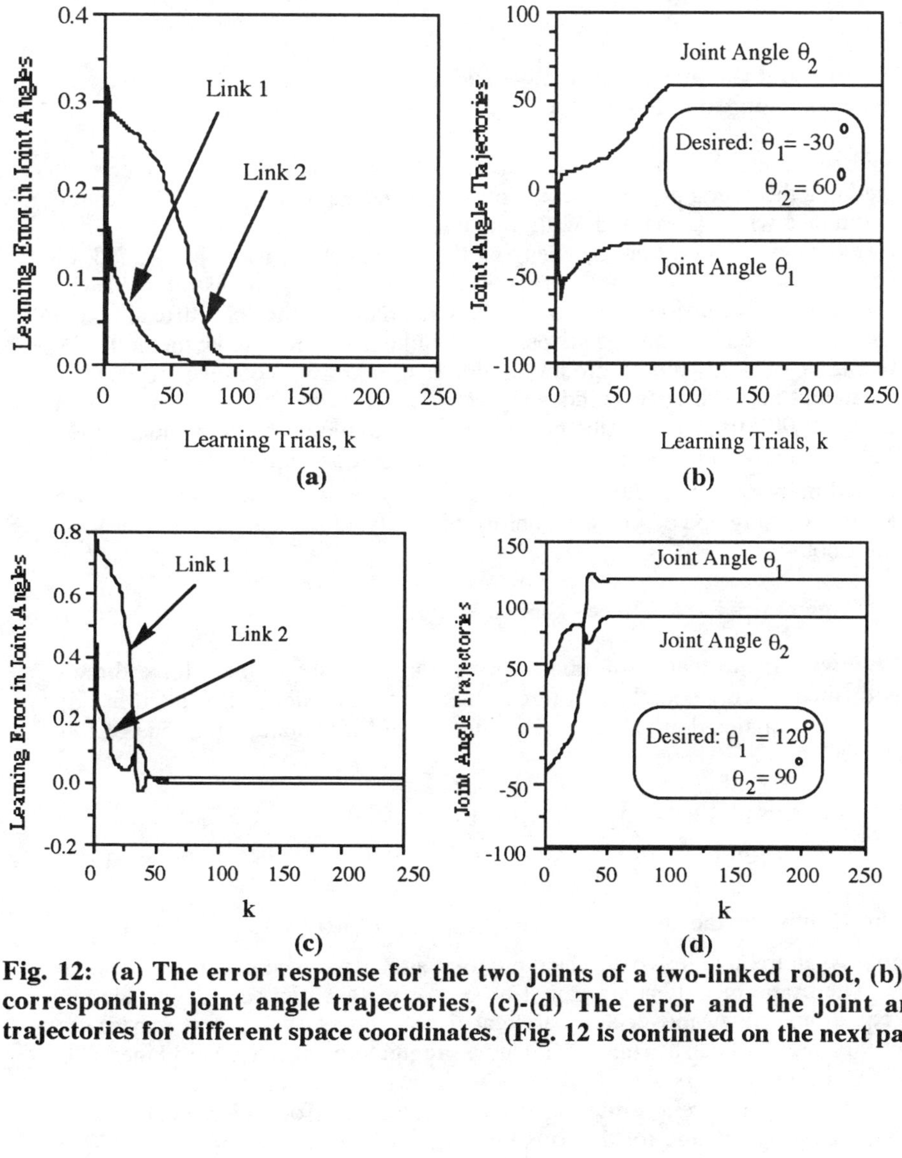

(a) (b)

(c) (d)

Fig. 12: (a) The error response for the two joints of a two-linked robot, (b) the corresponding joint angle trajectories, (c)-(d) The error and the joint angle trajectories for different space coordinates. (Fig. 12 is continued on the next page.)

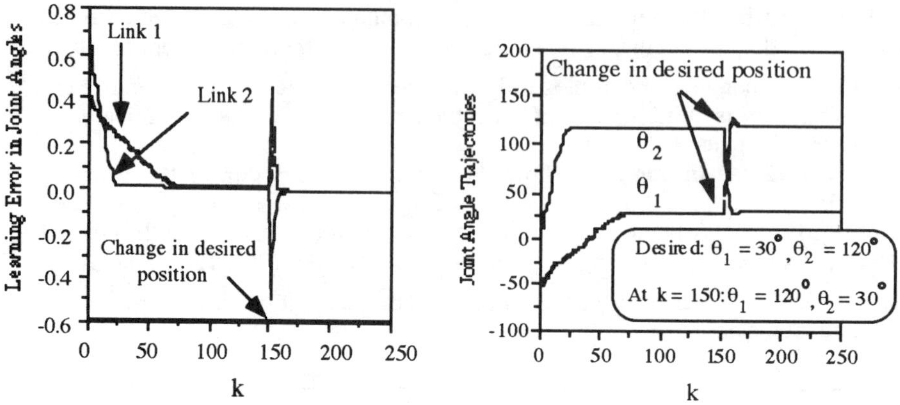

(e)-(f) The error and the joint angle trajectories for a case when the desired space coordinates were changed at instant k = 150.

Several simulation studies have been carried out for this robotic structure. It was observed that the neural processor could make the end-effector move to the desired space coordinates when presented with the untrained data pattern. The above simulation studies were carried out by changing the lengths of the robot joints, and the processor could adapt to the changes in robotic structure. This demonstrates that the dynamic neural processor possesses the two important characteristics of artificial neural networks, namely generalization and adaptation. Unlike the conventional methods, the major advantage of this neural model is the fast learning convergence. The feedforward neural network of two hidden layers with twenty neurons each is reported to have taken 120,000 trials before the output error decreased to the above mentioned value (0.05). A smaller number (ten) of the hidden neurons (with both one and two layers) resulted in much longer duration (236,000 trials).[22] In the next subsection we demonstrate the learning and adaptive capability of the dynamic neural processor for a three-linked robot.

5.2 Inverse Kinematics For a Three - Linked Robot

Now we will present a solution to inverse kinematics problem for a three - linked robot which is more complex. A three - linked robot model is illustrated in Fig. 13. With the joint angles defined as shown, the forward kinematic equations of this model are as follows:

$$x = L_1 \cos(\theta_1) + L_2 \cos(\theta_1 + \theta_2) + L_3 \cos(\theta_1 + \theta_2 - \theta_3), \text{ and}$$
$$y = L_1 \sin(\theta_1) + L_2 \sin(\theta_1 + \theta_2) + L_3 \cos(\theta_1 + \theta_2 - \theta_3). \quad (31)$$

The lengths of the three links were (L_1, L_2, L_3) 0.5, 0.5 and 0.2 meters respectively. As in the two link case, the three joint angles were constrained as shown in Fig. 13. The usage of a dynamic neural processor made it unnecessary to derive, from the equations given above, an analytical derivation of the inverse kinematic relations of this three-linked structure. All that was required was the forward kinematic equations given above. This shows the ease with which a neural processor can be adapted to include a different structure whose inverse transformation needs to be formed. The learning scheme for this robotic structure involved an additional DNP module in parallel with the existing scheme for the two-linked robot. This modification in the learning scheme is to take care of the increased number of joints and the corresponding joint angles. Only the output signal of excitatory neuron of the second DNP module is utilized for learning the third joint angle, θ_3. Similar to the learning scheme discussed for a two-linked robot, the input and the output units have appropriately scaled activation ranges. The processor outputs were suitably transformed to the Cartesian space using the forward kinematic equations.

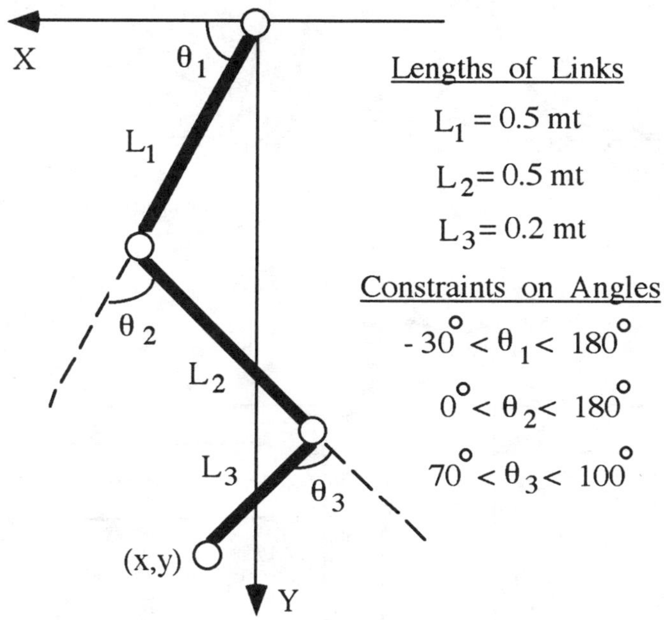

Fig. 13: An illustration of a two-dimensional, three-linked robot with joint angles and their ranges.

The correspondence between the desired and the obtained trajectories for different space coordinates is illustrated in Fig. 14(a). Also shown in this figure is desired coordinate point P_1 to which the neural processor could not learn because it was out of task space, even though it may be observed that the link orientations were in that direction. Figures 14 (b), (c) and (d) illustrate the convergence of the joint angles to the desired values, and, therefore, to the targeted Cartesian coordinates.

The simulation results reported in this section indicate that the dynamic neural processor, by virtue of its functional mapping capabilities, can be suitably employed for learning the coordinate transformations. In contrast to the conventional approaches for the inverse kinematics problem, the dynamic neural units provide convergence of the required transformation very fast compared to the conventional schemes which, on the other hand, need intensive computing. In the event that it is impossible for the model leg or robot to reach points within its task space, due to some physical constraints, the network can be made to learn to move to the nearest point still within its domain. Any modifications occurring in the structure of the robot can easily be taken care of by continuous learning. This reflects the adaptive nature of the neural processor. In the case of a robot, a conventional approach would have involved solving new potentially difficult transformations theoretically, and/or generating new software to implement the new transformations, and then installing them in the robot controller.

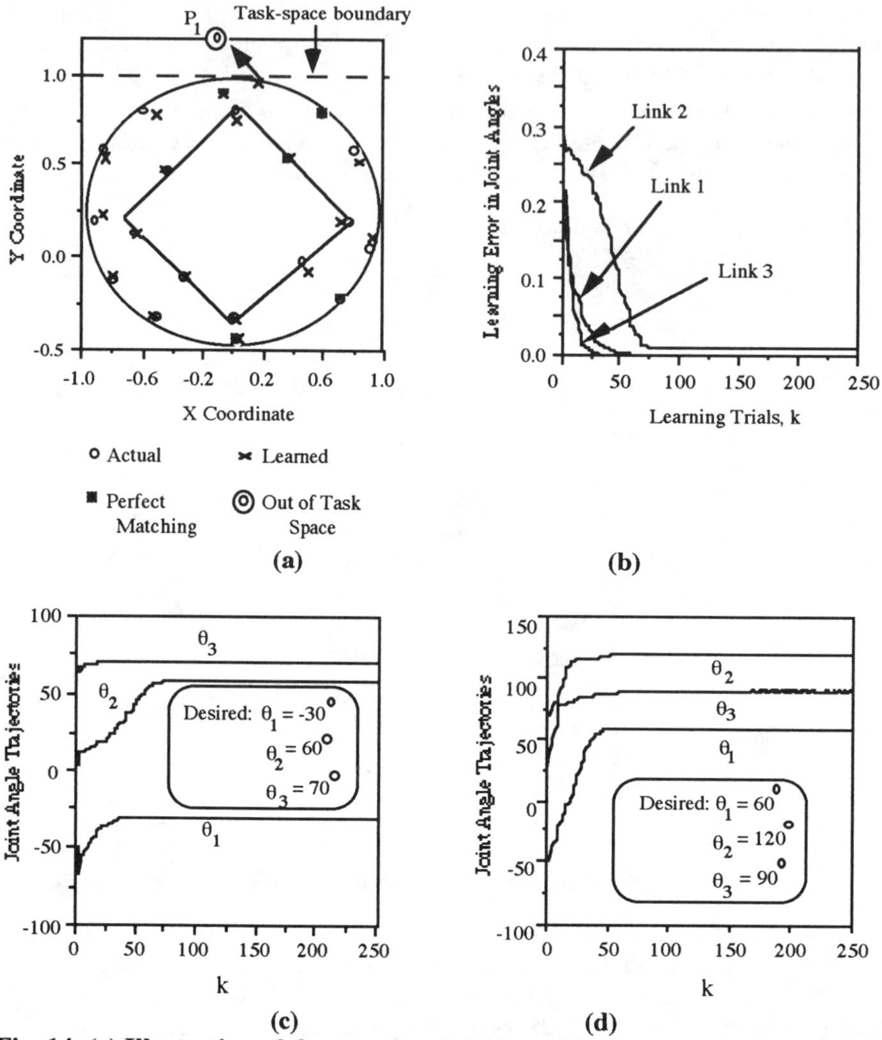

Fig. 14: (a) Illustration of the actual and the learned space coordinates; (b)-(d) The error responses and the joint angle trajectories of the three-linked robot for different desired space coordinates.

6. CONCLUSIONS

In this chapter, we have presented a new neural network architecture called the *dynamic neural processor* (DNP) for robotic applications. This neural structure emphasizes the aggregate dynamic properties of two subpopulations of neurons, namely excitatory (positive) and inhibitory (negative). The DNP, however, utilizes fixed inter-neural connections. The extension of this neural model would involve adaptable inter connections between neurons. It is envisaged that this neuronal morphology would lead to a general dynamic model of the biological neural population.

The DNP uses two *dynamic neural units* (DNUs) which are coupled to function as excitatory (positive) and inhibitory (negative) neurons. The DNU performs synaptic and somatic operations separately in more analogous to the biological neuron.

The learning and adaptive algorithm for the adjustable parameters of the DNU, namely the feedforward and feedback weights which contribute to the synaptic learning, and the slope of the activation function which contributes to the somatic learning, have been derived.

The architecture of the DNU based dynamic neural processor (DNP) is very general and can be used for various computational tasks in machine vision and control applications. However, in this chapter we have addressed a specific problem- that is the neural solution of inverse kinematics problem for a multi-linked robotic system. The DNP has been implemented to 'learn' the inverse kinematics transformations. The advantage of this approach, over the traditional inverse kinematics algorithms, is that it avoids time consuming computations and provides, more or less, instant recall. Furthermore, it is very easy to modify the learned associations upon changes in the structure of the mechanism, as it is demonstrated for two - and three - linked robots.

REFERENCES

1. A.D. Handelman, H.L. Stephen and J.J. Gelfand, " Integrating Neural Networks and Knowledge -Based Systems for Intelligent Robotic Control ", *IEEE Control System Magazine*, pp. 77-87, April 1990.
2. M.W. Spong and M. Vidyasagar, Robot Dynamics and Control, *John Wiley and Sons,* Inc., 1989.
3. W.A. Wolovich, Robotics: Basic Analysis and Design, *Holt, Rinehart and Winston,* New York, 1987.
4. A. Guez and Z. Ahmad, "Solution to the Inverse Kinematics Problem in Robotics by Neural Networks", *IEEE Int. Conf. on Neural Networks*, San Diego, Calif., pp. 617-624, March 1988.
5. J. Barhen, S. Gulati and M. Zak, "Neural Learning of Constrained Nonlinear Transformations", *IEEE Computer*, pp. 67-76, June 1989.
6. D.H. Rao and M.M. Gupta, "A Neural Processor for Coordinating Multiple Systems with Dynamic Uncertainties", *International Symposium on Uncertainty and Management* (ISUMA), Maryland , pp. 633-640, April 25-28, 1993.
7. D.H. Rao and M.M. Gupta, "A Multi-Functional Dynamic Neural Processor for Control Applications", *American Control Conference,* San Francisco, June 2-4, 1993.
8. H.R. Wilson and J.D. Cowan, "Excitatory and Inhibitory Interactions in Localized Populations of Model Neurons", *Biophysical Journal,* Vol. 12, pp. 1-24, 1972.
9. M.M. Gupta and D.H. Rao, " Dynamic Neural Units in the Control of Linear and Nonlinear Systems", *The International Joint Conference on Neural Networks (IJCNN),* pp. 100-105, Baltimore, June 9-12, 1992.
10. M.M. Gupta and D.H. Rao, "Dynamic Neural Units With Applications to the Control of Unknown Nonlinear Systems", *The Journal of Intelligent and Fuzzy Systems,* Vol. 1, No. 1, pp. 73-92, January 1993.
11. T. Yamada and T. Yabuta, "Neural Network Controller Using Autotuning Method for Nonlinear Functions", *IEEE Trans. on Neural Networks* , Vol. 3, No. 4, pp. 595-601, July 1992.
12. M.M. Gupta and D.H. Rao, "Synaptic and Somatic Adaptations in Dynamic Neural Networks", *Second Int. Conf. on Fuzzy Logic and Neural Networks,* Fukuoka, Japan, July 17-22, pp. 173-177, 1992.
13. D.H. Rao and M.M. Gupta, "Dynamic Neural Controller with Somatic Adaptation" *IEEE Conf. on Neural Networks*, pp. 558-563, San Francisco, March 28- April 1, 1993.

14. W.J. Freeman, *Mass Action in the nervous System*, Academic Press, New York, 1975.
15. J.J. Hopfield, "Neurons with Graded Response Have Collective Computational Properties Like of Those Two-State Neurons", *Proceedings of the National Academy of Sciences*, Vol. 81, pp. 3088- 3092, 1984.
16. W.J. Freeman, "Dynamics of Image Formation by Nerve Cell Assemblies", in E. Basar, H. Flohr, H. Haken and A.J. Mandell (Eds), Synergitcs of the Brain, Berlin, *Springer-Verlag*, 1983.
17. M.M. Gupta and G.K. Knopf, "A Multitask Visual Information Processor with a Biologically Motivated Design", *Journal of Visual Communication and Image Representation*, Vol. 3, No. 3, pp. 230-246, Sept. 1992.
18. G.E. Stelmach (Ed.), Motor Control: Issues and Trends, *Academic Press*, New York, 1976.
19. W.T. Powers, "The Nature of Robots", *Byte*, Vol. 82, pp. 96-111, 1979.
20. M. Kawato, K. Furukawa and R. Suzuki, "A Hierarchical Neural-Network Model for Control and Learning of Voluntary Movement", *Biological Cybernetics*, 57, pp. 169-185, 1987.
21. H. Asada and J.J.E. Slotine, Robot Analysis and Control, *John Wiley and Sons*, Inc., 1986.
22. S. Shankar, "A Rhythmic Movement Control Scheme Using Artificial Neural Networks", *Ph.D. Thesis*, University of Saskatchewan, Saskatoon, Canada, Jan. 1992.

TETHER OPERATIONS IN SPACE USING FUZZY LOGIC BASED LENGTH CONTROL

Robert N. Lea Ph.D.
NASA/Johnson Space Center
Houston, Texas 77058

Yashvant Jani Ph.D.
Togai InfraLogic Inc.,
Houston, Texas 77058

Abstract

The Tethered Satellite System (TSS) consisting of the Shuttle, tether, and an Italian satellite, has complex and non-linear dynamics because of the spring-like behavior of the tether and its interaction with two six degrees-of-freedom end-bodies. The TSS has natural librational and longitudinal oscillations, whose characteristics change as the tether is deployed and retrieved. When the finite mass of the tether is properly modeled, the TSS also exhibits 'skip rope' motion, particularly if the motion is excited by the interaction between the geomagnetic field and electric current in the tether. A conventional control system based on the linearization of the system dynamics is employed in the onboard software. Its inputs are sensed length, length rate, and tension parameters and it is expected to maintain not only the length but also the in-plane librational angle.

Due to the complexity of this problem and the recent success of fuzzy logic techniques in controlling systems with non-linear dynamics using imprecise measurements as input from sensors, a fuzzy logic based tether length controller has been developed and implemented into the TSS simulation to investigate the usefulness and robustness that can be achieved with fuzzy control. In this chapter, a concept which utilizes fuzzy membership functions for length, length rate, and voltage parameters, and the rulebase used in generating the control voltage signal is described. Initially, a simple proportional controller was created based on length error only. Later, the length rate error was included as input, and the rulebase was expanded. In order to have a fair comparison between the conventional and fuzzy controllers, tension feedback was not included as an input to the controllers since it is not an input to the conventional controller.

Test results indicate that the fuzzy controller keeps the length error smaller throughout the mission, and the amplitude of the tension oscillations during the retrieval phase is also smaller. One surprising benefit, in addition to better length control, is smaller in-plane librational oscillations during the retrieval phase. Overall the results of the study indicate that fuzzy logic based control can provide the simplicity and robustness desired for complex tethered operations in space.

1. Introduction

The recent successes of fuzzy logic control [1,2] in the late eighties and early nineties have provided momentum to look at this new emerging field introduced originally by Professor Lotfi Zadeh [3] about 25 years ago. Based on the industrial applications being built in Japan [4] it is evident that fuzzy logic provides new methods for control and decision making in many areas. It is a framework where heuristics and experiential knowledge about the system can be implemented easily. Fuzzy logic techniques have also shown utility in handling imprecise measurements from sensors and have been successful in controlling systems with non-linear dynamics [5,6]. Research experience gained so far in the Software Technology Laboratory at NASA Johnson Space Center [7,8], promises that control algorithms based on fuzzy logic can handle non-linearities of systems very well without any requirement to fully understand the dynamics of the system. However, it should be noted that an understanding of the system dynamics helps very much in tuning the membership functions and generating rules.

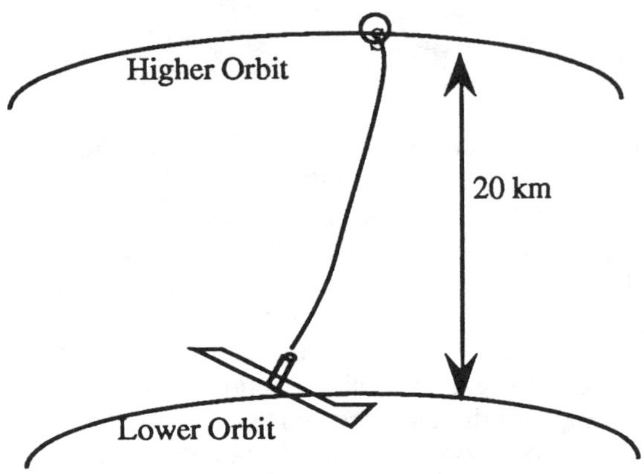

Figure 1. Tethered Satellite System Deployment in orbit

Fuzzy logic techniques have been successful in controlling systems with non-linear dynamics using imprecise measurements as input from sensors. A fuzzy logic based tether length controller has been developed and implemented into the simulation of Tethered Satellite System (TSS), as shown in figure 1, to investigate the usefulness and robustness that can be achieved with fuzzy control. A concept which utilizes fuzzy membership functions for length error, length rate error, and delta voltage parameters, and the rulebase used in generating the control delta voltage signal is described. Initially, a simple proportional controller was created based on length error only. Later, the length rate error was included as input, and the fuzzy rulebase was expanded. In order to have a fair comparison between the conventional and fuzzy controllers, tension feedback was not included as an input to the controllers since it is not used as an input to the conventional controller for the simulated mission scenario. The primary purpose of this analysis has been to evaluate the feasibility of such a fuzzy logic based controller and to understand the usefulness and advantages provided by such a technique.

The first tethered satellite system mission (TSS-1) tests for the comparisons of the fuzzy logic algorithm [9] with the conventional controller [10] were performed in a high fidelity simulation environment [11] of the space shuttle orbiter and the tethered Italian satellite. The comparison results utilizing the massless tether model were reported at the NAFIPS '91 conference [9]. At the FUZZ-IEEE conference [12] in 1992, the performance comparison results with the bead model were reported and benefits derived from the fuzzy controller were discussed.

In this chapter, a comprehensive summary of the tether study results and the performance of the fuzzy controller in the simulation of the tethered satellite system with the massless as well as the bead dynamics model of the tether, also known as finite element model, are given. First, a brief explanation of the nature of the physical control system and some of the behavioral characteristics of the tethered system is given. Then, a description of the algorithm based on fuzzy logic for controlling tether length is given and details of the membership functions and rulebase are provided. Results with massless and bead model are presented comparing the performance of both controllers in terms of length error, tension error and librational oscillations and, finally, a summary of the work and conclusions are given.

2.0 Tether Dynamics

Theoretical applications of long tethers in space have shown promise of substantial benefits for a variety of space operations. However, control of these tethers in low Earth orbit is a complex process because of the non-linear dynamics resulting from elasticity and mass distribution. NASA has gained some experience with short tethers during the Mercury and Gemini space programs. However, there is no experience base for long tethers and their behavior in space.

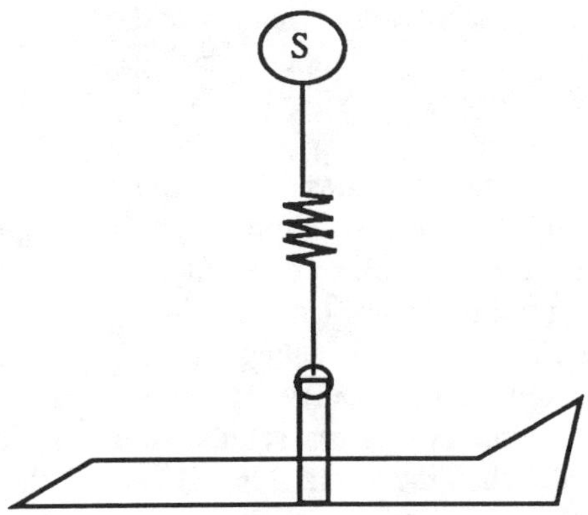

Figure 2. Tether behaves as a spring in massless Tether model

The Tethered Satellite System (TSS) consisting of the Shuttle, tether, and an Italian satellite, is a current NASA project being conducted in cooperation with Italy. The TSS center of mass is close to the shuttle center of mass because the satellite mass of 500 kg is very small compared with the shuttle mass of 102,000 kg. As a result, the orbit of the TSS is primarily determined by the shuttle orbit. The satellite, when fully deployed at 20 km tether length, has a substantially different orbit in which the magnitude of the gravity pull and the centrifugal force is smaller. The orbital time period in this higher orbit is larger if the satellite remains free (not tied to the tether) in that orbit. In other words, its orbital velocity is smaller, and therefore it will fall behind in relative position compared to the shuttle. Since it is attached to the shuttle via the tether, there will be exchange of forces and interaction between the three components of the TSS.

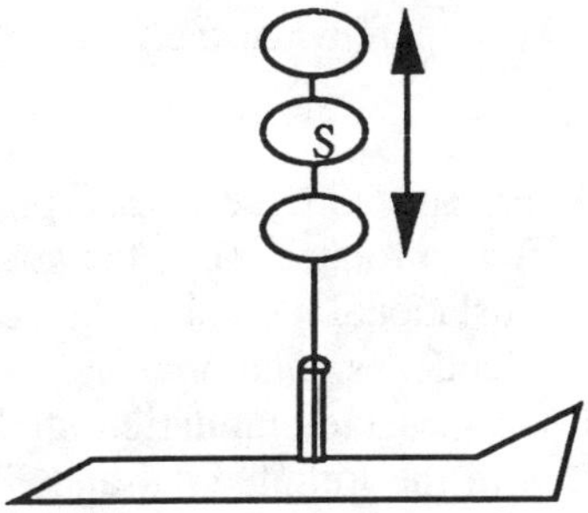

Figure 3. Due to spring like nature of tether, satellite moves up and down along orbital radial vector

Since the tether is a flexible string, it behaves like a spring with certain damping characteristics as shown in figure 2. When the satellite falls behind, it stretches the tether inducing tension force that will pull the satellite towards the shuttle. Thus, the satellite will have an 'up and down' motion along the tether line similar to spring oscillations. These oscillations known as longitudinal oscillations (figure 3) are small in magnitude and for the TSS have a time period of about 80 seconds. The nominal stretch in the tether at a length of 20 km is of the order of 15 meters. The time period and stretch in the tether are functions of tether length and thus will change continuously during the deployment and retrieval.

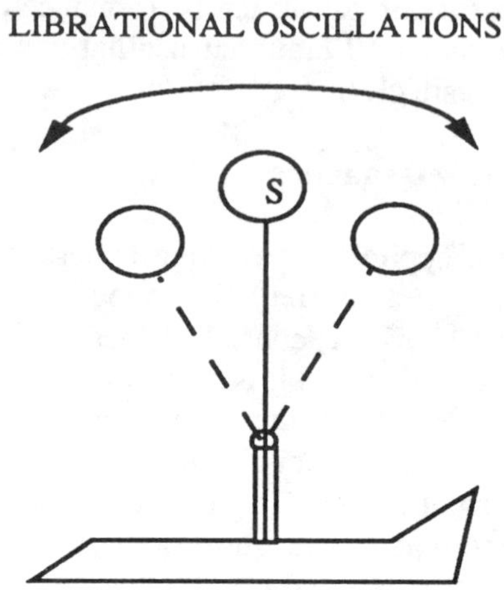

Figure 4. Due to interaction between gravity gradient, centrifugal force and tether tension, satellite moves back and forth in the orbital plane

The satellite oscillates along the tether line, but it also continues to fall behind in relative position slowly until the tether tension really pulls it forward with a large pull. The satellite's velocity is then increased such that it passes the shuttle in its relative position and continues to move forward. When the satellite is right above the shuttle, the tether tension is minimum and it continues to increase as the satellite moves forward. The tension again pulls it back when it is too far in front. Just like a pendulum, the satellite moves back and forth in the orbital plane. This back and forth relative motion is called in-plane librational oscillations as shown in figure 4. It has a time period of about 53 minutes for the TSS. Amplitude of these oscillations in terms of angle depends on how the deployment/retrieval procedures were executed and what other perturbations the system experiences. The satellite can similarly move perpendicular to the orbital plane. The motion called out-of-plane librational oscillations has a time period of about 45 minutes for the TSS.

In summary, the TSS has natural librational and longitudinal oscillations, whose characteristics change as the tether is deployed and retrieved. Longitudinal oscillations are due to the elasticity of tether and the gravity gradient forces acting on the satellite. In-plane libration is the motion of the tethered satellite along the velocity vector as it pivots at the tether boom tip attach point. In other words, the satellite moves in a line from the nose to the tail of the orbiter. Out-of-plane libration occurs when the tethered satellite moves from the port to starboard side of the shuttle.

The satellite is a 6 degree-of-freedom space vehicle, and its tether attach point is not located at the center of mass. Thus, the satellite is free to rotate around the attach point, when the tether tension induces torque on it. The translational motion of the satellite (in terms of longitudinal and librational oscillations) induces rotational motion, called satellite attitude oscillation as shown in figure 5. Extensive simulation studies have shown that the attitude oscillations are coupled with the longitudinal and librational motion (figure 6), and thus it is expected that a proper control of longitudinal and librational motion will induce less attitude oscillations. Currently, the Italian satellite has its own control jets for yaw and pitch axes.

SATELLITE ATTITUDE OSCILLATIONS

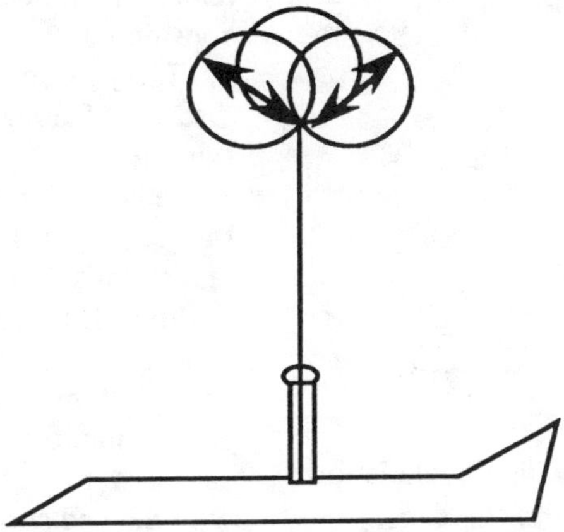

Figure 5. Satellite has three rotations at the attach point primarily induced by longitudinal and librational oscillations

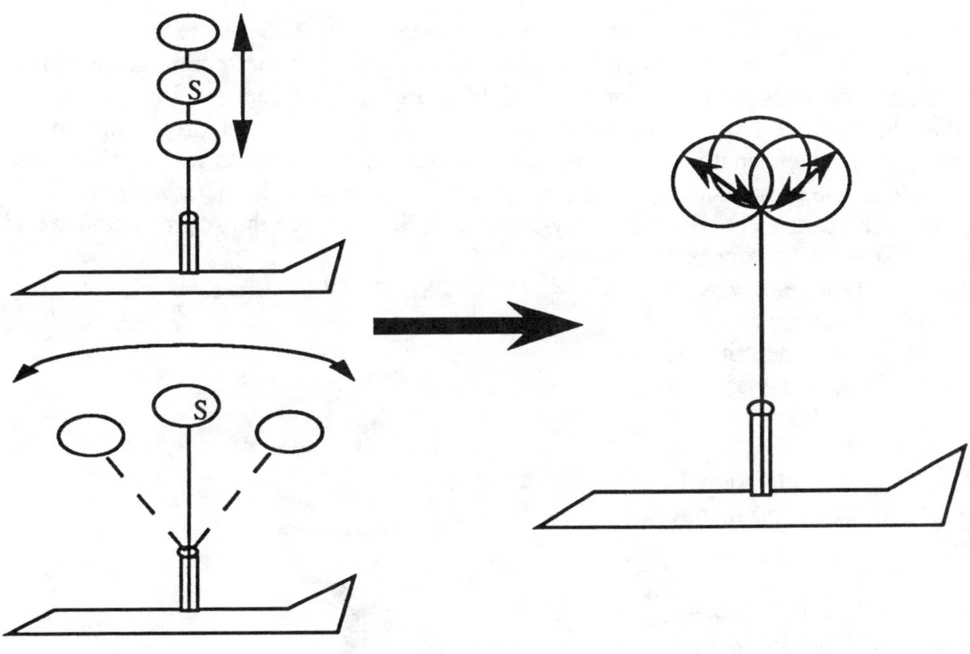

Figure 6 There is a strong coupling between longitudinal-librational motion and satellite attitude oscillations

BEAD MODEL OF TETHER

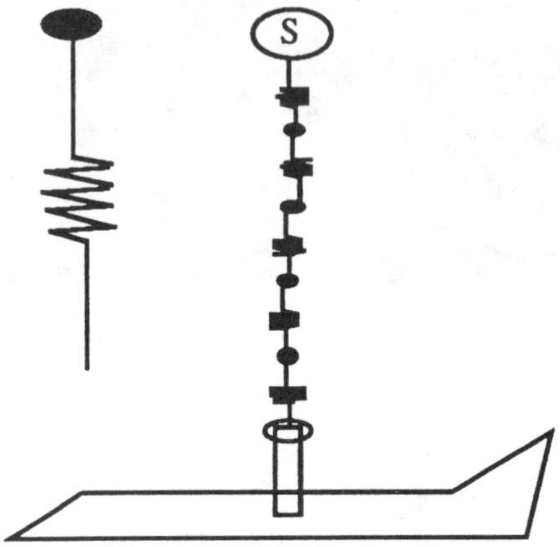

Figure 7. Tether mass is distributed in form of beads connected by springs.

In the TSS dynamics described above, the tether may be modeled as a spring without mass. Thus, it is called a massless tether model. However, the tether has a finite mass and it should be considered properly. When the satellite is fully deployed, this mass is distributed along the line connecting the satellite and the shuttle. In simulations, the tether mass can be represented as a lumped mass in a bead. More realistically, the entire tether length is divided into several segments thus making several beads with three degrees of freedom motion in the orbital environment. These beads are further assumed to be connected by springs with the end springs being connected to the shuttle and the satellite respectively (figure 7). In this way, the spring like characteristic as well as tether mass are accounted for in the TSS dynamics.

When the finite mass of the tether is properly modeled, the TSS also exhibits a special motion known as 'skip rope' motion (figure 8), particularly if the motion is excited by the interaction between the geomagnetic field and electric current passing through the tether. Parameters like the amplitude and time period for the skiprope motion are functions of tether length, current and the geomagnetic field around it. Magnitude and direction of the geomagnetic field depends on the TSS position in the orbit, and thus, is mission time dependent. There is a very strong coupling between the skiprope motion and the tether modes described earlier. Detailed analysis of the bead model simulations have shown that the skiprope motion induces a very low frequency characteristic satellite attitude oscillation (figure 9) that could possibly help measure skiprope motion.

"SKIP ROPE" Effect

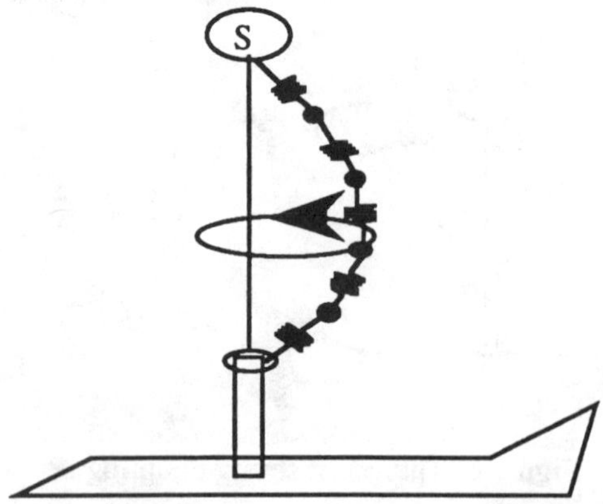

Figure 8. Tether circular transverse motion occurs naturally in bead model, particularly when geomagnetic field interacts with the current passing through tether.

The initial tether mission named TSS-1 will involve an upward deployment of the Italian satellite weighing approximately 500 kg. on a 20 km conducting tether. The mission time line as shown in figure 10 consists of five phases : Deployment phase lasting about 5 hours to reel out the tether and stop at 20 km tether length, On-station 1 phase of about 11 hours to perform scientific experiments and investigate the tether dynamics in orbit, Retrieval phase of about 7 hours to reel in the tether from 20 km to 2.4 km, On-station 2 phase of about 5 hours to damp out the unwanted motion of the tether and satellite, and the Final Retrieval phase lasting about 2 hours to reel in the tether to a length of 200 meters where the crew takes over for manual operations. The total mission duration is about 40 hours. As the tether is dragged through the magnetic field of the Earth, electric current is induced in this conducting tether creating another force on the tethered system. Due to the interaction between the orbital environment and tether dynamics as described above, tether length control during the mission is a complex problem.

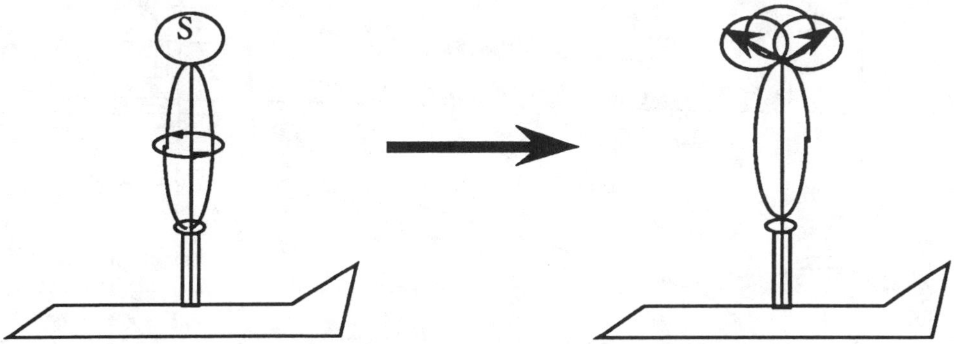

Figure 9 The tether "SKIPROPE" effect induces highly characteristics satellite attitude oscillations

The control system for the TSS consists of a reel powered by an electric motor, satellite thrusters, and the orbiter attitude control system. Due to the physical nature of the TSS and design of the hardware, only the deployed tether length and tether tension at the reel end can be directly measured and controlled. In-plane and out-of-plane libration amplitude cannot be directly measured or controlled, but are indirectly controlled through tether length maintenance [10].

Based on the simulation analysis, it was determined that the librational angles can be maintained within a range if particular mission profiles are selected for deployment and retrieval phase and the tether length, length rate and tension are maintained within an error envelope. The mission timeline shown in figure 10 is one particular profile converted into length as a function of mission time. Thus, a definite length value is required/desired at a definite mission time and if the actual measured length is different, then there is a length error that needs to be corrected. Since the speed of the reel/motor drive is a function of the motor characteristics and voltage applied, the desired length profile can be combined with motor specifications to compute a reference voltage profile for the mission. When the reference voltage is applied to the motor according to the profile, then the motor will generate a length rate that will deploy and retrieve the satellite according to the length profile selected. The motor will also maintain the desired tension in the tether. The algorithm that generates the reference voltage profile based on the length profile is called tether guidance. A reliable motor model has been created using the experimental data from laboratory testing, and is included in this algorithm.

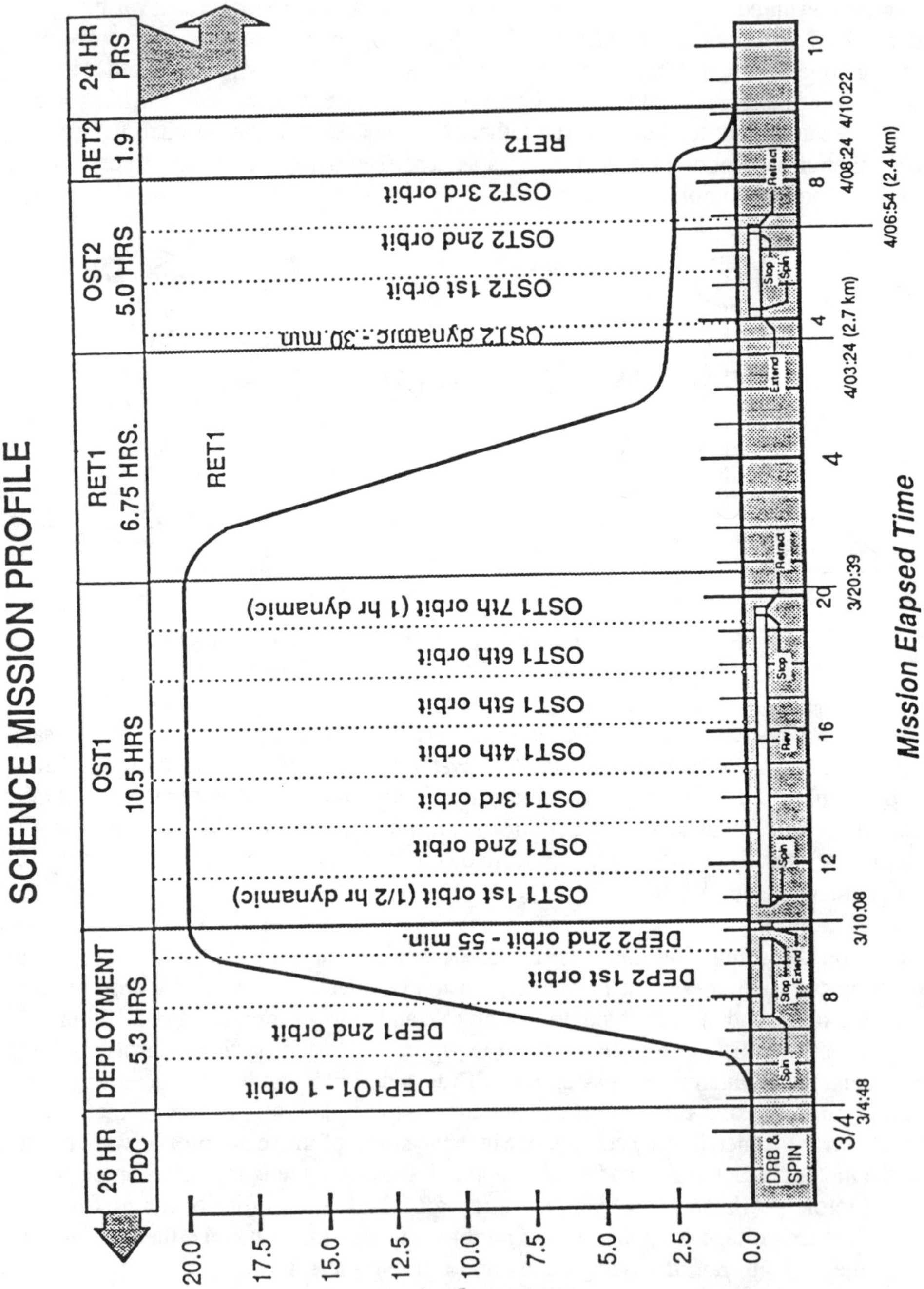

Fig. 10 TSS-1 mission time line.

Under the nominal situation in the orbit for TSS, the satellite can be deployed and retrieved using reference guidance voltage. If there are no perturbations in the orbital environment and the tether behaves as it is modeled, then, the length profile as well as tension profile will be generated properly by this reference voltage. However, there are perturbations (small and large) in the orbital environment, and tether behavior is going to be different resulting in small deviations in length profile. Thus, when there is a length error, then a delta voltage can be combined with this reference voltage to null the error and achieve the desired length. If length measurements show that the length profile is not followed properly, a delta voltage must be generated using length error. Similarly, there could be length rate error and tension error due to the orbital environment, and a delta voltage should be generated to null these errors. The algorithm that generates the delta voltage as function of length error, length rate error and tension error is called the tether length controller.

A conventional control system based on the linearization of the dynamics has been developed to control the deployed length and length rate during mission phases [10]. This conventional controller employed in the onboard software is based on the sensed length, length rate, and tension parameters and uses length error, length rate error and tension error to compute the delta voltage that is added to the reference voltage generated by the guidance. Due to the particular mission profile selected it is expected to maintain not only the length but also the in-plane librational angle within reasonable limits.

3. Fuzzy Logic Based Controller Design

A fuzzy logic controller for the tethered system can be designed by writing a few common sense, linguistic rules about how the tether should be controlled. For example, a designer might say that if the actual length of the tether is much less than the desired length, then the voltage to the reel mechanism must be increased. The assumption here is that the increase in voltage increases length rate. If the increase in voltage decreases the rate, then, of course the rule must be changed. Another way to describe this action is to say that the delta voltage should be positive if the length error is negative. A designer can relate the length error with the delta voltage in descriptive terms like positive_small, positive_large, and negative_large. The control system designer can use as many symbolic values as are necessary to quantify the length error and delta_voltage, and a number of rules can be written to relate these symbolic values. Fuzzy logic provides a framework for defining the symbolic values (called membership functions) and relate the input and output parameters through fuzzy rules.

If there is a plan to utilize length rate error as well as length error then the designer can reason that if the actual length of the tether is much less than the desired length and the actual length rate is greater than the desired length rate (that is, the length rate is compensating for the shortness of the tether), then the change in voltage (or delta_voltage) should be very small. Through this type of reasoning process, the designer of a fuzzy system generates a matrix (such as Table I or Table II), which describes the entire control system logic.

Table I Rules for proportional fuzzy controller

LE	DV
ZE	ZE
NL	NL
NM	NM
NS	NS
PS	PS
PM	PM
PL	PL

From this rule matrix, the designer can use a high level tool, such as Togai InfraLogic's TIL Shell [13] to generate the fuzzy rule base and compile this rule base using the Fuzzy-C compiler [14] into C Language source code. The main components of the fuzzy tether controller are the linguistic variables Length Error, Length Rate Error, and Delta Voltage. Using the TIL Shell, membership functions for each linguistic variable are defined in terms of its range and its belief value as shown in figures 11 and 12. So far, the design of the system has required intuitive knowledge of the physical system, rather than the understanding of the complex dynamics.

At this point, however, to define reasonable values for each membership function, some specific knowledge of the system is required. For example, the linguistic variable Length Error contains the linguistic values Negative Big, Negative Medium, Negative Small, Zero, Positive Small, Positive Medium, and Positive Big. Some familiarity with the tethered satellite system and its behavior during the mission is required to provide reasonable values for these membership functions. If a designer does not have this familiarity, then, a simple proportional controller can be tested to gain more knowledge about the system, and a new set of membership functions can be implemented.

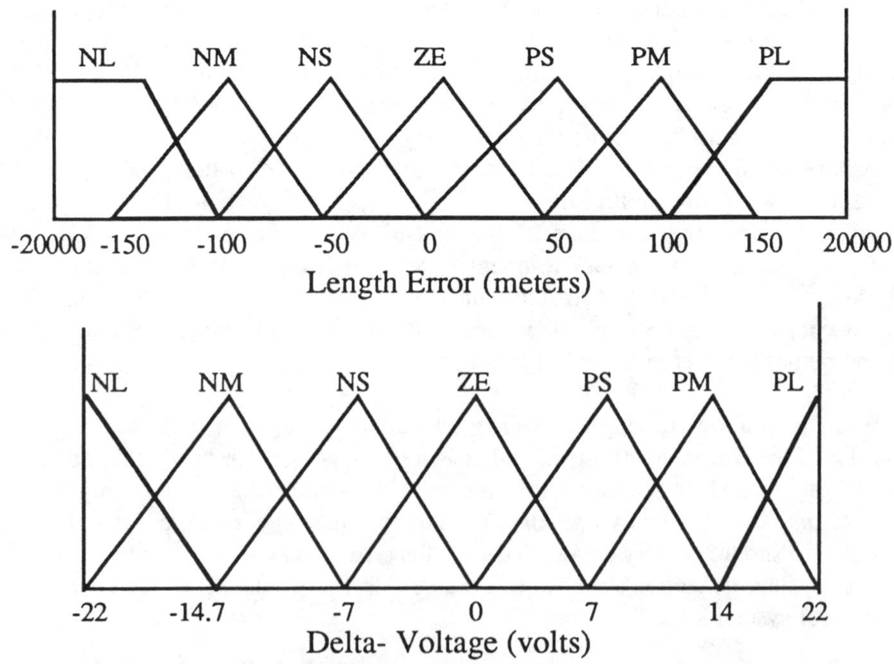

Figure 11 Membership Functions for simple proportional type fuzzy controller with length error and delta-voltage

The design of the fuzzy logic based tether controller was influenced by the tools available in the Software Technology Laboratory such as Togai InfraLogic's TIL Shell [13], and the Fuzzy-C development system [14]. A validated tether simulation based on the Orbital Operations Simulator (OOS)[11] with the base-lined control system [10] implemented for the TSS-1 mission is also available in the software technology laboratory. The fuzzy controller was designed to be compatible with the UNIX based OOS.

The first step in the development process was a simple linear proportional fuzzy controller (figure 11) created using the TIL shell. This proportional fuzzy controller consisted of seven membership functions for length error input, seven membership functions for the delta voltage output, and seven fuzzy rules (Table I). All rules are very simple and result from applying common sense reasoning. The full tether length that can be deployed is 20 km, so the universe of discourse for the length error was set from -20000 to 20000 meters, and the universe of discourse for delta voltage was set to its full range of -22 to 22 volts. The membership functions for both parameters are very

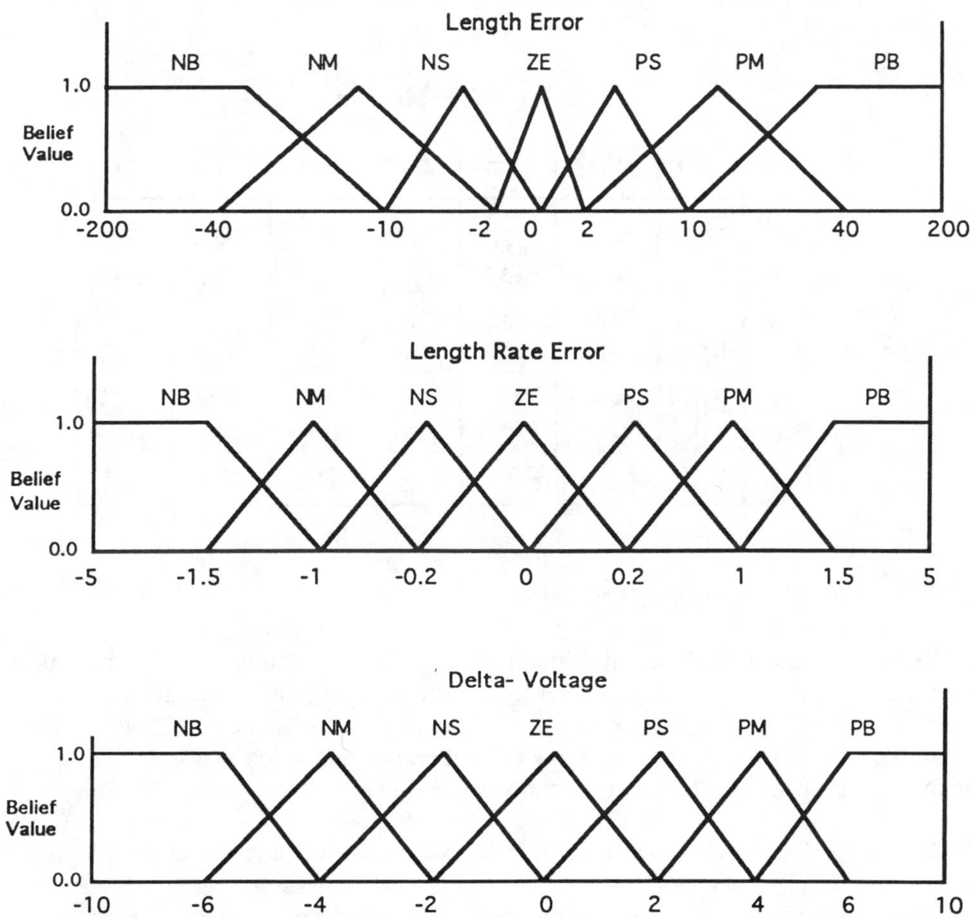

Figure 12. Membership functions for length error, length rate error and delta_voltage.

symmetric and no attempt is made to tune the performance. In fact, this simple proportional controller was designed and tested to assess the feasibility of using fuzzy logic for complex system such as tethers.

Having proven the feasibility of combining a fuzzy controller with the OOS, the fuzzy controller was improved by adding rules and membership functions for length rate error (figure 12). The membership functions were tuned, and the universe of discourse of the linguistic variables modified. For instance, the universe of discourse for Length Error was -20,000 to +20,000 meters in the linear proportional fuzzy controller. As shown in Fig. 12, it was reduced to allow for length errors from -200 to +200 meters because the analysis indicated that the length error would never exceed these values. The rulebase for delta voltage based on length error and length rate error consists of 49 rules as shown in Table II. Please note the asymmetry in the rulebase required to maintain tight control over the length. Since this controller is based on error and error rate, it is called a proportional-derivative type of controller in the classical control terminology.

Table II Rulebase for proportional-derivative type fuzzy controller for tether length.

Length Rate Error \ Length Error	NB	NM	NS	ZE	PS	PM	PB
NB	NB	NB	NB	NB	NS	NS	ZE
NM	NB	NM	NM	NM	NS	ZE	PS
NS	NB	NM	NM	NS	ZE	PS	PM
ZE	NB	NM	NS	ZE	PS	PB	PB
PS	NM	NS	ZE	PS	PM	PM	PB
PM	NS	ZE	PS	PM	PM	PM	PB
PB	ZE	PS	PS	PB	PB	PB	PB

4. Simulation Testing and Results

The tether simulation includes high fidelity models for the Shuttle Orbiter, the Italian Satellite, a massless tether, and a bead or finite element tether model. The simulation also contains a model of the reel motor mechanism and the conventional control law baselined for the TSS-1 Mission. To test the fuzzy controller, the simulation was modified to call the fuzzy controller instead of the conventional control law.

The TSS-1 mission simulation profile (as described earlier) includes upward deployment of the tether to 20 kilometers, followed by an On-station phase during which scientific experiments are conducted. Upon completion of the scientific experiments, the satellite is retrieved to a 2.4 km. length by commanding the reel motor drive. The OOS initializes the simulation from initial conditions defined in input data files, and then propagates the states of each of the vehicles in the simulation. Separate test cases were set up for the conventional controller and the fuzzy controller both running finite element tether models. We conducted experiments first with the massless tether simulation to gain confidence in the fuzzy controller before running with the more time consuming finite element tether simulation.

The performance of the linear proportional controller for the massless tether simulation (figure 13) was very encouraging, resulting in a length error of around 16 meters during deployment compared to 10.5 meters for the conventional controller. During the on-station phase, the fuzzy controller keeps length error at about 2.5 meters compared to 2.2 meters for the conventional controller. During the retrieval phase, the fuzzy controller reached a maximum length error of 8 meters while the conventional controller reached 6.5 meters length error. Tension error was about the same throughout the mission for both controllers.

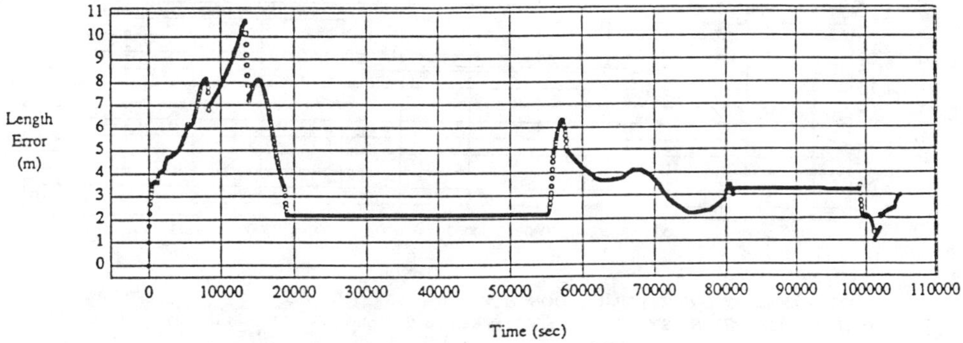

SIMULATION APPLICATION: TSS SIMULATION WITH MASSLESS TETHER
RUN IDENTIFICATION: TSS-1 FULL MISSION ***NEW*** PROFILE (ICLB97)
MODEL: TETHER_ML
DATE: Thu Sep 27 1990 08:20:07 AM
NUMBER OF DATA POINTS: 1751
DATA SAMPLING FREQUENCY: 0.017 Hz

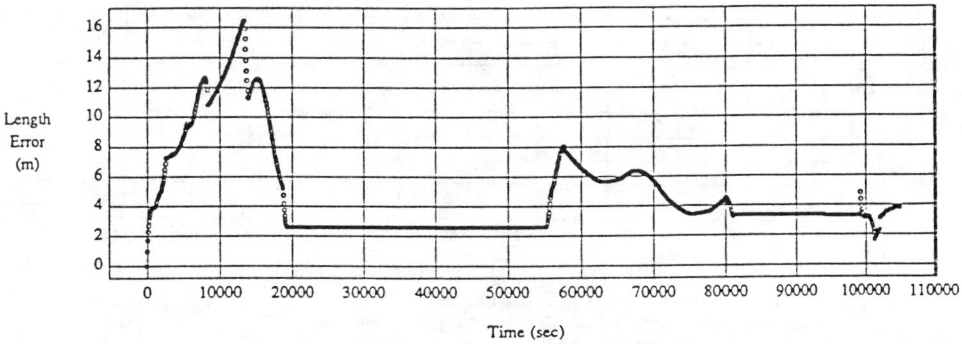

SIMULATION APPLICATION: TSS SIMULATION WITH MASSLESS TETHER
RUN IDENTIFICATION: Fuzzy Tether Delta Voltage Controller (Filtered)
MODEL: TETHER_ML
DATE: Wed Oct 03 1990 01:06:58 PM
NUMBER OF DATA POINTS: 1751
DATA SAMPLING FREQUENCY: 0.017 Hz

Fig. 13 Comparison of length error between simple proportional fuzzy and conventional controller in terms .

When length rate error was included in the fuzzy controller (figure 12) for the massless tether, the overall length error was reduced to within 5 meters (figure 14), while the conventional controller resulted in a length error of about 7.5 meters during the deployment phase[8]. During the on-station phase, the fuzzy controller (PD type) maintained a maximum length error of 2.1 meters as compared to 4.2 meters length error for the conventional controller. During the retrieval phase, the conventional controller exhibited a length error spike from 4.2 to 5.4 meters and then decreased to 1.5 meters length error. The fuzzy controller provided very smooth behavior during this phase and did not exhibit any length error spike. The fuzzy controller started the retrieval phase with a length error of 2 meters and decreased it to less than 1 meter. In general, the fuzzy controller provided smoother length control during the TSS-1 mission, when the massless tether model was used in the simulation.

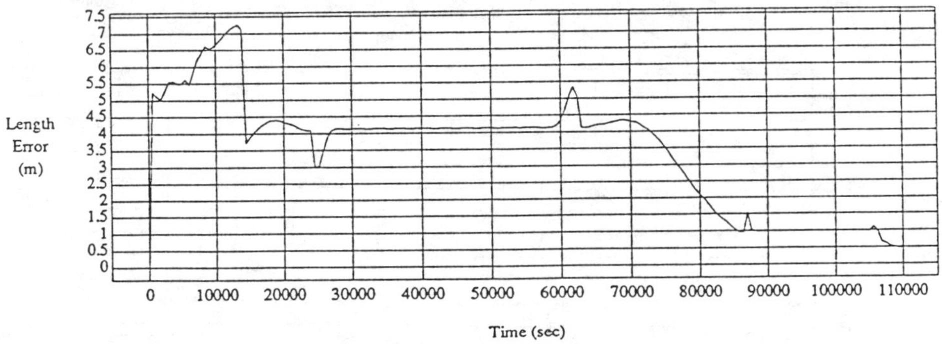

SIMULATION APPLICATION: TSS SIMULATION WITH MASSLESS TETHER
RUN IDENTIFICATION: SYS-17: Deployer CDMS Reel Control Software Model (proportional only)
MODEL: TETHER_ML
DATE: Tue Feb 05 1991 10:27:07 AM
NUMBER OF DATA POINTS: 184
DATA SAMPLING FREQUENCY: 0.002 Hz

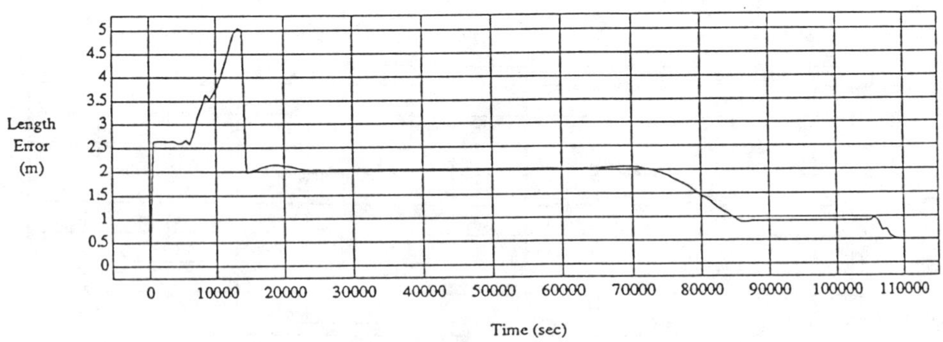

SIMULATION APPLICATION: Tether With Fuzzy Reel Control
RUN IDENTIFICATION: SYS-17: Deployer CDMS Reel Control Software Model
MODEL: TETHER_ML
DATE: Tue Feb 05 1991 02:12:25 PM
NUMBER OF DATA POINTS: 184
DATA SAMPLING FREQUENCY: 0.002 Hz

Fig. 14 Comparison of length error using massless model between Proportional-Derivative type fuzzy and conventional controllers.

Since the massless tether simulation provided a fair estimation of the performance of the fuzzy controller, we decided to utilize a bead or finite element model of the tether system in order to accurately model the complex dynamics of the tethered system. The finite element simulation allowed us to study the dynamic effects that can lead to a transverse oscillatory motion known as the skiprope phenomenon. The finite element simulation models the tether as a string of beads, each bead being a 3 degree of freedom object connected to the next bead by a spring. The first bead is connected to the shuttle and the last bead is connected to the satellite. In our simulation, we used ten beads to model the tether. The bead model requires proper initialization of all of its beads. As a result, the test case was set up to begin at 8 km. deployed tether length and the mission time line was set to retrieve up to 2.4 km. only. Since the deployment phase begins with so much deployed length, we expected that both, the fuzzy as well as the conventional controller, would not perform well and both would get saturated. Therefore, in evaluating the performance in terms of length error and other parameters we concentrated only on the retrieval phase.

As we expected, the length error saturates beyond 15 meters and in fact it goes over 100 meters during deployment. Then during on-station phase the length error reduces to a reasonable value. Both controllers maintain nearly the same value of the length error. However, during the retrieval phase of the mission, the fuzzy controller results in 2 to 4 meters less length error than the conventional controller as shown in figure 15. The tension error plots shown in figure 16 indicate that the fuzzy controller provides better overall tension control than the conventional controller. From 12,000 to 45,000 seconds of On-station 1 mission time, the tension error in the fuzzy controller varies from 3 to slightly over 9 Newtons, with most of the activity between 6 and 8 Newtons as indicated by the dark band. The fuzzy controller compared to the conventional controller shows a smoother tension error control from about 45,000 to 65,000 seconds during the retrieval of the tether from 20 km. to 2.4 km. length.

While it is important to control length error and tension error as tightly as possible, it may be more important to control the in-plane and out-of-plane libration amplitudes, especially during the retrieval. If the in-plane librations increase beyond 45 degrees, the satellite motion can result in a situation known as wrap-around due to orbital forces. It is very important to avoid such situations for the safety of the crew and the shuttle. The fuzzy control shows real advantage over conventional control in this case as shown in the plots (figure 17) of in-plane libration amplitude. During the on-station phase, the fuzzy control results in about half the in-plane libration amplitude as compared to the conventional control. During retrieval, between 70,000 and 90,000 seconds, the conventional control results in a librational amplitude of -12 to +12 degrees. The fuzzy control during the same time shows a libration amplitude of only -4.5 to +4.5 degrees. The out of plane librational angle remains nearly the same in both cases without much appreciable difference.

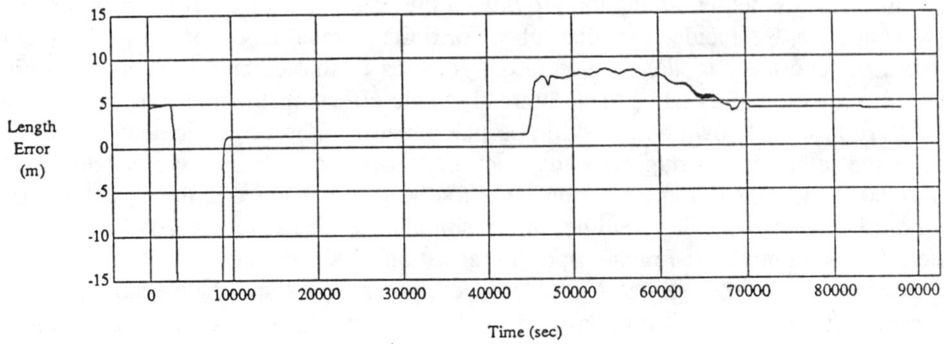

SIMULATION APPLICATION: TSS SIMULATION WITH FINITE ELEMENT TETHER
RUN IDENTIFICATION: DRM: friction, rip/cog, rate fdbk, 10 bd , 0.01dt
MODEL: TETHER_BD
DATE: Tue Sep 10 1991 03:58:31 PM
NUMBER OF DATA POINTS: 1765
DATA SAMPLING FREQUENCY: 0.020 Hz

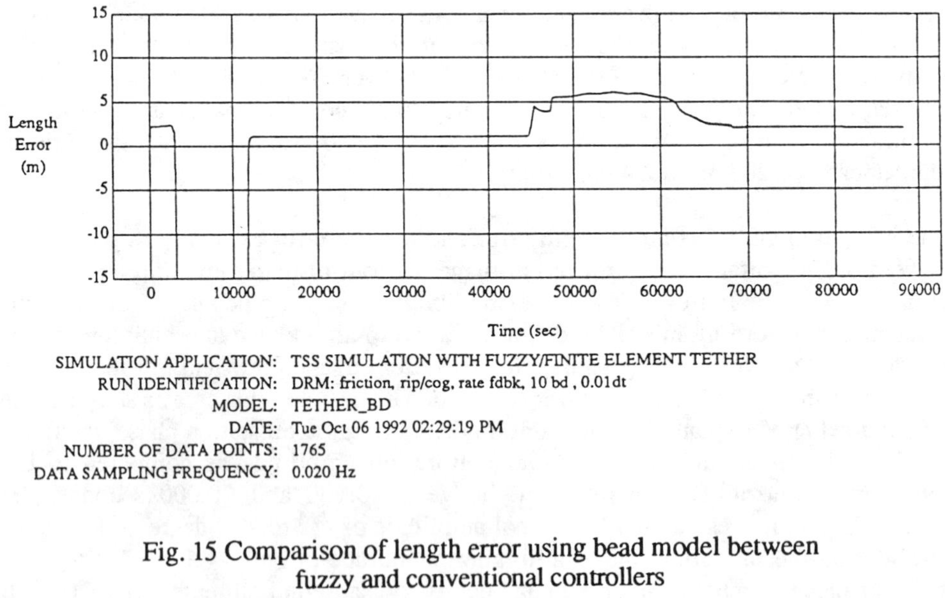

SIMULATION APPLICATION: TSS SIMULATION WITH FUZZY/FINITE ELEMENT TETHER
RUN IDENTIFICATION: DRM: friction, rip/cog, rate fdbk, 10 bd , 0.01dt
MODEL: TETHER_BD
DATE: Tue Oct 06 1992 02:29:19 PM
NUMBER OF DATA POINTS: 1765
DATA SAMPLING FREQUENCY: 0.020 Hz

Fig. 15 Comparison of length error using bead model between fuzzy and conventional controllers

CONTROL ERROR TIME HISTORY

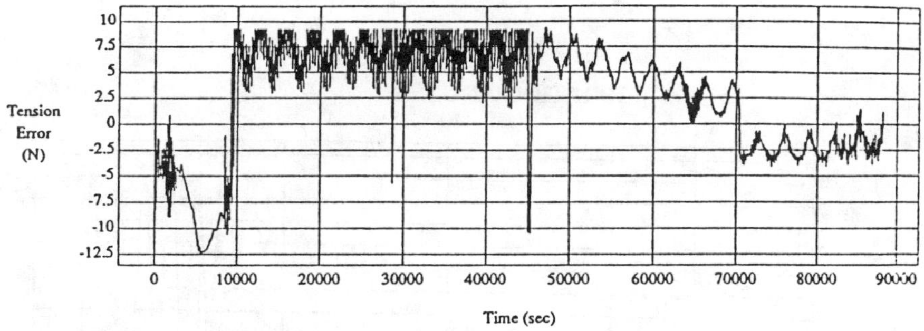

SIMULATION APPLICATION: TSS SIMULATION WITH FINITE ELEMENT TETHER
RUN IDENTIFICATION: DRM: friction, rip/cog, rate fdbk, 10 bd , 0.01dt
MODEL: TETHER_BD
DATE: Tue Sep 10 1991 03:58:31 PM
NUMBER OF DATA POINTS: 1765
DATA SAMPLING FREQUENCY: 0.020 Hz

CONTROL ERROR TIME HISTORY

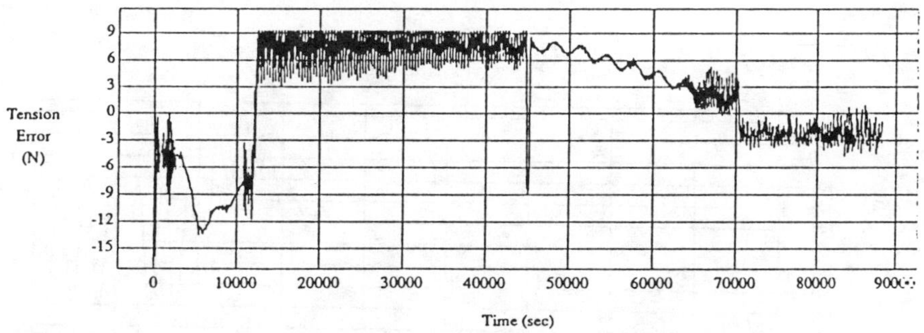

SIMULATION APPLICATION: TSS SIMULATION WITH FUZZY/FINITE ELEMENT TETHER
RUN IDENTIFICATION: DRM: friction, rip/cog, tens fdbk, rate fdbk, 10 bd , 0.01dt
MODEL: TETHER_BD
DATE: Tue Sep 10 1991 03:45:14 PM
NUMBER OF DATA POINTS: 1765
DATA SAMPLING FREQUENCY: 0.020 Hz

Fig. 16 Comparison of tension error using bead model between fuzzy and conventional controllers.

LIBRATION ANGLES

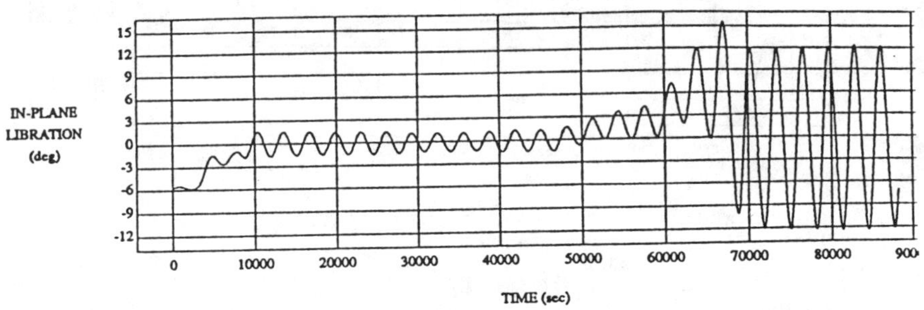

SIMULATION APPLICATION: TSS SIMULATION WITH FINITE ELEMENT TETHER
RUN IDENTIFICATION: DRM: friction, rip/cog, rate fdbk, 10 bd, 0.01dt
DATE: Thu Sep 12 1991 03:13:13 PM
NUMBER OF DATA POINTS: 1765
DATA SAMPLING FREQUENCY: 0.020 Hz

LIBRATION ANGLES

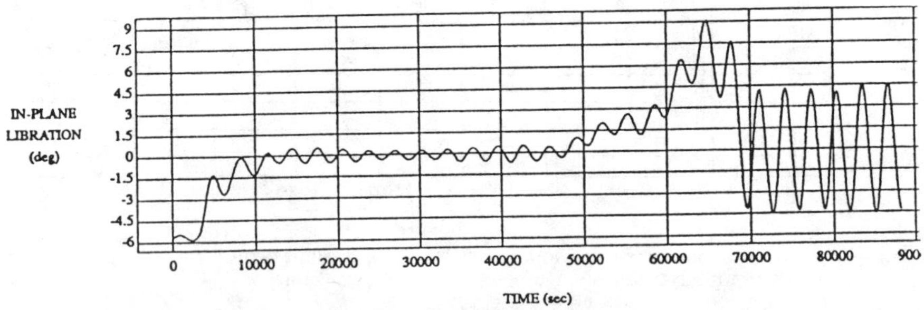

SIMULATION APPLICATION: TSS SIMULATION WITH FUZZY/FINITE ELEMENT TETHER
RUN IDENTIFICATION: DRM: friction, rip/cog, tens fdbk, rate fdbk, 10 bd, 0.01dt
DATE: Thu Sep 12 1991 02:37:23 PM
NUMBER OF DATA POINTS: 1765
DATA SAMPLING FREQUENCY: 0.020 Hz

Fig. 17 Comparison of librational oscillations using bead model between fuzzy and conventional controllers.

5. Summary

Our results based on the massless as well as finite element bead models of the tethers show promise regarding the applicability of fuzzy logic based length control to space based tethers. Deployed length of tethers can be easily controlled using length error and length rate error as input and imposing the output delta voltage in addition to the guidance command on the reel motor drive. Better length control by the fuzzy controller has the additional advantage of small tension oscillation during on-station and retrieval phases. Our simulation results also indicate that the fuzzy controller outperforms the conventional controller in terms of in-plane and out-of-plane libration amplitudes.

In our experience, fuzzy controllers can be easily designed using heuristics and experience, especially with tools like the TIL shell and Fuzzy-C compilers. Such commercially available tools simplify the development of fuzzy control systems significantly. Furthermore, the tuning of membership functions and modifications to the rulebase is also simplified to a point where rapid implementation and testing is possible.

The examination of whether librational oscillations, particularly in-plane oscillations, can also be controlled with the fuzzy logic approach will continue. It is planned to study the fuzzy controller's ability to handle different mission profiles such as a 100 kilometer downward deployment. Studies thus far have dealt with upward deployments. If a fuzzy controller can be easily modified to handle multiple mission profiles, overall life cycle costs for space tethered operations can be drastically reduced, allowing space tethers to be a useful technology for future space activities. One problem that neither conventional nor fuzzy controllers can address with respect to TSS-1 is the problem of controlling the skiprope motion. We plan to investigate the possibility of using data collected during such crew operations to train a hybrid fuzzy logic/neural network based control system that would possibly damp out oscillations and control any "skiprope" activity.

As mentioned earlier, the skiprope motion of the tether resulting from the interaction of the Earth's magnetic field with the current pulsing through the tether is very important to study for overall mission success. The skiprope dynamics is highly non-linear and cannot be minimized by optimizing a current pulsing profile. If the skiprope effect is not controlled, it can result in mission failure and serious damage to the orbiter. Since neither conventional nor fuzzy controllers can damp out the skiprope effect (again, due to the physical nature of the tethered system), manual damping methods must be used. The methods which will be used for TSS-1 involve the orbiter crew performing attitude maneuvers by firing attitude control thrusters. The effectiveness of these maneuvers depends heavily on the ability to accurately detect the parameters of skiprope motion. Our efforts are currently focused on applying neural network techniques toward rapid and accurate detection of the skiprope motion.

Acknowledgment

The authors would like to express their appreciation to William L. Othon and Jeffrey Hoblit of LinCom Corporation, Charles Copeland of Loral Space Information Systems, and James Villarreal of Software Technology Branch all of whom participated enthusiastically in this project and provided technical support in developing the software and analysis help.

References :

1. M. Rogers and Y. Hoshiai, "The Future Looks *'Fuzzy'* ", *Newsweek*, 46-47, May 28, 1990.
2. L. Armstrong and N. Gross "Why 'Fuzzy Logic' beats black-or-white thinking", Science & Technology section, *Business Week*, 92-93, May 21, 1990.
3. L. Zadeh, "Fuzzy Sets",*Information and Control*, vol. 8, 338-353 (1965).
4. R. C. Johnson, "Clear Leader Emerges : Japan at fuzzy fore", *EETimes*, Technology Section, Sept. 11, 1989.
5. G.J. Klir and T.A. Folger, *Fuzzy Sets, Uncertainty, and Information*, Prentice-Hall, New Jersey (1988).
6. C. C. Lee, "Fuzzy logic in control systems: fuzzy logic controller", *IEEE transactions on Systems, Man and Cybernetics*, **20**(2), 404-435 (1990).
7. R. N. Lea and Y. Jani, " Applications of Fuzzy Logic to Control and Decision Making", *Technology 2000*, NASA Conference Publication no. 3109, vol. 2, 67-75 (1990).
8. R. N. Lea, J. Hoblit, J. and Y. Jani, "Performance Comparison of a Fuzzy Logic based Attitude Controller with the Shuttle On-orbit Digital Auto Pilot", *North American Fuzzy Information Processing Society 1991 Workshop Proceedings*, 291-295 (1991).
9. R. N. Lea, J. Villarreal, Y. Jani and C. Copeland, "Fuzzy Logic Based Tether Control", *North American Fuzzy Information Processing Society 1991 Workshop Proceedings*, 398-402 (1991).
10. C. Bodley, "Tethered Satellite System Deployer Control System Design Review", vol. IV. (1989).
11. H. C. Edwards and R. Bailey, *The Orbital Operations Simulator User's Guide*, LinCom corporation, ref. LM85-1001-01 (1987).
12. R. N. Lea, J. Villarreal, Y. Jani and C. Copeland, "Tether Operations Using Fuzzy Logic Based Length Control", *Proceedings of the IEEE International Conference on Fuzzy Systems FUZZ-IEEE '92*, 1335-1342, 1992.
13. G. Hill, E. Horstkotte and J. Teichrow, *TIL Shell User's Manual*, release 1.1, Togai InfraLogic Inc. (1990).
14. C. Perkins, J. Teichrow and E. Horstkotte, "Fuzzy-C development system : A complete overview", SOAR-89 conference held at Johnson Space Center, Houston, 25-27 (1989).

Fuzzy and Possibilistic Clustering Methods for Computer Vision

Raghu Krishnapuram and James M. Keller

Department of Electrical and Computer Engineering
University of Missouri-Columbia
Columbia, MO 65211

Abstract

Uncertainty abounds in all aspects of computer vision. As a result, methods which explicitly model and manage that uncertainty have a better chance in producing meaningful results in such complex situations. Fuzzy set theory provides a framework to initially model the uncertainty in a vision problem, and many methods to exploit it in producing realistic results. One very important tool which has been used extensively in pattern recognition and computer vision is that of objective function based clustering. In this chapter, we will review classical and novel clustering methods as they apply to computer vision and show examples of their utility. In particular, we will focus on the use of clustering to detect and recognize regular boundaries of objects from images of edge magnitudes. The problem of fitting an unknown number of boundary curves to edge magnitude data is one of the major challenges of computer vision. We will show that fuzzy clustering can be readily adapted to the problem of curve detection, and that a new possibilistic clustering method introduced by the authors can produce significantly more accurate results than either crisp or fuzzy clustering in noisy situations.

1. Introduction

Clustering methods have been used extensively in pattern recognition and computer vision. Objective function based clustering methods are one particular class of clustering methods in which a criterion function is iteratively minimized until a global or local minimum is reached. Objective function based clustering can be either hard (crisp) or fuzzy, depending on whether each feature vector belongs exclusively to one cluster or to all clusters to different degrees. In general, the performance of fuzzy algorithms is superior to that of the corresponding hard versions, and they have a lower tendency to get stuck in local minima[1].

While clustering methods have been used for region segmentation for years, that will not be the main focus of application here. The detection and recognition of boundaries in two dimensional images and surfaces in three dimensional scenes from a sparse set of data points is one of the major problems in computer vision. If one assumes that the boundary segments come from a parametrized family of curves or surfaces, for example, linear subspaces, then the problem of both detection and recognition reduces to that of finding an "optimal" fit of a number of such functions to the edge or surface point data generated from the original image. This is a difficult problem which is exasperated by the presence of noise in the image. Thus the problem can be stated as *simultaneously* fitting a possibly unknown number of parametrized curves/surfaces to an unsegmented data set. There is a plethora of techniques to fit parametrized curves such as conics to segmented edge pixels and to fit parametrized surfaces to segmented range data. Segmentation of edge and range data is difficult in the case of jagged edges and noisy or sparse range data, since features such as gradients and curvatures cannot be computed reliably. A better approach in such situations would be to perform segmentation and boundary/surface fitting simultaneously on the data, without making use of features that assume continuity and smoothness of the edges and surfaces.

The generalized Hough transform[2] (GHT) can be used when the boundaries/surfaces are noisy or sparse. One disadvantage of the GHT is that its computational complexity is $O(N \times N_{p_1} \times N_{p_2} \cdots \times N_{p_{s-1}})$, where N is the total number of points in the image to be processed, N_{p_i} is the number of quantization levels of the i-th parameter, and s is the total number of parameters. The memory requirement of the GHT is $O(N_{p_1} \times N_{p_2} \cdots \times N_{p_s})$. In the case of a general second-degree curve in 2D, we need five parameters to describe the curve. The speed of the GHT can be improved only if we make certain assumptions about the curve, (i. e., if the curve is circular, elliptic etc.), and if the gradient information is available[3,4]. Also, if the edge points are somewhat scattered around the ideal curve (or surface), then peak detection is very difficult in the multidimensional Hough space due to bin splitting. Moreover, the detection of small segments is virtually impossible, since small peaks in the GHT are lost in the bias. The GHT also suffers from a high probability of spurious peaks[5].

Objective function based clustering has been traditionally used to detect "compact" or "filled" clusters in feature spaces, whose prototypes are typically represented by cluster centers. The Fuzzy C Means (FCM) algorithm[1] and its derivatives may be used to find clusters that resemble filled hyperspheres or filled hyperellipsoids. Lately this approach has been extended to the case of hollow or shell-like clusters by using shells (manifolds) for prototypes and measuring the distances to the shells rather than to the cluster centers. Since shell clustering algorithms can perform segmentation and fitting simultaneously, they are ideally suited for boundary detection and surface approximation when the boundaries/surfaces are ill-defined. They also require far less computations and memory compared to the GHT methods. Since they look for global structures and do not use edge following or region growing, they are insensitive to local aberrations and deviations in shape. They do not use features such as gradients and curvature and hence are not sensitive to noise and sharp discontinuities at the boundaries.

Determining the correct number of clusters present in a data set is an important aspect of the application of clustering to unknown situations. This determination is usually based on some measure of "cluster validity", which is represented by some metric on the final results of the clustering process. While this is clearly a crucial issue in boundary/surface detection, we will not devote much discussion to this point because of space limitations. The reader is referred elsewhere[1,6,7] for in-depth treatments of cluster validity.

In this chapter, we will review classical and novel clustering methods as they apply to computer vision and show examples of their utility. In what follows, “Linear subspace clustering” refers to the process of identifying linear or hyperplanar subspace clusters in a multidimensional feature space. More generally, “shell clustering” refers to the process of identifying clusters that resemble curves in 2-D or thin sheets (hypersurfaces) in a multidimensional feature space. All the algorithms that are discussed in this chapter are based on iteratively minimizing a least-square-type objective function, which can also be called the "sum of fuzzy intra-cluster distances".

2. Objective Function Based Fuzzy Clustering

Let $\boldsymbol{X} = \{\boldsymbol{x}_j \mid j = 1...N\}$ be a set of feature vectors in $\Re^n$ with coordinate-axis labels $x_1, x_2, \ldots, x_n$. Let $\boldsymbol{x}_j = [x_{j1}, x_{j2}, \ldots, x_{jn}]^{\mathrm{T}}$, and let $B = (\beta_1,\ldots,\beta_C)$ represent a C-tuple of prototypes each of which characterizes one of the C clusters. Each β_i consists of a set of parameters that define the prototype. In the following, we use β_i to denote both cluster i and its prototype. Let u_{ij} represent the grade of membership of feature point $\boldsymbol{x}_j$ in cluster β_i. The $C{\times}N$ matrix $U = [u_{ij}]$ is called a fuzzy C-partition matrix if it satisfies the following conditions:

$$u_{ij} \in [0,1] \text{ for all } i, \quad 0 < \sum_{j=1}^{N} u_{ij} < N \text{ for all } i, j \qquad \text{and}$$

$$\sum_{i=1}^{C} u_{ij} = 1 \text{ for all } j. \tag{1}$$

The Fuzzy C-partition is a generalization of the hard C-partition[1,8]. The problem of fuzzily partitioning the feature vectors into C clusters can be formulated as the minimization of an objective function $J(B,U;X)$ of the form

$$J(B,U;X) = \sum_{i=1}^{C} \sum_{j=1}^{N} (u_{ij})^m \, d^2(\mathbf{x}_j,\beta_i). \tag{2}$$

The objective function $J(B,U;X)$ is simply a generalization of the classical sum of intracluster distances criterion. In the above equation, $m \in [1,\infty)$ is a weighting exponent called the fuzzifier. When m approaches 1, the memberships are hard, and when $m \rightarrow \infty$, they are maximally fuzzy. A value of 2.0 works best for m in most applications. In (2), $d^2(\mathbf{x}_j,\beta_i)$ represents the distance from a feature point $\mathbf{x}_j$ to the prototype β_i. Minimization of the objective function in (2) can be done in two stages. In the first stage, the objective function is minimized with respect to B. However, this procedure varies according to the choice of the prototypes and the distance measure. Each choice leads to a different algorithm, as will be explained in subsequent sections. In the second stage[1], the objective function is minimized with respect to U subject to the constraints in (1). This gives us

$$\left.\begin{array}{ll} u_{ij} = \dfrac{1}{\displaystyle\sum_{k=1}^{C}\left(\dfrac{d^2(\mathbf{x}_j,\beta_i)}{d^2(\mathbf{x}_j,\beta_k)}\right)^{\frac{1}{m-1}}} & \text{if } I_j = \varnothing \\ \left.\begin{array}{ll} u_{ij} = 0 & i \notin I_j \\ \displaystyle\sum_{i\in I_j} u_{ij} = 1 & i \in I_j \end{array}\right\} & \text{if } I_j \neq \varnothing \end{array}\right\} \tag{3}$$

where $I_j = \{i \mid 1 \le i \le C,\ d^2(\mathbf{x}_j,\beta_i) = 0\}$. Thus, the membership of a point $\mathbf{x}_j$ in cluster β_i is inversely proportional to its *relative* distance from the prototype, except when the distance of $\mathbf{x}_j$ to a prototype β_i is zero. The hard (crisp) versions of these algorithms are easily obtained by changing the updating rule for the memberships so that they are always binary. In other words, use

$$u_{ij} = \begin{cases} 1 & \text{if } d^2(\mathbf{x}_j,\beta_i) < d^2(\mathbf{x}_j,\beta_k) \text{ for all } k \neq i \\ 0 & \text{otherwise} \end{cases}$$

Ties are broken arbitrarily. Objective function based clustering algorithms consist of Picard iterations through the two stages for minimizing the objective function. The general form of such clustering algorithms is given below.

OBJECTIVE FUNCTION BASED FUZZY CLUSTERING:

Fix the number of clusters C; fix m, $m \in [1,\infty)$;
*Initialize the fuzzy C-partition **U**;*
REPEAT
  Update the parameters of each cluster prototype;
  *Update the partition matrix **U** by using* (3);
UNTIL ($\|\Delta U\| < \varepsilon$);

In the next sections, we discuss various possible choices for the distance measure and the prototypes, and the minimization with respect to the prototype parameters in each case.

3. Classical Clustering Algorithms

3.1 The Fuzzy *C*-Means (FCM) Algorithm

The FCM uses points (also known as cluster centers) c_i for prototypes[1]. It uses the A-norm distance given by

$$d^2(x_j,\beta_i) = \| x_j - c_i \|_A^2 = (x_j - c_i)^T A (x_j - c_i),$$

where A is any positive-definite $n \times n$ matrix. In practice, A is usually the $n \times n$ identity matrix I_n, and hence the above distance reduces to the Euclidean norm

$$d^2(x_j,\beta_i) = d_{Eij}^2 = \| x_j - c_i \|^2. \tag{4}$$

When the membership values are fixed and the objective function is minimized with respect to the centroids c_i, we obtain the following equation for updating the centroids:

$$c_i = \frac{1}{N_i} \sum_{j=1}^{N} (u_{ij})^m x_j , \tag{5}$$

where

$$N_i = \sum_{j=1}^{N} (u_{ij})^m . \tag{6}$$

The FCM algorithm looks for spherical clusters when A is the identity matrix, or ellipsoidal clusters when A is a general positive definite matrix. However there is no preset way to select A, although the choice of A is very important and it should depend on the shape of the clusters. Moreover, when the clusters are of different shapes, a single matrix A cannot be used for all clusters. These considerations have motivated researchers to look for modifications to the FCM.

3.2 The Gustafson-Kessel (GK) Algorithm

Let us consider an ellipsoidal cluster whose covariance matrix C_i is diagonal. (If C_i is not diagonal, one can always rotate the coordinate system such that it becomes diagonal, and this process will not affect the distances.) Let us now consider the linear transformation

$$y_j = (\det C_i)^{1/2n} C_i^{-1/2} x_j,$$

where $C_i^{-1/2}$ denotes a matrix with all diagonal terms replaced by the square-roots of their reciprocals. If W_i denotes the covariance matrix of the cluster in the transformed space, then it is easy to show that $W_i = (\det C_i)^{1/n} C_i^{-1/2} C_i (C_i^{-1/2})^T = (\det C_i)^{1/n} I_n$. Thus, this transformation makes the cluster spherical. We may also note that the hypervolume of the cluster in the transformed domain is $(\det W_i)^{1/2} = (\det C_i)^{1/2}$, which is the same as the hypervolume of the cluster before the transformation. Thus, this transformation preserves the volume. Let the cluster center in the original space be c_i, and in the transformed space be w_i. The Euclidean distance of a point y_i to the center in the transformed space is $(y_j - w_i)^T (y_j - w_i) = (\det C_i)^{1/n} (x_j - c_i)^T (C_i^{-1/2})^T C_i^{-1/2} (x_j - c_i) = (\det C_i)^{1/n} (x_j - c_i)^T C_i^{-1} (x_j - c_i)$. Thus, using the Euclidean distance in the transformed space is equivalent to using a scaled Mahalanobis distance in the original space. It can also be shown that this transformation minimizes the total intra-cluster distance of the cluster among all possible such transformations[9]. Diday and Simon[10] devised the Adaptive Distance Dynamic Clusters (ADDC) algorithm based on this idea. Gustafson and Kessel seemingly independently derived a similar algorithm for the fuzzy case[11]. Thus, the Gustafson-Kessel algorithm seeks clusters that are hyperellipsoidal in shape, and it allows the norm inducing matrix $A = (\det C_i)^{1/n} C_i^{-1}$ to be different for each cluster. The distance used is

$$d^2(x_j, \beta_i) = d^2_{Gij} = (\det C_i)^{1/n} (x_j - c_i)^T C_i^{-1} (x_j - c_i). \qquad (7)$$

The prototypes β_i in this algorithm consist of the pairs (c_i, C_i). The centers c_i are updated using (5) as in the C-Means algorithm, where as the covariance matrices C_i are updated using

$$C_i = \frac{1}{N_i} \sum_{j=1}^{N} (u_{ij})^m (x_j - c_i)(x_j - c_i)^T. \qquad (8)$$

The FCM algorithm may be used to provide a reasonably good initialization for this algorithm.

3.3 The Gath and Geva (GG) Algorithm

This algorithm assumes that the feature points in each cluster have a Gaussian distribution. The distance of a point to a cluster can be interpreted as being inversely proportional to the *a posteriori* probability (or likelihood) of the point belonging to the cluster. Since the parameters of the Gaussian distribution and the *a priori* probabilities can vary for each cluster, this algorithm has the advantage of being able to handle clusters of different sizes, shapes, and densities. Let cluster β_i be Gaussian distributed with center c_i and covariance matrix C_i. Let the *a priori* probability of a point belonging to a cluster β_i be P_i. It can be easily verified[12] that the *a posteriori* probability of the point belonging to the cluster is proportional to $P_i [\det(C_i)]^{-1/2} \exp\{-\frac{1}{2}(x_j - c_i) C_i^{-1} (x_j - c_i)^T\}$. Therefore, the distance measure is given by

$$d^2(x_j, \beta_i) = d^2_{Xij} = \frac{[\det(C_i)]^{1/2}}{P_i} \exp\left[\frac{(x_j - c_i) C_i^{-1} (x_j - c_i)^T}{2}\right], \qquad (9)$$

If we assume that the *a priori* probability P_i of selecting a cluster is proportional to its size, then P_i can be updated in every iteration as

$$P_i = \frac{N_i}{N}. \qquad (10)$$

Thus, in this algorithm, the prototypes β_i consist of the triplets (P_i, c_i, C_i), and the center and the covariance matrix may be updated using (5) and (8). One needs to be careful while implementing this algorithm since the distance is exponential and computer precision is limited. Gath and Geva[13] also developed an unsupervised version of the algorithm known as the Unsupervised Fuzzy Partition-Optimal Number of Classes (UFP-ONC) algorithm. This algorithm evaluates the partitions of the data for a range of C values. The optimal partition is the one that optimizes the given validity criterion. The validity criteria are based on the hypervolume and density of the clusters. This algorithm can be used when the number of clusters is not known.

3.4 Examples with Classical Clustering Algorithms

Fig. 1 shows some examples involving classical clustering techniques. In all the examples, the hard partition obtained by assigning each feature point to the class with the highest membership is shown. (The points assigned to different classes are plotted with different symbols.) Fig. 1(a) shows the final hard partition obtained by the FCM algorithm with the Euclidean distance measure. It can be seen that the small cluster is influenced by the presence of the large cluster and moves closer to it, thereby taking away a few points from it. Figs. 1(b) shows the final hard partition obtained by the GK algorithm. One may notice the tendency of the GK algorithm to find elongated clusters. This can be a disadvantage. Fig. 1(c) shows the results of the GG algorithm on the same data set. The GG algorithm takes into account the size of the clusters, and its performance is thus excellent. Figs. 1(d), (e) and (f) show similar results on a data set with three clusters of different sizes and shapes.

These results seem to indicate that the performance of the GG algorithm is generally superior. However, it is to be borne in mind that the GG algorithm is computationally and implementationally more complex. Also, its performance is good only in a local region of the solution space since it is highly sensitive to initialization when compared with the other algorithms. It is best to initialize the GG algorithm with a partition obtained either by the FCM or the GK algorithm, which are less prone to get stuck in local minima.

4. Clustering in Linear Subspaces

4.1 The Fuzzy C-Varieties (FCV) Algorithm

The FCV was developed by Bezdek et al. to detect clusters that resemble lines or planes[1]. It retains the global metric structure induced by a fixed matrix $\boldsymbol{A}$, but allows the C prototypes β_i to be r-dimensional linear varieties, $0 \leq r \leq n\text{-}1$, rather than just points (cluster centers) in $\mathfrak{R}^n$. The "linear variety" of dimension r, $0 \leq r \leq n\text{-}1$ through the point $\boldsymbol{c}_i \in \mathfrak{R}^n$, spanned by the linearly independent vectors $\{\boldsymbol{e}_{i1}, \boldsymbol{e}_{i2},...,\boldsymbol{e}_{ir}\} \subset \mathfrak{R}^n$ is the set

$$L_{ri} = (\boldsymbol{c}_i; \boldsymbol{e}_{i1}, \boldsymbol{e}_{i2},...,\boldsymbol{e}_{ir}) = \{\boldsymbol{y} \in \mathfrak{R}^n \mid \boldsymbol{y} = \boldsymbol{c}_i + \sum_{k=1}^{r} t_k\, \boldsymbol{e}_{ik},\ t_k \in \mathfrak{R}\}. \quad (11)$$

Thus, L_{ri} can be thought of as "flat" sets in $\mathfrak{R}^n$, and r is the number of directions in which this "flatness" extends. The FCV uses the distance from the data points to the prototypical linear varieties. The perpendicular distance from $\boldsymbol{x}_j$ to L_{ri}, when $\{\boldsymbol{s}_{ij}\}$ are an orthonormal basis for their span, is

$$d^2(\boldsymbol{x}_j,\beta_i) = d^2_{Lij} = d^2(\boldsymbol{x}_j,L_{ri}) = \left[\|\boldsymbol{x}_j - \boldsymbol{c}_i\|^2 - \sum_{k=1}^{r} ((\boldsymbol{x}_j - \boldsymbol{c}_i)\bullet \boldsymbol{s}_{ik})^2 \right]. \quad (12)$$

The second term in (12) amounts to projecting $(\boldsymbol{x}_j - \boldsymbol{c}_i)$ onto the span formed by $\{\boldsymbol{e}_{ik}\}$. Note that for $r = 0$, the FCV reduces to the FCM. For $r = 1$, the FCV becomes the Fuzzy C-Lines, for $r = 2$, the Fuzzy C-Planes, and for $r = n - 1$, the Fuzzy C-Hyperplanes. In these algorithms, the fuzzy C-partition matrix $\boldsymbol{U}$ is updated the same way as in the FCM

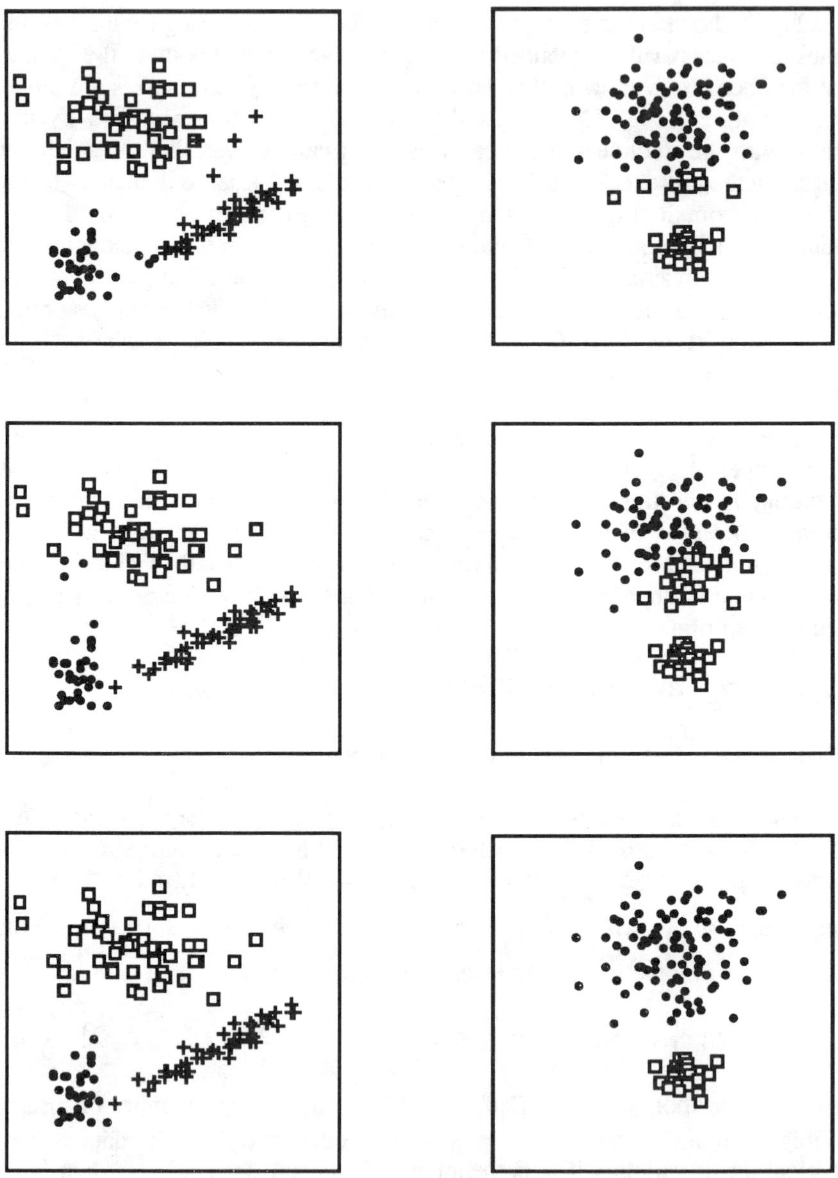

Fig. 1. Effect of different distance metrics on the clustering of two data sets. Top row: Euclidean distance; Middle Row: scaled Mahalanobis distance (GK); Bottom Row: maximum likelihood distance (GG). The results shown in all cases are the closest hard partitions, identified by different symbols.

except that the distance metric in (12) is used. For fixed U, the centers c_i are updated as in (5) and the spanning vectors are updated by using

$$e_{ik} = \text{the } k\text{th unit eigenvector of } C_i. \tag{13}$$

The eigenvectors are assumed to be arranged in descending order of the corresponding eigenvalues. Since the FCV uses the perpendicular distance to the linear variety clusters being sought, it does not take in account the length or extent of the linear cluster. For example, if $r = 1$, the FCV seeks infinitely long thin lines, and thus can lump approximately collinear clusters together, even when they are very far apart. One solution[1] to this problem is to choose a distance measure given by

$$d^2(x_j,\beta_i) = d^2_{Cij} = \alpha\, d^2_{Lij} + (1 - \alpha) d^2_{Eij} \tag{14}$$

which is a convex combination of the perpendicular and the Euclidean distances. Depending on the cluster shapes, α can vary from 0 for spherical clusters to 1 for linear clusters. The resulting algorithm is called the Fuzzy C Elliptotypes (FCE) algorithm. The FCE can seek clusters of ellipsoidal shapes in the data set, where the shape can vary all the way from spheres to lines. However, one must know the shape of the clusters in the data set *a priori* to be able to select a proper value for α. Also, the clusters have to be of similar shapes since one single value of α is used for all the clusters.

4.2 The Adaptive Fuzzy Clustering (AFC) Algorithm

Davé offered a solution to the problem of selecting α in the FCE algorithm[14]. The distance metric used in his AFC is identical to the one in (14) except that the α is different for each cluster, reflecting an individualized distance measure that best represents the shape of each different cluster. The choice for the α value for cluster β_i, denoted by α_i, is heuristically set to

$$\alpha_i = 1 - \frac{\min\{\lambda_{ik}\}}{\max\{\lambda_{ik}\}}, \quad i = 1, \ldots, C, \tag{15}$$

where λ_{ik} is the kth largest eigenvalue of the generalized covariance matrix C_i of cluster β_i. The performance of the AFC algorithm has been shown to be good for finding linear clusters in 2-D.

4.3 The Compatible Cluster Merging (CCM) Algorithm

Although the GK and GG algorithms are meant for ellipsoidal clusters, they can also be used to find linear or planar clusters. This is because lines and planes can be viewed as special cases of ellipsoids for which one or more of the radii are zero. Krishnapuram and Freg[15] developed a Compatible Cluster Merging (CCM) technique which finds the optimal number of line/plane segments in a data set using this idea. The CCM algorithm performs the GK algorithm with a large value C_{max} for the number of clusters C, and after convergence of the GK algorithm, merges compatible clusters among the C_{max} clusters to obtain the final result. Let the cluster centers be c_i and c_j, their eigenvalues be $\{\lambda_{i1},\ldots\lambda_{in}\}$ and $\{\lambda_{j1},\ldots\lambda_{jn}\}$, and their eigenvectors be $\{e_{i1},\ldots,e_{in}\}$ and $\{e_{j1},\ldots,e_{jn}\}$. Assuming that the eigenvalues and their eigenvectors are arranged in descending order, the compatibility conditions of clusters β_i and β_j are given below.

$$|e_{in} \bullet e_{jn}| \geq C_1,\ C_1 \text{ close to } 1 \tag{16}$$

$$\left| \frac{e_{in} + e_{jn}}{2} \bullet \frac{c_i - c_j}{\|c_i - c_j\|} \right| \leq C_2, \qquad C_2 \text{ close to 0, and} \qquad (17)$$

$$\|c_i - c_j\| \leq C_3 \left(\sqrt{\lambda_{i1}} + \sqrt{\lambda_{j1}}\right), \qquad C_3 \approx \sqrt{3}. \qquad (18)$$

Eq. (16) ensures that the hyperplanes are parallel, (17) ensures that the line joining the cluster centers be approximately orthogonal to the normals of the two lines (planes), and finally (18) verifies that the cluster centers are relatively close. This algorithm can therefore detect an "unknown" number of linear (2-D) or planar (3-D) clusters in a data set. A value of 1.5 is recommended for the fuzzifier m. The compatible cluster merging algorithm is given below.

THE COMPATIBLE CLUSTER MERGING (CCM) ALGORITHM

$C = C_{max}$;
Run the GK algorithm with $C = C_{max}$;
Find eigenvalues and eigenvectors of clusters;
Form groups of compatible clusters using (16-18);
IF *number of groups* $< C$
  Merge each group into one cluster;
  $C =$ *number of groups;*
  Update centers and covariance matrices of each cluster;
END IF;
Re-assign samples to clusters based on updated centers and covariance matrices.

The clustering and merging can be done either in one pass or iteratively. The one pass algorithm is given above. The iterative merging algorithm consists of repeatedly performing the CCM algorithm, and stopping when no more clusters can be merged. Krishnapuram and Freg[15] provide several examples of line and plane fitting using this approach.

4.4 Examples of Clustering in Linear Subspaces

As an example of the effect of the distance metric on the results of clustering linear subspaces, consider the two situations in the top row of Fig. 2. The left side depicts three intersecting noisy linear clusters of different sizes, whereas the right side shows three noisy linear clusters out of which two are collinear. The FCV algorithm does a poor job in both cases, for example, in the colinear situation, it finds two almost identical clusters and interprets the two collinear line segments as parts of one long line. Therefore, the points belonging to these two clusters are arbitrarily assigned to two of the prototypes. On the other hand, the GK (middle), AFC (bottom), and GG (not shown) algorithms correctly identify the two line segments as two different clusters, and hence the assignment of the points to their respective clusters is much better. This shows the advantage of the AFC, GK, and GG algorithms which are able to detect compact ellipsoidal clusters of finite size. The tendency of the FCV to disregard compactness completely and lump points that are widely separated into the same cluster can easily be seen in the cluster denoted by the +'s in the top left image of the three intersecting lines.

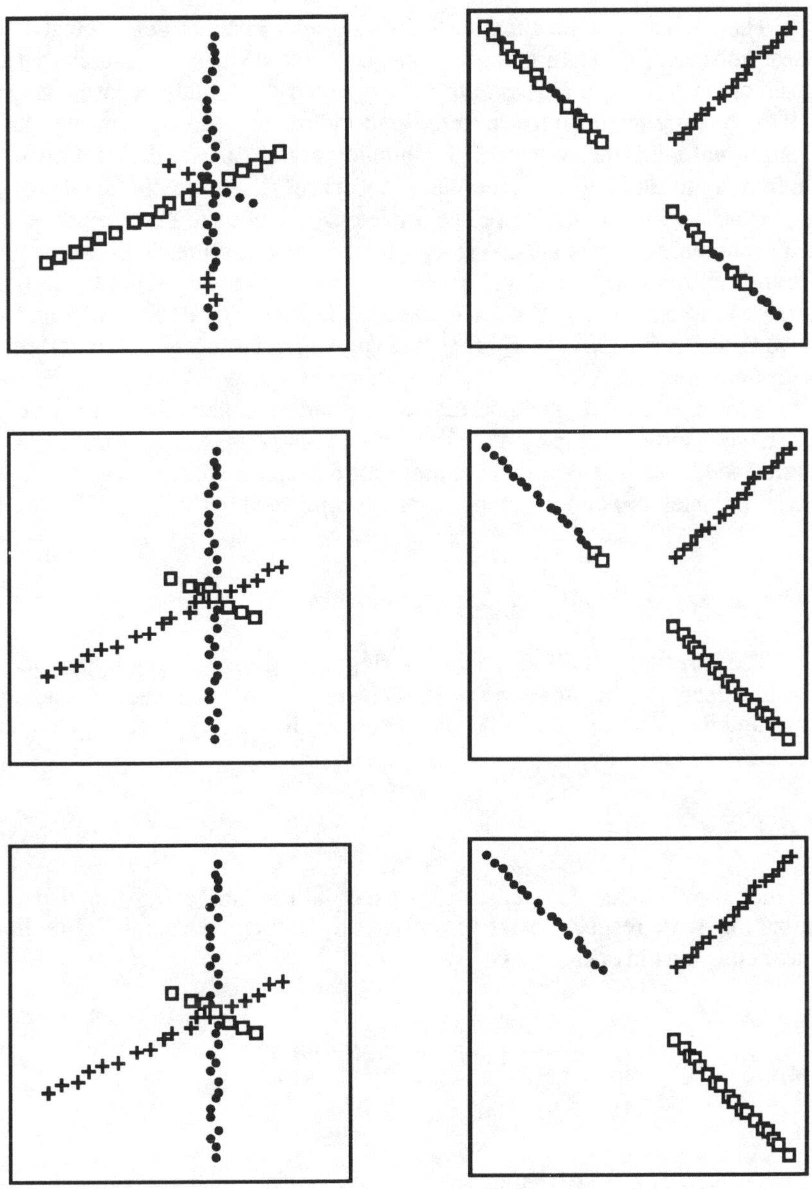

Fig. 2. Effect of different distance metrics on linear subspace clustering of two data sets. Top Row: perpendicular distance (FCV); Middle Row: scaled Mahalanobis distance (GK); Bottom Row: combined distance (AFC). The results shown in all cases are the closest hard partitions, identified by different symbols.

5. Shell Clustering

The clustering algorithms in Sections 3 and 4 cannot detect clusters that lie in nonlinear subspaces of feature space because all the distance measures that have been discussed assume that the clusters are compact (or filled), and except in the case of the FCV, they measure the distance from the centroid of the clusters. However, linear/planar clusters are not sufficiently general for boundary and surface description in computer vision. In order to find clusters in nonlinear subspaces that resemble "shells" or patches of (hyper)surfaces with no interior points, different distance measures must be developed, i.e., ones that measure the distance from a feature vector to the shell of the cluster rather than to the centroid of the cluster. Coray suggested the use of this idea to find circular clusters[16]. More recently, Davé's Fuzzy *C* Shells (FCS) algorithm[17] and the Fuzzy Adaptive C-Shells (FACS) algorithm[18] have proven to be successful in detecting circular and elliptical shapes. A computationally simpler Fuzzy *C* Spherical Shells (FCSS) algorithm for clustering hyperspherical shells, and extensions to more general quadric shapes, have also been proposed[19,20]. In shell clustering, the membership updates are still performed by using (3) with an appropriate distance measure. However, updating the prototype parameters is considerably more complicated, as will be seen the following sections.

5.1 The Fuzzy C-Shells (FCS) Algorithm

The Fuzzy C-Shells (FCS) clustering algorithm[17] seeks hyperspherical shell clusters in a data set. The prototype of each cluster is a spherical shell characterized by its center $\mathbf{c}_i$ and its radius r_i. The FCS uses the exact distance from feature vector $\mathbf{x}_j$ to the shell of cluster i given by

$$d^2(\mathbf{x}_j,\beta_i) = (\| \mathbf{x}_j - \mathbf{c}_i \| - r_i)^2. \tag{19}$$

When the above distance measure is used in the objective function in (2), its minimization with respect to the prototypes $\mathbf{c}_i$ and r_i yields the following coupled nonlinear equations in $\mathbf{c}_i$ and r_i :

$$\sum_{j=1}^{N} u_{ij}^{m}\left[1 - \frac{r_i}{\| \mathbf{x}_j - \mathbf{c}_i \|}\right](\mathbf{x}_j - \mathbf{c}_i) = 0, \text{ and} \tag{20}$$

$$\sum_{j=1}^{N} u_{ij}^{m} (\|\mathbf{x}_j - \mathbf{c}_i \| - r_i) = 0. \tag{21}$$

The above equations are not explicit in r_i and $\mathbf{c}_i$. Therefore a technique such as Newton's method that solves a set of coupled nonlinear equations must be used. When the prototype parameters r_i and c_i are updated for the very first time, the starting value for the centers $\mathbf{c}_i$ may be obtained through (5). The following estimate may be used as the starting values for r_i :

$$r_i = \frac{1}{N_i}\sum_{j=1}^{N} u_{ij}^{m} \| \mathbf{x}_j - \mathbf{c}_i \|^2.$$

In the subsequent updates, the values of the prototype parameters from the previous iteration can be used to initialize Newton's method. The FCS is quite successful in detecting spherical shell clusters. However it is computationally somewhat expensive because it needs to use a numerical method to solve two coupled nonlinear equations in every iteration. Bezdek and Hathaway[21] suggested a method to accelerate the convergence of the above algorithm. They showed that an exact solution of the system of equations (20), (21) is not required at every step of the fixed point iteration. Instead, they suggested using only one step of Newton's method in each iteration. However, this may increase the overall number of iterations required for the algorithm to converge.

5.2 The Fuzzy C Spherical Shells (FCSS) Algorithm

A computationally simpler algorithm to detect spherical shells can be devised if one uses an alternative distance measure known as the algebraic distance[22]. The algebraic distance from x_j to a prototype $\beta_i = (c_i, r_i)$ is given by

$$d^2(x_j,\beta_i) = d^2_{Sij} = (\| x_j - c_i \|^2 - r_i^2)^2. \tag{22}$$

In order to minimize the resulting objective function with respect to the prototype parameters, we rewrite the distance in (22) as

$$d^2_{Sij} = p_i^T M_j p_i + v_j^T p_i + b_j ,$$

where

$$b_j = (x_j^T x_j)^2, \quad v_j = 2(x_j^T x_j) y_j, \qquad y_j = \begin{bmatrix} x_j \\ 1 \end{bmatrix}, \tag{23}$$

$$M_j = y_j\, y_j^T, \quad \text{and} \quad p_i = \begin{bmatrix} -2\, c_i \\ c_i^T c_i - r_i^2 \end{bmatrix}. \tag{24}$$

It is easy to show[23] that when the distance measure in (22) is used, vectors p_i that minimize (2) must satisfy

$$p_i = -\frac{1}{2} (H_i)^{-1} w_i , \tag{25}$$

where

$$H_i = \sum_{j=1}^{N} (u_{ij})^m M_j , \quad \text{and} \quad w_i = \sum_{j=1}^{N} (u_{ij})^m v_j . \tag{26}$$

Equation (26) requires the inversion of the matrix H_i. This is quite trivial when the feature space is two-dimensional or three-dimensional. In theory the inverse will always exist as long as $N > n+1$, and the feature vectors are not all collinear. Here N is the number of feature vectors and n is the dimensionality. In practice, there is very little difference between the results obtained by this algorithm and the FCS algorithm. Krishnapuram et al.[23] also developed an unsupervised version of this algorithm which can be used when the number of clusters is not known.

5.3 The Adaptive Fuzzy C-Shells Algorithm : Unconstrained A_i (AFCS-U)

Davé and Bhaswan[18] proposed an algorithm for elliptical shells. This algorithm is based on the fact that the equation of an ellipsoid with a center c_i may be written as

$$(x_j - c_i)^T A_i (x_j - c_i) = 1, \tag{27}$$

where c_i is the center, and A_i is a positive definite symmetric matrix which determines the major and minor axes lengths as well as the orientation of the ellipse. Let us consider the distance D_{ij} defined by

$$(D_{ij})^2 = ([(x_j - c_i)^T A_i (x_j - c_i)]^{1/2} - 1)^2. \tag{28}$$

The above distance can be interpreted as a normalized radial distance, as will be explained in the next section. Davé and Bhaswan[18] showed that the minimization of the objective function in (2), with D_{ij} as the distance measure results in the following equations for updating the prototypes $\beta_i = (c_i, A_i)$.

$$\sum_{j=1}^{N} (u_{ij})^m \frac{D_{ij}}{d_{ij}} (x_j - c_i) = 0, \text{ and} \tag{29}$$

$$\sum_{j=1}^{N} (u_{ij})^m \frac{D_{ij}}{d_{ij}} (x_j - c_i)(x_j - c_i)^T = 0. \tag{30}$$

where

$$d_{ij}^2 = (x_j - v_i)^T A_i (x_j - v_i). \tag{31}$$

The above system of equations must be solved using numerical techniques in each iteration. Several examples are given[18].

5.4 The Fuzzy C Ellipsoidal Shells Algorithm

Let the equation of an ellipse be defined by (27). Consider the radial distance from a point x_j to the ellipse. This distance can be computed as

$$d^2(x_j, \beta_i) = d_{Rij}^2 = \|x_j - z\|^2, \quad 32)$$

where z is the point of intersection of the ellipse and the line joining the center c_i and the point x_j, i. e., z satisfies the following two equations:

$$(z - c_i)^T A_i (z - c_i) = 1, \text{ and} \tag{33}$$

$$z - c_i = \alpha (x_j - c_i). \tag{34}$$

In (34), $\alpha \in [0, 1]$. Substituting (34) in (33), we can solve for α. Then we can use (32) to obtain

$$d_{Rij}^2 = \| x_j - z \|^2 = (1 - \alpha)^2 \|x_j - c_i\|^2,$$

or

$$d_{Rij}^2 = \frac{(\sqrt{(x_j - c_i)^T A_i (x_j - c_i)} - 1)^2 \|x_j - c_i\|^2}{(x_j - c_i)^T A_i (x_j - c_i)}$$

$$= (\sqrt{(x_j - c_i)^T A_i (x_j - c_i)} - 1)^2 \|z - c_i\|^2$$

$$= (D_{ij})^2 \|z - c_i\|^2 \tag{35}$$

The distance measure defined in equation (35) is a good approximation to the exact (perpendicular) distance for points located close to the axes, and has the advantage of being simpler to compute when compared with the exact distance (see section 5.6). Minimization of the resulting objective function with respect to A_i and c_i yields the equations

$$\sum_{j=1}^{N} (u_{ij})^m \frac{d_{Rij}}{d_{ij}^3} \|x_j - c_i\| (x_j - c_i)(x_j - c_i)^T = 0, \text{ and} \tag{36}$$

$$\sum_{j=1}^{N} (u_{ij})^m \left[\frac{d_{Rij}\|x_j - c_i\|}{d_{ij}^3} (x_j - c_i) + (1 + d_{ij}) \|x_j - c_i\| A_i (x_j - c_i)\right] = 0, \tag{37}$$

where d_{ij} is as defined in (31). Equations (36) and (37) can be solved numerically by using the Levenberg-Marquardt algorithm.

From (35) we see that the distance used in AFCS-U algorithm is simply a normalized radial distance. The normalization factor depends not only on the size of the ellipse, but also on the position of the point with respect to the ellipse. Thus, the distances, given by equation (27), from large ellipses will be relatively small, and distances near the major axis will be divided by larger values than distances near the minor axis. This type of normalization may not be desirable when the data points are scattered and when the ellipses are of widely varying sizes.

5.5 The Fuzzy C Quadric Shells Algorithm

This algorithm is a generalization of the shell clustering algorithms discussed above, in the sense that it allows each cluster to be any second-degree curve or hyperquadric surface, and the prototypes β_i consist of parameter vectors p_i which define the equations of the hyperquadric surfaces. The general equation for such a hyperquadric surface is

$$p_i^T q = 0, \tag{38}$$

where

$$p_i^T = [p_{i1}, p_{i2}, \ldots, p_{in}, p_{i(n+1)}, \ldots, p_{ir}, p_{i(r+1)}, p_{i(r+2)}, \ldots, p_{i(r+n)}, p_{is}].$$

$$q^{\mathrm{T}} = [x_1^2,\ x_2^2, \ldots, x_n^2, x_1x_2, \ldots, x_{n-1}x_n, x_1, x_2, \ldots, x_n, 1], \text{ and}$$

$$s = \frac{n(n+1)}{2} + n + 1 = r + n + 1.$$

We may define the algebraic (or residual) distance from a point x_j to a prototype β_i as

$$d^2(x_j,\beta_i) = d^2_{Qij} = p_i^{\mathrm{T}} q_j q_j^{\mathrm{T}} p_i = p_i^{\mathrm{T}} M_j p_i\,, \tag{39}$$

where

$$M_j = q_j q_j^{\mathrm{T}}, \tag{40}$$

with $q_j^{\mathrm{T}} = [x_{j1}^2,\ x_{j2}^2, \ldots, x_{jn}^2, x_{j1}x_{j2}, \ldots, x_{j(n-1)}x_{jn}, x_{j1}, x_{j2}, \ldots, x_{jn}, 1]$.

In order to obtain a fuzzy C-partition of the data, we may minimize the objective function in (2) with d^2_{Qij} as the underlying distance measure. However, since the objective function is homogeneous with respect to p_i, we need to constrain the problem in order to avoid the trivial solution. One of the possibilities is:

$$\|p_{i1}^2 + p_{i2}^2 + \ldots + p_{in}^2 + \tfrac{1}{2}p_{i(n+1)}^2 + \tfrac{1}{2}p_{i(n+2)}^2 + \ldots + \tfrac{1}{2}p_{ir}^2\|^2 = 1. \tag{41}$$

This constraint was used by Bookstein[24] and has the advantage that the resulting distance measure is invariant to rigid transformations of the prototype. However, it does not allow the solution to be linear or planar. Many other constraints are also possible[25]. In our experience, the above constraint is the best compromise between computational complexity and performance in the 2-D case. If we define

$$a_{ik} = \begin{cases} p_{ik} & 1 \le k \le n \\ \dfrac{p_{ik}}{\sqrt{2}} & n+1 \le k \le r \end{cases}, \quad \text{and} \quad b_{ik} = p_{i(r+k)} \quad 1 \le k \le s-r,$$

then the constraint in (41) simplifies to $\|a_i\|^2=1$. It is easy to show[20] that the solution is given by

$$a_i = \text{eigenvector of } (F_i - G_i^{\mathrm{T}} H_i^{-1} G_i) \text{ with the smallest eigenvalue, and}$$

$$b_i = -H_i^{-1} G_i a_i, \tag{42}$$

where

$$F_i = \sum_{j=1}^{N} (u_{ij})^m R_j, \qquad G_i = \sum_{j=1}^{N} (u_{ij})^m S_j\,, \qquad H_i = \sum_{j=1}^{N} (u_{ij})^m T_j\,,$$

$$R_j = r_j r_j^{\mathrm{T}}\,, S_j = r_j t_j^{\mathrm{T}}\,, T_j = t_j t_j^{\mathrm{T}},$$

$$r_j^{\mathrm{T}} = [x_{j1}^2,\ x_{j2}^2, \ldots, x_{jn}^2, \sqrt{2}x_{j1}x_{j2}, \ldots, \sqrt{2}x_{jk}x_{jl}, \ldots, \sqrt{2}x_{j(n-1)}x_{jn}],$$

and $t_j^{\mathrm{T}} = [x_{j1}, x_{j2}, \ldots, x_{jn}, 1]$.

It is to be noted that H_i^{-1} exists as long as there are at least n+1 non collinear feature points in the data set. Thus, in the FCQS algorithm, the parameters are updated using (42), and the memberships are updated using (3) except that $d^2(x_j,\beta_i)$ is replaced by d^2_{Qij}. The Fuzzy C Spherical Shells (FCSS) algorithm[23] discussed in Section 5.2 provides a good initialization for the FCQS algorithm.

Since the FCQS algorithm uses the algebraic distance given by (39) which is highly nonlinear in nature, the membership assignments are not very meaningful. It can be shown that, when there are curves (surfaces) of highly-varying sizes, the algebraic distance is biased toward smaller curves (surfaces), and for a particular curve (surface) the distance measure is biased towards points inside the curve (surface) as opposed to points outside.

5.6 The Modified Fuzzy C Quadric Shells Algorithm

To alleviate the problem due to the nongeometric nature of d^2_{Qij}, one may use the geometric (perpendicular) distance denoted by d^2_{Pij} between the point x_j and the shell β_i. To compute d^2_{Pij} we first rewrite (39) as

$$x^T A_i x + x^T b_i + c_i = 0.$$

In the 2-D case, A_i, b_i and c_i are given by

$$A_i = \begin{bmatrix} p_{i1} & \frac{1}{2} p_{i3} \\ \frac{1}{2} p_{i3} & p_{i2} \end{bmatrix}, \; b_i = \begin{bmatrix} p_{i4} \\ p_{i5} \end{bmatrix}, \text{ and } c_i = p_{i6}.$$

In order to simplify the computation of d^2_{Pij}, we first rotate the cluster prototype β_i and the point x_j so that the matrix A_i becomes diagonal. This does not change the distance. In the 2-D case, the rotation matrix is given by

$$R_i = \begin{bmatrix} \cos\alpha_i & \sin\alpha_i \\ -\sin\alpha_i & \cos\alpha_i \end{bmatrix}, \text{ where } \alpha_i = \frac{1}{2} \tan^{-1}\left(\frac{p_{i3}}{p_{i1} - p_{i2}}\right)$$

Let x_j' denote the point x_j after rotation. The distance d^2_{Pij} can be obtained by minimizing $\|x_j' - z'\|^2$ subject to

$$z'^T A'_i z' + z'^T b'_i + c'_i = 0 \qquad (43)$$

where z' is a point on the rotated quadric β_i. It is easily verified that

$$x_j = R_i x_j' \text{ and } z' = R_i z'$$

$$A' = R_i^T A R_i = \begin{bmatrix} p'_{i1} & 0 \\ 0 & p'_{i2} \end{bmatrix}, \; b' = R_i b = \begin{bmatrix} p'_{i4} \\ p'_{i5} \end{bmatrix}, \text{ and } c'_i = c_i.$$

By using a Lagrange multiplier λ, the solution is found to be

$$z' = \frac{1}{2}(I - \lambda A'_i)^{-1} (\lambda b'_i + 2x'_j) \tag{44}$$

Since A'_i is a diagonal matrix, $(I - \lambda A'_i)^{-1}$ can be easily inverted, and from (44) we obtain

$$z'^T = [\frac{\lambda p'_{i4}+2x'_{j1}}{2(1-\lambda p'_{i1})}, \frac{\lambda p'_{i5}+2x'_{j2}}{2(1-\lambda p'_{i2})}] \tag{45}$$

Substituting (45) into (43) yields a quartic equation in λ, which has at most four real roots λ_k, k =1,...,4. The four roots can be computed using the standard closed-form solution. For each real root λ_k so computed, we calculate the corresponding z' vector z'_k. Then, we compute d^2_{Pij} using

$$d^2_{Pij} = \min_k ||x_j' - z_k'||^2 \tag{46}$$

Minimization of the objective function in (2) with respect to p_i when d^2_{Pij} is used as the underlying distance measure can be achieved only by using numerical techniques. To reduce the computational burden, we may assume that we can obtain approximately the same values for p_i by using (42), which will be true if all the feature points lie reasonably close to the hyperquadric shells. This leads to the Modified Fuzzy C Quadric Shells (MFCQS) algorithm, in which the memberships are computed using d^2_{Pij}, but the parameters are computed using d^2_{Qij}. Our simulations indicate that the performance of the MFCQS algorithm is adequate for most computer vision applications.

The MFCQS algorithm can also be used to find linear clusters, even though the constraint forces all prototypes to be of second degree. This is because in practice this algorithm fits a pair of coincident lines for a single line, a hyperbola for two intersecting lines, and a very elongated ellipse, or a flat hyperbola, or a pair of lines for two parallel lines. When the data set contains many linear clusters, the MFCQS algorithm may characterize them variously as hyperbolas, an extremely elongated ellipses, etc. In this case, we can group all the points belonging to such pathological clusters into a data set and then run a line finding algorithm such as the GK algorithm on this data set with an appropriate initialization.

The procedure described above to solve for the exact distance is practical only in the 2-D case. In higher dimensions, this procedure results in a sixth (or higher) degree polynomial equation. To overcome this problem, one could use what is known as the "approximate distance" (first-order approximation of the exact distance)[25]. Krishnapuram et al.[26] developed an alternative algorithm based on this idea. They apply this algorithm to fit quadric surfaces to range data[27].

5.7 Examples of Shell Clustering

Figure 3 shows some examples that are typical of boundary detection problems. The data sets in images of size 200×200 were artificially generated, and had between 50 and 200 points. Uniformly distributed noise with an interval of 3 to 5 was added to the feature point locations so that they do not always lie on ideal quadric curves. Fig. 3(a) consists of three overlapping circles, Fig. 3(b) shows two overlapping circles enclosing another ellipse, Fig. 3(c) shows three differently oriented parabolas, and Fig. 3(d) shows three different types of quadric curves: a parabola, a circle and an ellipse, all in the same data set. In all cases the clusters criss-cross one another, and conventional methods cannot separate them. The different clusters obtained using the FCQS algorithm are shown with different symbols. As can be seen, the results are excellent.

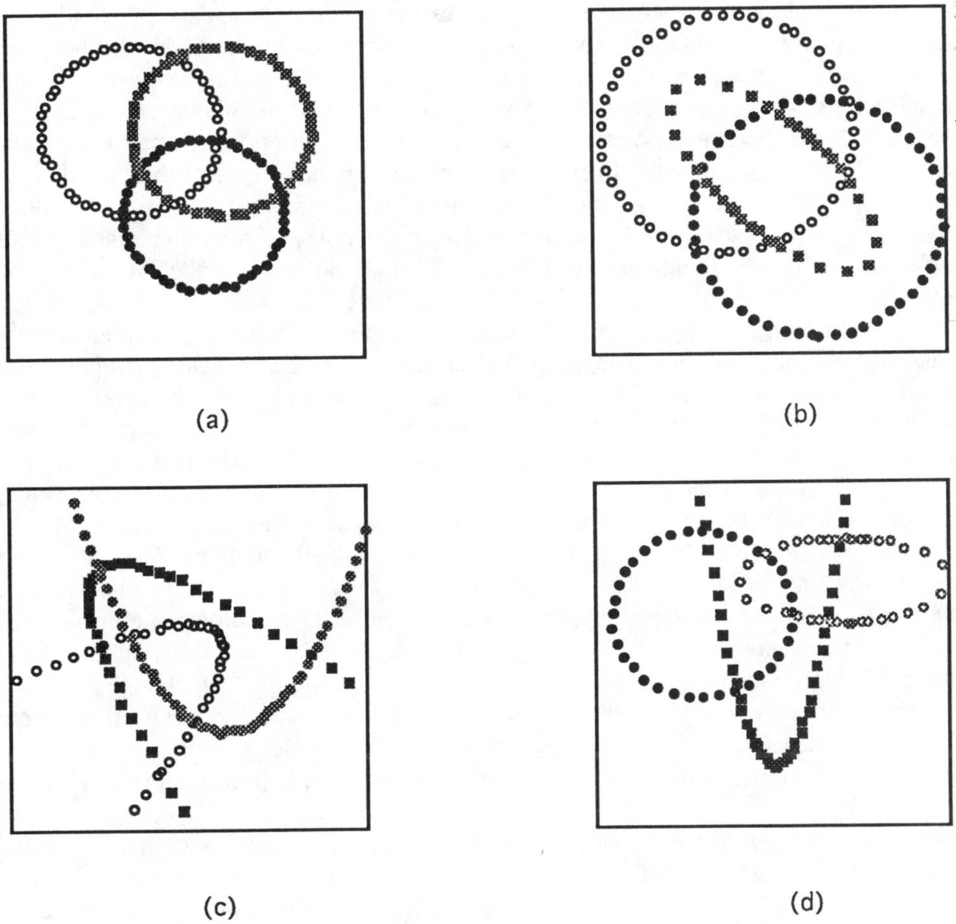

(a) (b) (c) (d)

Figure 3: Examples of clustering using the fuzzy C Quadric Shells algorithm. (a) three overlapping circles, (b) an ellipse enclosed by two overlapping circles, (c) three parabolas with different orientations, and (d) a mixture of three types of quadrics: a circle, an ellipse and a parabola.

6. Possibilistic Memberships for Robust Clustering

6.1 Possibilistic Clustering

Fuzzy clustering algorithms do not always estimate the parameters of the prototypes accurately. The main source of this problem is the probabilistic constraint used in fuzzy clustering, which states that the memberships of a data point across all clusters must sum to one. (See Eq.(1)). This problem has been discussed in detail elsewhere[28]. Here we present two simple examples to illustrate its drawbacks as related to this probabilistic constraint.

Figure 4(a) shows a situation containing two linear clusters. The FCM would produce very different membership values in cluster 1 for the points A and B, even though they are equally typical (i. e., they are equidistant from the prototypical member - the fuzzy linear prototype) of this class. This problem arises from the constraint on the memberships, which forces point B to give up some membership in cluster 1 in order to increase its membership in cluster 2. Similarly, point A and point C may have equal membership values in cluster 1, even though point C is more typical than point A. In other words, in the FCM algorithm, the membership of a point in a class is a relative number, and it depends on the membership of the point in all other classes, thus indirectly on the total number of classes itself. That is, the final membership of point B reflects a "sharing" of B between the two clusters. This affects the estimate of the cluster center. While this may not be a problem in some applications such as pattern classification, this is not always appropriate for many fuzzy set applications. Figure 4(b) presents a situation with two intersecting circular shell clusters. In this case, point A is a "good" member of both clusters, whereas point B is a "poor" member, and point C is an outlier. Here again, the constraint in the fuzzy clustering updates would force a membership of 0.5 in the two clusters on points A, B, and C. The membership values cannot distinguish between a moderately atypical member and an extremely atypical member, because the membership of a point in a class a relative number. Therefore noise points, which are often quite distant from the primary clusters, can drastically influence the estimates of the class prototypes, and hence the final partition. One established technique for reducing the effect of noise on least-squares fits is to use weights that are inversely related to the distance of the point to the prototype[29]. Although one may interpret memberships as weights, the memberships generated by fuzzy clustering are not inversely related to the distance, since they are relative numbers. Several ways to deal with noise in clustering (but not the relative membership problem) have been explored[30,31].

If there is only one cluster in the data, there is no difference between the crisp and fuzzy cases. Otherwise, the effect of noise points on the final partition is more drastic in the crisp case, since the membership of noise points tends to distributed among all classes in the fuzzy case. However, since the sum of the memberships of a point is constrained to be equal to one, this difficulty still remains with a lesser degree in the fuzzy case.

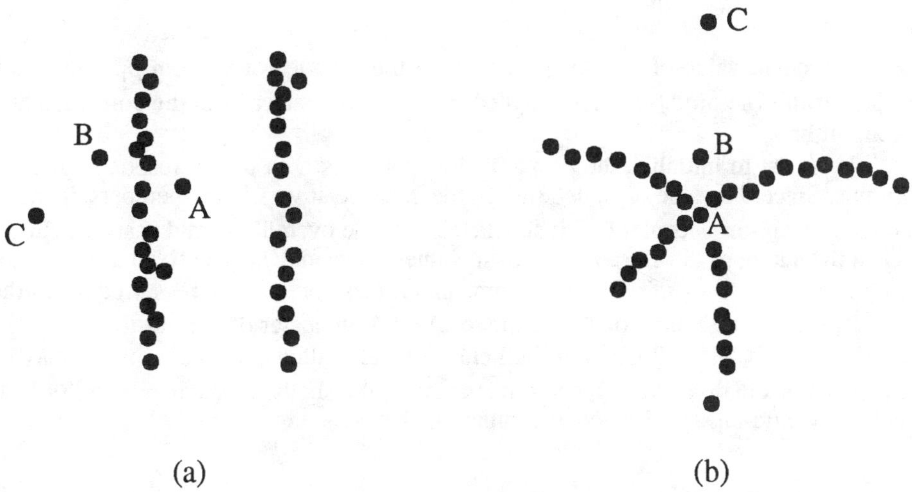

Figure 4: Disadvantages of fuzzy memberships with the probabilistic constraint. (a) A data set with two linear clusters. Membership values in cluster 1 for points A and B will be different, even though they are equidistant from the prototype. Point A and point C may have equal membership values in cluster 1, even though point C is far less typical of cluster 1 than point A. (b) A data set with two intersecting circular shell clusters. The memberships of points A, B, and C are all 0.5 in the two clusters.

One can cast the clustering problem into the framework of possibility theory[32,33] by relaxing the constraint in (1) and reformulating the objective function in (2) as:

$$J_m(B,U;X) = \sum_{i=1}^{C} \sum_{j=1}^{N} (u_{ij})^m d^2(x_j,\beta_i) + \sum_{i=1}^{C} \eta_i \sum_{j=1}^{N} (1 - u_{ij})^m , \qquad (47)$$

where η_i are suitable positive numbers[28]. The first term in (47) requires that the distance from the feature vectors to the prototypes be as low as possible, whereas the second term forces u_{ij} to be as large as possible, thus avoiding the trivial solution. It is easy to show[28] that U may be a global minimum of $J_m(U, V; X)$ only if the memberships are updated by

$$u_{ij} = \frac{1}{1 + \left(\frac{d^2(x_j,\beta_i)}{\eta_i}\right)^{\frac{1}{m-1}}}. \tag{48}$$

Thus, the updated value of u_{ij} depends only on the distance of x_j from β_i, which is a desirable result. The prototypes are updated in the same manner as in the corresponding fuzzy algorithms.

It is best to initialize the possibilistic algorithms with the corresponding fuzzy algorithms. Since the value of η_i determines the distance at which the membership value of a point in a cluster becomes 0.5, it should relate to the overall size and shape of cluster β_i. When the nature of the clusters is known, values for η_i may also be fixed *a priori*. For example, in the case of the GK algorithm, the values for η_i may be set equal to the expected value of the square of the (geometric) mean diameter of the clusters (i. e., the expected value of $(\det C_i)^{1/n}$), and in shell clustering algorithms, the values for η_i may be set equal to the square of the expected thickness of the shells. Another possibility is to use the fuzzy intra-cluster distance to compute η_i, i. e.,

$$\eta_i = \frac{K}{N_i}\sum_{j=1}^{N} u_{ij}^m\, d^2(x_j,\beta_i). \tag{49}$$

Typically K is chosen to be 1. Krishnapuram and Keller[28] also describe several other ways to estimate the η_i. Although the fits themselves are not very sensitive to the exact values of η_i, some of the cluster validity measures do depend on the choice of η_i. For this reason, the best approach is to compute approximate values for the η_i from the initial fuzzy partition using (49), and after the possibilistic algorithm converges, run a few more iterations of the algorithm with a reestimated value of η_i. The resulting possibilistic shell clustering algorithm is summarized below.

THE POSSIBILISTIC CLUSTERING ALGORITHM:

> *Fix the number of clusters C; fix* m, $1 < m < \infty$;
> *Initialize C-partition* U *using the corresponding fuzzy algorithm*;
> *Estimate* η_i;
> ***REPEAT***
>   *Update the prototypes using* U;
>   *Compute* U *using* (48);
> ***UNTIL*** $(\|\Delta U\| < \varepsilon)$;
> *Reestimate the values of* η_i;
> ***REPEAT***
>   *Update the prototypes using* U;
>   *Compute* U *using* (48);
> ***UNTIL*** $(\|\Delta U\| < \varepsilon)$;

Possibilistic versions of all the algorithms discussed in the prevision sections can be derived very easily by merely changing the membership updating equation to (48).

6.2 Example of Possibilistic Clustering

The top left of Fig. 5 shows a noisy data set containing two ellipses. The top right of Fig. 5 depicts the result of the hard Modified C Quadric Shells algorithm with the prototypes superimposed on the data. It can be seen that the fits are very poor. The bottom left of Fig. 5 shows the result of the MFCQS algorithm. This is an improvement over the result of the hard case, however, the effect of the noise points on the fits is still significant. Finally, in the bottom right of Fig. 5 the result of the possibilistic Modified C Quadric Shells algorithm is displayed demonstrating its superior performance in the presence of noise.

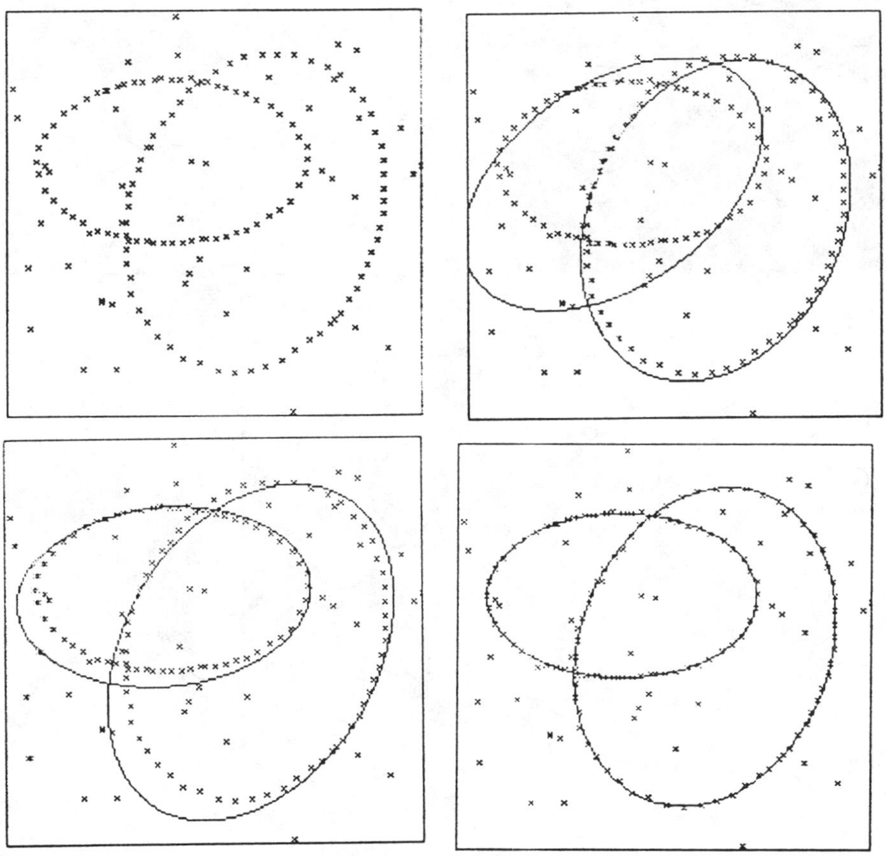

Figure 5: Examples of crisp, fuzzy and possibilistic shell clustering in the presence of noise. Top Left: a noisy data set containing two ellipses; Top Right: the result of the hard Modified C Quadric Shells algorithm with the prototypes superimposed on the data; Bottom Left: the result of the MFCQS algorithm; Bottom Right: the result of the possibilistic Modified C Quadric Shells algorithm.

Fig. 6 gives an example of the possibilistic Modified C Quadric Shells algorithm used to detect linear and elliptical boundaries from the edge map of a space shuttle image. Here again we see the robustness of the algorithm in detecting the salient boundaries in spite of the clutter. Krishnapuram and Keller[28,34] give several other examples of the use of possibilistic clustering in computer vision.

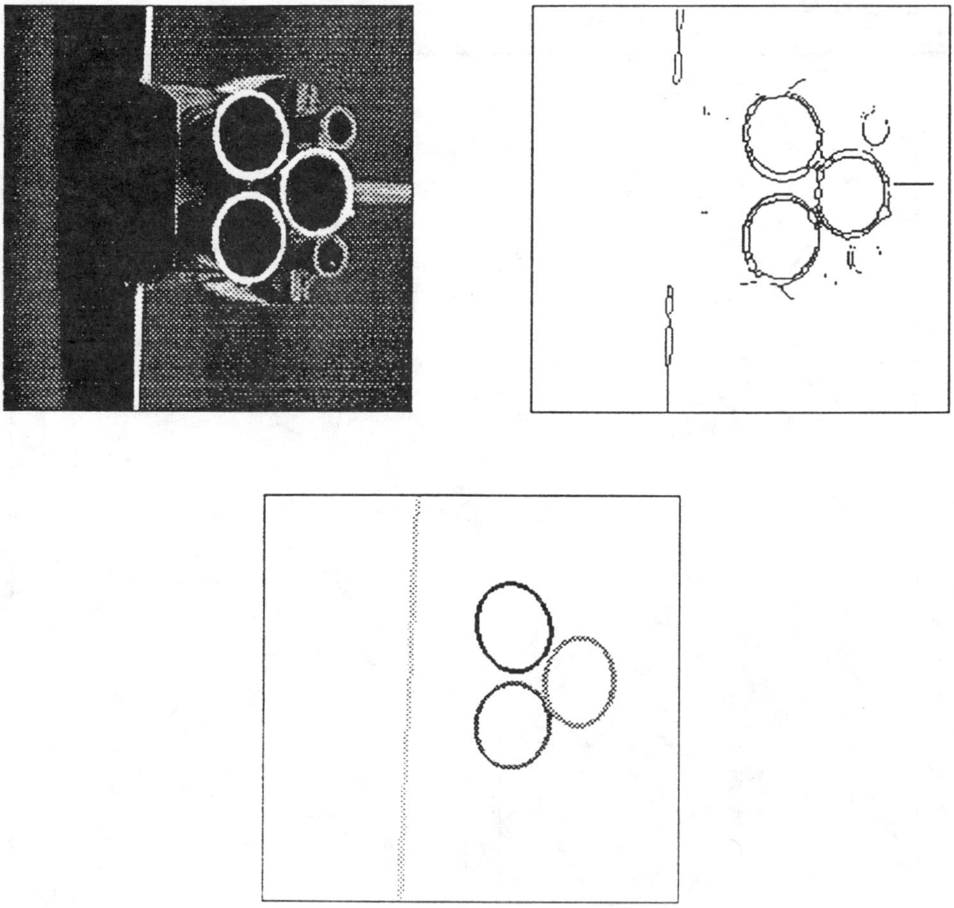

Figure 6: Example of the possibilistic Modified C Quadric Shells algorithm used to detect linear and elliptical boundaries from the edge map of a space shuttle image. Top Left: original shuttle image; Top Right: noisy edges extracted from shuttle image; Bottom: results of the possibilistic Modified C Quadric Shells algorithm applied to the edge image.

7. Conclusions

In earlier work[35,36], we have considered the more general application of fuzzy set theoretic approaches to all areas of computer vision. In this chapter, we have described existing clustering techniques with respect to their use in computer vision. We have also discussed modifications to these approaches so that they can be used for the detection and simultaneous recognition of curves and surfaces. Since shell clustering algorithms can perform segmentation and fitting simultaneously, they are ideally suited for boundary detection and surface approximation when the boundaries/surfaces are ill-defined. They also require far less computations and memory compared to the GHT methods. Since they look for global structures and do not use edge following or region growing, they are insensitive to local aberrations and deviations in shape. They do not use features such as gradients and curvature and hence are not sensitive to noise and sharp discontinuities at the boundaries. We have also shown that by refining the objective function and thereby relaxing the constraint that the memberships sum to one across clusters, we can construct clustering-based shape segmentation algorithms which are much less affected by noise in the edge image. Hence, these approaches provide more accurate estimates of the curve or surface parameters than either crisp or fuzzy c-means type techniques in that situation.

8. Acknowledgment

Dr. Krishnapuram would like to acknowledge the partial support for this research by the Humboldt Foundation, Germany.

9. References

1. J. C. Bezdek, *Pattern Recognition with Fuzzy Objective Function Algorithms*, Plenum Press, New York, 1981.
2. D. H. Ballard, "Generalizing the Hough transform to detect arbitrary shapes." *Pattern Recognition* Vol. 13, No. 2, 1981, pp. 111-122.
3. D. Casasent and R. Krishnapuram, "Curved object location by Hough transformations and inversions", *Pattern Recognition*, Vol. 20, No. 2, 1987, pp. 181-188.
4. J. Illingworth and J. Kittler, "A survey of Hough transforms", *Computer Vision, Graphics and Image Processing*, Vol. 44, No. 1, October 1988, pp. 87-116.
5. W. E. L. Grimson and D. P. Huttenlocher, On the sensitivity of the Hough transform for object recognition. *IEEE Transactions on Pattern Analysis and Machine Intelligence.*, Vol. PAMI-12, No. 3, 1990, pp. 2555-2574.
6. R. N. Davé, "New measures for evaluating fuzzy partitions induced through c-shells clustering," *Proceeding of the SPIE Conference on Intelligent Robot and Computer Vision X*: SPIE Vol. 1607, Boston, Nov. 1991, pp. 406-414.
7. R. Krishnapuram, O. Nasraoui, and H. Frigui, "The surface density criterion and its application to linear/circular boundary detection and planar/spherical surface approximation", *Proceedings, Second IEEE Conference on Fuzzy Systems*, San Francisco, CA, March 1993, pp. 725-730.
8. E. H. Ruspini, "A new approach to clustering," *Inform. and Control*, Vol. 15, 1969, pp. 22-32.

9. G. S. Sebestyen and Roemder, *Decision-Making Processes in Pattern Recognition*, Macmillan Company, New York, 1962.
10. E. Diday and J. C. Simon, "Clustering analysis", in *Digital Pattern Recognition*, K. S. Fu (ed.), Springer, New York, 1976, pp. 47-94.
11. E. E. Gustafson and W. C. Kessel, "Fuzzy clustering with a fuzzy covariance matrix," *In Proceedings, IEEE CDC*, San Diego, Calif., 1979, pp. 761-766.
12. J. C. Bezdek and J. C. Dunn, "Optimal fuzzy partition: A heuristic for estimating the parameters in a mixture of normal distributions," *IEEE Trans*actions on Computers, Vol. C-24, 1975, pp. 835-838.
13. I. Gath and A. B. Geva, "Unsupervised optimal fuzzy clustering", *IEEE Transactions on Pattern Analysis and Machine Intelligence*, Vol. PAMI-11, No. 7, 1989, pp. 773-781.
14. R.N. Davé, "Use of the adaptive fuzzy clustering algorithm to detect lines in digital images, "*Proceedings, Intelligent Robots and Computer Vision VIII*, Vol. 1192, No. 2, 1989, pp. 600-611.
15. R. Krishnapuram and C.-P. Freg, "Fitting an unknown number of lines and planes to image data through compatible cluster merging," *Pattern Recognition*, Vol. 25, No. 4, 1992, pp. 385-400.
16. C. Coray, "Clustering Algorithms with Perspective Selection", *Proceedings of the Hawaii International Conference on System Sciences*, January 1981, pp. 945-955.
17. R. N. Davé, "Fuzzy shell-clustering and application to circle detection in digital images," *International J. of General Systems*, vol. 16, 1990, pp. 343-355.
18. R. N. Davé and K. Bhaswan, "Adaptive fuzzy C-shells clustering and detection of ellipses," *IEEE Transactions on Neural Networks*, Vol. 3, No. 5, 1992, pp. 643-662.
19. R. Krishnapuram, H. Frigui, and O. Nasraoui, " New fuzzy shell clustering algorithms for boundary detection and pattern recognition," *Proceedings of the SPIE Conf. on Intelligent Robots and Computer Vision X*: Algorithms and Techniques, Boston, Nov. 1991, pp 458-465.
20. R. Krishnapuram, H. Frigui, and O. Nasraoui, "Quadratic shell clustering algorithms and their applications," to appear in *Pattern Recognition Letters*.
21. J. C. Bezdek and R. H. Hathaway, "Numerical convergence and interpretation of the fuzzy C-shells clustering algorithm," *IEEE Transactions on Neural Networks*, Vol. 3, No. 5, 1992, pp. 787-793.
22. V. Pratt, "Direct least squares fitting of algebraic surfaces," *Computer Graphics*, Vol. 21, No. 4, 1987, pp. 145-152.
23. R. Krishnapuram, O. Nasraoui, and H. Frigui, "The fuzzy C spherical shells algorithms: A new approach," *IEEE Transactions on Neural Networks*, Vol. 3, No. 5, 1992, pp. 663-671.
24. F. L. Bookstein, "Fitting conic sections to scattered data", *Computer Vision, Graphics, and Image processing*, vol. 9,, 1979 pp. 56-71.
25. G. Taubin, "Estimation of planar curves, surfaces, and nonplanar space curves defined by implicit equations with application to edge and range image segmentation", *IEEE Transactions on Pattern Analysis and Machine Intelligence*, Vol. PAMI-13, No. 11, 1991, pp. 1115-1138.
26. R. Krishnapuram, H. Frigui, and O. Nasraoui, "Fuzzy and possibilistic shell clustering algorithms and their application to boundary detection and surface approximation", submitted to *IEEE Transactions on Fuzzy Systems*.

27. R. Krishnapuram, H. Frigui, and O. Nasraoui, "Surface approximation through fuzzy shell clustering", *Proceedings, International Fuzzy Systems Association Congress*, Seoul, Korea, July 1993.
28. R. Krishnapuram and J. Keller, “A possibilistic approach to clustering,” *IEEE Transactions on Fuzzy Systems*, Vol. 1, No. 2, 1993, pp. 98-110.
29. R. M. Haralick and L.G. Shapiro, *Computer and Robot Vision*. vol. I, Addison Wesley, Reading, MA, 1992.
30. J. Jolion and A. Rosenfeld, “Cluster detection in background noise,” *Pattern Recognition*, Vol. 22, No. 5, 1989, pp. 603-607.
31. R. N. Davé, “Characterization and detection of noise in clustering,” *Pattern Recognition Letters*, Vol. 12, No. 11, 1991, pp. 657-664.
32. L. A. Zadeh, “Fuzzy sets as a basis for a theory of possibility,” *Fuzzy Sets and Systems*, Vol. 1, 1978, pp. 3-28.
33. D. Dubois and H. Prade, *Possibility Theory: An Approach to Computerized Processing of Uncertainty*. Plenum Press, New York, 1988.
34. J. Keller and R. Krishnapuram, "Possibilistic clustering for shape description", *Proceedings, Third International Workshop on Neural Networks and Fuzzy Logic*, NASA/JSC, Houston, TX, June, 1992, pp. 227-236.
35. J. Keller and R. Krishnapuram, "Fuzzy set methods in computer vision", in *An Introduction to Fuzzy Logic Applications in Intelligent Systems*, R. Yager and L. Zadeh (eds), Kluwer Academic Press, 1992, pp. 121-146.
36. R. Krishnapuram and J. Keller, “Fuzzy set theoretic approach to computer vision: An overview", invited paper, *Proceedings, IEEE International Conference on Fuzzy Systems*, San Diego, March, 1992, pp. 135-142.

NEURO-FUZZY IMAGE PROCESSING: RELEVANCE AND FEASIBILITY

Sankar K. Pal and Ashish Ghosh

Machine Intelligence Unit
Indian Statistical Institute
203 B. T. Road
Calcutta 700 035

ABSTRACT

A brief review of the applications of neural networks in various aspects of *Image Processing* (e.g., image segmentation, image restoration, texture segmentation, image coding etc.) along with their parallelism and fault tolerance characteristics will be provided first. Relevance of various fuzzy tools and methodologies for handling uncertainties and in providing soft decisions in image processing will be described. A discussion on making fusion of the merits of fuzzy logic and neural network technologies will then be made. Finally, a neuro fuzzy system for extracting objects from noisy images and its effectiveness will be described. The system consist of a *multi-layer neural network* with back-propagation of error and with feed back connections. In each layer there are $M \times N$ neurons (for an $M \times N$ image). Each neuron corresponds to a single pixel. Neurons in the same layer do not have any connection among themselves. Each neuron in a layer is connected to the corresponding neuron in the previous layer and to its neighbors over some window. Input is given as fuzzy set *'brightness'*. The membership value for *'brightness'* involves both global and local information of an image. Various measures of fuzziness of a set (like index of fuzziness, entropy, fuzzy correlation etc.) are used to model the error in the output layer. The status of the neurons in the output layer is also considered to be a fuzzy set '*object pixels*'.

1. INTRODUCTION

For any image analysis/vision system, one desires to achieve robustness of the system with respect to random noise and failure of components, and to obtain output in real time. Moreover, a system can be made artificially intelligent if it is able to emulate some aspects of human information processing system. Neural network (*NN*)[1-11] based approaches are attempts to achieve these goals. A neural network can formally be defined as: *massively parallel interconnected network of simple (usually adaptive) processing elements which are intended to interact with the objects of the real world in the same way as biological systems do.* The architecture of the network depends on the goal one is trying to achieve. The massive connectivity among the neurons usually makes the system fault tolerant (with respect to noise and component failure) while the parallel processing capability enables the system to produce output in real time. In addition to this, most of the image processing/vision operations use neighborhood information to reduce local ambiguity and to attain global consistency, i.e., the operations are distributed and parallel in nature. The parallel and distributed processing characteristics of neural networks suggests that many image processing algorithms are candidates for neural network implementation .

Practical systems for image analysis or computer vision are expected to work with real life data which are not always precisely defined and fully reliable. Besides that, the basic concepts of image analysis, (e.g., the concepts of an edge, a boundary, a corner and a relation between regions) do not lend themselves well to precise definition.[12-17] Thus uncertainty may arise in every phase of decision making of such a system (as the input data obtained is not certain). Fuzzy set theories[18-22] are reputed to handle uncertainties,[12,23,24] to a reasonable extent, arising from deficiencies of information available from a situation (the deficiency may result from incomplete, ill-defined, not fully reliable, vague and contradictory information). Therefore the incorporation of fuzzy set theory in image processing and recognition systems[19-28] can handle real life data more efficiently.

Integration of the merits of these two technologies therefore promises to provide, to a great extent, more intelligent systems to handle real life processes. A large number of researchers are now concentrating on exploiting these modern concepts in the field of pattern recognition and machine vision. The present article deals with the utility and feasibility of such an integrated system. Rest of the article is organized as follows. In section 2, relevance of fuzzy set theoretical methods for image processing tasks is described. Relevance of neural network based techniques in image processing operations is described in section 3. In section 4 various ways of fusion of the fuzzy sets and neural network models are provided. Suggestions for further research are given in section 5. Concluding remarks can be found in section 6.

2. RELEVANCE OF FUZZY SET THEORETICAL METHODS

Fuzzy sets were introduced in 1965 by Lotfi Zadeh[18] as a new way to represent vagueness in everyday life. They are generalizations of conventional (crisp) set theory. Conventional sets contain objects that satisfy precise properties required for membership. Fuzzy sets, on the other hand, contain objects that satisfy imprecisely defined properties to varying degrees. A fuzzy set A of the universe X is defined as a collection of ordered pairs

$$A = \{(\mu_A(x), x), \forall x \in X\} \tag{1}$$

where $\mu_A(x)$, $(0 \le \mu_A(x) \le 1)$ gives the degree of belonging of the element x to the set A. Different aspects of fuzzy set theory including membership functions, basic operations, and uncertainty measures, can be found in.[19-23]

A gray tone image possesses ambiguity within each pixel because of the possible multi-valued levels of brightness. If the gray levels are scaled to lie in the range[0,1], we can regard the gray level of a pixel as its degree of belonging (membership) in the set of high-valued ('bright') pixels - thus a gray tone image can be viewed as a fuzzy set. Regions, features, primitives, properties, and relations among them that are not crisply defined can similarly be regarded as fuzzy subsets of images.

An L level image X $(M \times N)$ can therefore be considered as an array of fuzzy singletons, each having a value of membership denoting its degree of possessing some property (e.g., brightness, darkness, edginess, blurredness, texture etc.). In the notation of fuzzy sets one may therefore write $X = \{\mu_X(x_{mn}): m = 1,2, \ldots, M; n = 1,2.,N\}$; where $\mu_X(x_{mn})$ denotes the grade of possessing such a property μ by the (m,n)th pixel.

This property μ of an image may be defined using global information or local information or positional information or a combination of them depending on the problem. Basic principles and operations of image processing and recognition in the light of fuzzy set theory are available in.[21]

Incertitude in an image pattern may be explained in terms of grayness ambiguity or spatial (geometrical) ambiguity or both. Grayness ambiguity means 'indefiniteness' in deciding whether a pixel is white or black. Spatial ambiguity refers to 'indefiniteness' in the shape and geometry of a region within the image. For example, grayness ambiguity measures are reflected by index of fuzziness[29] and entropy,[30-33] whereas the spatial ambiguity measures are represented by fuzzy geometrical properties.[14,17,34-37]

Conventional approaches to image analysis and recognition[38,39] consist of segmenting the image into meaningful regions, extracting their edges and skeletons, computing various features/properties (e.g., area, perimeter, centroid etc.)and primitives (e.g., line, corner, curve etc.) of and relationships among the regions, and finally, developing decision rules/grammars for describing, interpreting and/or classifying the image and its sub-regions. In a conventional system each of these operations involves crisp decisions (i.e., yes or no, black or white, 0 or 1) to make regions, features, primitives, properties, relations and interpretations crisp.

Since the regions in an image are not always crisply defined, uncertainty can arise within every phase of the aforesaid tasks. Any decision made at a particular level will have an impact on all higher level activities. An image recognition (or computer vision) system should have sufficient provision for representing and manipulating the uncertainties involved at every processing stage; i.e., in defining image regions, features, matching, and relations among them, so that the system retains as much of the 'information content' of the data as possible. If this is done, the ultimate output (result) of the system will possess minimal uncertainty (and unlike conventional systems, it may not be biased or affected as much by lower level decision components).

For example, consider the problem of object extraction from a scene. Now, the question is 'how can one define exactly the target or object region in a scene when its boundary is ill-defined?' Any hard thresholding made for the extraction of the object will propagate the associated uncertainty to subsequent stages (e.g., thinning, skeleton extraction, primitive selection, etc.) and this might, in turn, affect feature analysis and recognition. Thus, it is convenient, natural and appropriate to avoid committing ourselves to a specific (hard) decision (e.g., segmentation / thresholding, edge detection and skeletonization), by allowing the segments or skeletons or contours to be fuzzy subsets of the image, the subsets being characterized by the possibility (degree) to which each pixel belongs to them. Prewitt[13] first suggested that the results of image segmentation should be fuzzy subsets, rather than ordinary subsets. Similarly, while describing relations among different components and features or classifying the subregions it is necessary to make the decision making algorithms flexible by providing soft decisions.

Some of the areas where the theory of fuzzy sets has been adequately applied are given below. Computation of fuzzy geometric properties is described in.[14,17,36,37,40,41]

Fuzzy segmentation is discussed in.[20,27,34,42-50] The problem of evaluation of image quality has also been dealt with fuzzy set theory.[21,34,42,51,52] Other image operations, like thinning and edge detection have been developed in.[21,34,35,53,54,55] There have been several attempts to extract fuzzy primitives (or features) from fuzzy edges and segmented outputs of the image regions for shape analysis, matching, and recognition. Readers may consult the articles.[21,24,57,58] A review showing the development of this area is provided recently in.[12]

3. RELEVANCE OF FUZZY NEURAL NETWORK BASED METHODS

Neural network *(NN)* models[1-11] try to emulate the biological neural network/nervous system with electronic circuitry. *NN* models have been studied for many years with the hope of achieving human like performance (artificially), particularly in the field of image and pattern recognition by capturing the key ingredients responsible for the remarkable capabilities of the human nervous system. The models are extreme simplifications of the actual human nervous system.

*NN*s are designated by the network topology, connection strength between pairs of neurons (called weights), node characteristics and the status updating rules. Node characteristics mainly specify the primitive types of operations it can perform, like summing the weighted inputs coming to it and then amplifying it or doing some fuzzy aggregation operations. The updating rules may be for weights and/or states of the processing elements (neurons). Normally an objective function is defined which represents the complete status of the network and the set of minima of it corresponds to the set of stable states of the network. The neurons operate asynchronously (status of any neuron can be updated at random times independent of the other) or synchronously (in parallel) thereby providing output in real time. Since there are interactions among the neurons the collective computational property inherently reduces the computational task and makes the system fault tolerant. Thus *NN* models are more suitable for tasks where collective decision making is required. Hardware implementations of neural networks are attempted in.[59-61]

A wide variety of models has been proposed for various purposes; they differ in structure and details, but share some common features. The common features are:

- Each node/neuron/processor is usually described by a single real variable (called its *state*), representing its output.

- The nodes are densely connected so that the state of one node affects the potential (total input) of many other nodes according to the weight or strength of the connection.

- The new state of a node is a non-linear function of the potential created by the firing activity of other neurons to it.

- Input to the network is given by setting the states of a subset of the nodes to specific values initially.

- The processing takes place through the evolution of the states of all the nodes of the network according to the dynamics of a specific network and continues until it stabilizes.

- The training (learning) of the net is a process whereby the values of the connection strengths are modified in order to achieve the desired output.

Neural network based systems are reputed to enjoy the following major advantages:

- *adaptivity*- adjusting the connection strengths to new data/information,

- *speed*- due to massively parallel architecture,

- *robustness*- to missing, confusing, ill-defined/noisy data,

- *ruggedness*- to failure of components,

- *optimality*- as regards error rates in performance.

Image processing and analysis systems mainly employ simple arithmetic operations at each pixel cite in parallel in the spatial domain. Operations involve neighboring information of pixels and they are independent of the other. An objective measure is then evaluated (representing the overall status of the system), the optimum of which represent the goal. The system thus involves collective decisions. One desires such a system to be fast and robust (i.e., result should not deteriorate even if the input is noisy/ill-defined or a part of it is missing). On the other hand, we notice that neural network models also based on parallel and distributed working principles (all neurons work in parallel and independently). The operations performed at each processor cite are also simpler. Overall status of a neural networks can also be measured. These direct analogies show that the *NN*s can be a better choice for implementing image processing and analysis algorithms. The review provided in this section further supports this claim.

3.1 Review on neural network based techniques

Major areas in which neural networks have been applied in order to exploit its high computational power and to make robust decisions are: supervised classification,[62,63] expert system design,[64,65] clustering,[66-68] text classification,[69] cloud classification,[70] pixel classification[71,72] histogram thresholding[73] image compression[74,75] image segmentation[76-79] object extraction from noisy scenes,[80-83] edge detection,[85,86] image restoration[87,88] object recognition,[89,90] scene labeling[91] and character recognition.[92-94] A few of them related to image analysis are described below.

3.1.1 Image segmentation

Several texture segmentation algorithms based on deterministic and stochastic relaxation[95] principles and their implementation on multi-layer neural networks are suggested by Manjunath et al..[78] The segmentation problem is posed as an optimization task and two different optimality criteria (namely maximum aposteriori probability estimate & maximization of posterior marginal distribution) are considered for its solution. The texture labels are modeled by a Gibbs Markov random field.[95] A neural network architecture having k layers (k = number of textures present in the image) and $M \times N$ neurons per layer (for an $M \times N$ image) is used. The dynamics of the network is the same as that of Hopfield's model.[3] An energy function is designed in such a way that its local minima correspond to different segmented versions. In the converged state, the set of ON neurons in different layers correspond to different textures.

An image segmentation system which uses a multi-layer perceptron *MLP* as a central building block is described in.[71] The complete system consists of two parts. First module is called feature extractor and the second one connectionist classifier. The feature extractor basically defines pixel descriptors which capture the distinguishing features of objects and background (may be in gray levels, texture or both). Different features such as local maxima, gradient, median, variance etc. are used. A set of important features are then selected. These feature values are then used as input to the input layer of the feed-forward multi-layer perceptron classifier. Labeled pixels (pixels with known classes) are used for learning the network parameters (weights and node biases) using back-propagation learning. After the network has learned (i.e., weights have settled), test images are used to measure the performance of classification. Demonstrations are given on 'real images' of combustion chamber, textures and printed circuit boards.

Lee[70] et al. used a 4-layer neural network as a supervised classifier for cloud classification. Textural features (like mean, standard deviation, contrast, angular second moment, entropy and homogeneity) obtained from a single channel high resolution visible data are used for learning. Classification accuracy of different types of clouds (stratocumulus, cumulus and cirrus) were found to be more compared to discriminant analysis and k-nearest neighbor algorithms.

In,[73] Babaguchi et al. used a three-layer perceptron architecture for image binarization. Binarization is basically a task of converting a gray level image to a bilevel image. In particular they applied it to document processing (news paper clipping, technical papers etc.) binarization. The input to the network is the gray level histogram (mapped on to [0,1]) and the output is the corresponding threshold. Number of nodes in the input layer is equal to the maximum number of gray levels available, and number of nodes in the output layer is two. Number of nodes in the hidden layer was variable. A set of images (with known thresholds) is used to adjust the parameters (weights and biases) of the feed forward multi-layer back-propagating network. The known thresholds are given in different forms (e.g., a single threshold, a number of thresholds with varying degree in an interval). The algorithm has been demonstrated on a set of images.

Silverman[79] suggested a method for using both back propagation neural networks and competitive learning networks in segmentation of ultrasonic images. A four layer back propagation neural network is used for learning the operation of a 'gradient operator' using the gradient image as teaching signal and median image as input for a number of images. After the learning is complete, if an unknown input image is presented the system provides a gradient version as output. They also used a competitive learning network in image segmentation using a single image as input. An extension to 3-dimensional image segmentation is also attempted using competitive learning scheme.

In,[77] image segmentation problem has been formulated as a constraint satisfaction problem *(CSP)*, and a class of constraint satisfaction neural network *(CSNN)* is proposed. A *CSNN* consists of a set of objects, a set of labels, a collection of constraint relations and a topological constraint describing the neighborhood relationships among various objects. The *CSNN* is viewed as a collection of interconnected neurons. The architecture is chosen in such a way that it represents constraints in the *CSP*.

A mapping method that makes the Hopfield neural network perform the probabilistic relaxation is suggested in.[72] In such a network the neurons/nodes are used to represent various possible hypotheses and pieces of evidence in the relaxation process; and the links embody the relationship among them. Experimental results demonstrating the feasibility of the proposed mapping process by performing a pixel classification task are also given.

Different types of neural network models have been designed in order to develop methodologies for object extraction or object background classification from noisy scenes by Ghosh et al..[80-83] In one of the algorithms[81] the object extraction problem has been viewed as a self-organizing task and is solved by a modified version of Kohonen's network model. Here a single neuron is assigned to each pixel. Unlike Kohonen's model, the inputs given to different neurons are different. The convergence of the algorithm has been proven mathematically. Statistical criteria such as square error minimization and correlation maximization have been suggested for choosing the optimal output.

In another case,[82] the task of image segmentation is viewed as an optimization problem and solved by a Hopfield type network. An objective function has been formulated for object background separation (which is similar to the expression of energy of a Hopfield type network) in such a manner that in a stable state of the network neurons corresponding to homogeneous regions would be in ON state. A modified version (topologically) of Hopfield's model is then proposed for this optimization task. Both the discrete and the continuous dynamics have been studied. A polynomial over a fixed domain is used as the input/output transfer function, instead of an asymptotic function. The problem of assignment of initial weights is tackled using the concept of probability of co-occurrence of pixels.

An approach is provided[80] for solving a computationally hard optimization problem using a neural network. The maximum a-posterior probability (*MAP*) estimate of a scene from a noise corrupted realization (modeled by Gibbs random field) is computationally exponential in nature.[95] A modified version (both topologically and neuron characteristic wise) of Hopfield's model has been designed for solving this problem. The energy function of the network is formulated in such a way that its minimum value corresponds to the *MAP* estimate of the scene. A possible hardware realization of the neuron is provided. The discrete and the continuous dynamics, the input/output transfer function used are the same as in the previous case.[82] The convergence of the algorithm is also proved.

Shah[85] formulated the problem of edge detection in the context of an energy minimizing model. The method is capable of eliminating weak boundaries and small regions. Cortes et al.[86] proposed a neural network to detect potential edges in different orientations. The performance of the system has been investigated through simulation studies using simulated annealing and mean field annealing.

3.1.2 Image restoration

In[87] the problem of image restoration using the maximum entropy method is formulated as a constrained optimization task in a primal-dual Lagrange multiplier frame-work. They defined an iterative scheme to solve the problem. The objective function for this solution scheme is shown to be similar to the energy function of a Hopfield type neural network and is solved by such a network. On the basis of this scheme the equations governing the dynamics of the network are derived, and some electronic circuits are discussed which could be effective in a practical implementation of the network.

3.1.3 Image compression

Luttrell[75] tried to demonstrate that Kohonen's network model[2] may be used to compress synthetic aperture image date. He suggested an improved version of the learning algorithm employed in Kohonen's network learning which converges more rapidly than Kohonen's original algorithm. He also extended the concept of Kohonen's learning algorithm to a multi-layer version of the same. The ability of a single network to compress a variety of synthetic aperture images are also suggested.

In[74] attempts are made by Cotterel et al. to use a three-layer feed forward neural network architecture with back-propagation of error for image data coding. Experiments are done on a three-layer network having 8 x 8 neurons in the input and output layers, and variable number of neurons in the hidden layer. Input to the network was patches of size 8 x 8 chosen randomly from the entire image. The square error was then calculated from the target output and was back-propagated for learning the weights. After sufficient amount of learning the weights between the layers specify the coded data. If an unknown patch is given as input, the output can easily be obtained.

3.1.4 Image matching

The work of Jamison and Schalkoff[91] concerns with the utility of neural networks in problems involving constraint satisfaction. Many symbolic constraint satisfaction problems involve instanciation of values to variables in an observed relational structures which are consistent with an underlying constraint-based semantic model. One such problem is image labeling, i.e., determining the types (labels) of regions in a scene under some constraints (like a region can have only one label, a car is on a road, grass is adjacent to a road etc.). They used a Hopfield type neural network with $N = R$ $x\ L$ neurons, (where R = number of regions to be labeled and L = number of labels) arranged in an $R \times L$ array. They formulated an energy function in the line suggested by Hopfield and Tank (for solving the traveling salesman problem[96]) in such a way that the minimum of it correspond to the solution which satisfy the constraints. The dynamics of the network are the same as that in Hopfield's model. In the converged state, if the ith region takes label j, then neuron at the (i, j)th location is ON, else it is OFF.

A two dimensional model-based object recognition technique to identify and locate isolated or overlapping 2-D objects using a co-operative feature matching technique is implemented by a binary Hopfield type neural network. For each object a model graph is constructed from its feature primitives and then integrated into a model-database graph. The feature primitives used were corners extracted from the polygonal approximation of object boundaries. In the model data-base graph each node represents a feature that has a numerical value and is connected to other nodes by an arc representing the relationship or compatibility between them. Object recognition problem is formulated as matching this model graph with the scene graph representing the constraints that the nodes in the two graphs has to satisfy in order to find the best match. A 2-D binary Hopfield type neural network is used to minimize this energy function. The synaptic interconnection weights between the neurons are designed such that matched features belonging to the same model receive excitatory supports, and matched features belonging to different models receive an inhibitory support or a manual support depending on whether the input scene consists of an isolated object or several overlapping objects. The co-ordinate transformation from mapping each pair of matched nodes from the model onto the input scene is calculated, followed by a simple clustering technique to eliminate any false match. The orientation and the position of objects in the scene are then calculated by averaging the transformation of correct matched nodes.

4. INTEGRATION OF THE THEORIES OF FUZZY SETS AND NEURAL NETWORKS

Fuzzy set theory provides an approximate but effective and more flexible means of describing the behaviors of the systems which are too complex or too ill-defined to admit of precise mathematical analysis by classical methods and tools. The approach has some drawbacks. The choice of a membership function to represent a fuzzy set is normally heuristic. Since the operations involved in fuzzy logic based systems are set operations, it becomes computationally prohibitive. Though the approach tries to model the human thought process in a decision making system, it has no relation with the architecture of the human neural information processing system, nor does it take into consideration the information storage technique of human beings.

Human intelligence and discriminating power are mainly attributed to the massively connected network of biological neurons in the human brain. We mentioned earlier that attempts have recently been made to emulate electronically the architecture and information representation scheme of human neural network under the name *neural network* models. The collective computational abilities of the densely interconnected nodes or processors may provide a material technique, at least to a great extent, for solving highly complex real life problems in a manner a human being does.

We see that the fuzzy set theoretical models try to mimic human reasoning and the capability of handling uncertainty, whereas the neural network models attempt to emulate the architecture and information representation schemes of human brain. The fusion or integration is mainly tried out in the following ways or in any combination of them.

i: Incorporating fuzziness into neural network frameworks. This includes assigning fuzzy labels to training samples, fuzzifying the input data and obtaining output in terms of fuzzy sets.

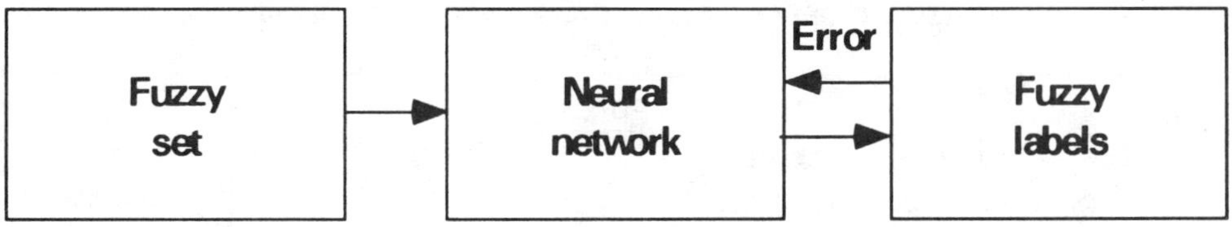

Figure 1: A neural network implementing a fuzzy classifier

ii: Making the individual neuron fuzzy (input to such a neuron is a fuzzy set and the output also is a fuzzy set). Activity of the networks involving fuzzy neurons is a fuzzy process.

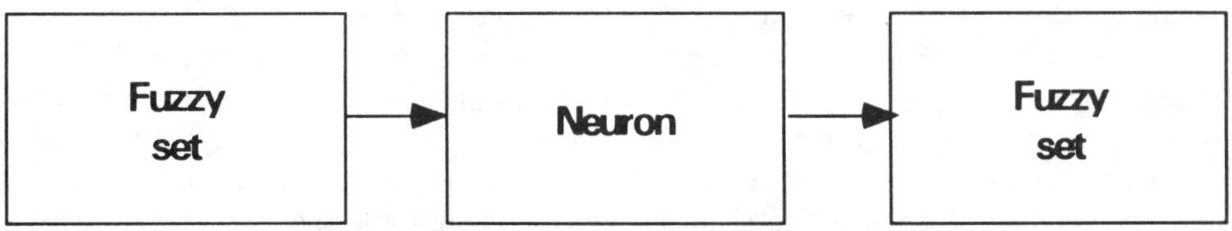

Figure 2: A block diagram of a fuzzy neuron

iii: Designing neural networks guided by fuzzy logic formalism (i.e., designing neural networks to implement fuzzy logic) and realization of membership functions representing fuzzy sets by neural networks.

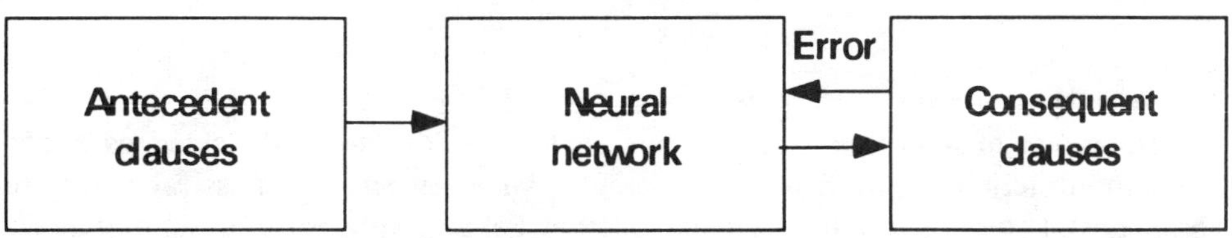

Figure 3: A neural network implementing fuzzy logic

iv: Changing the basic characteristics of the neurons so that they perform the operations used in the fuzzy set theory (like fuzzy union, intersection, aggregation) instead of doing the standard multiplication and addition

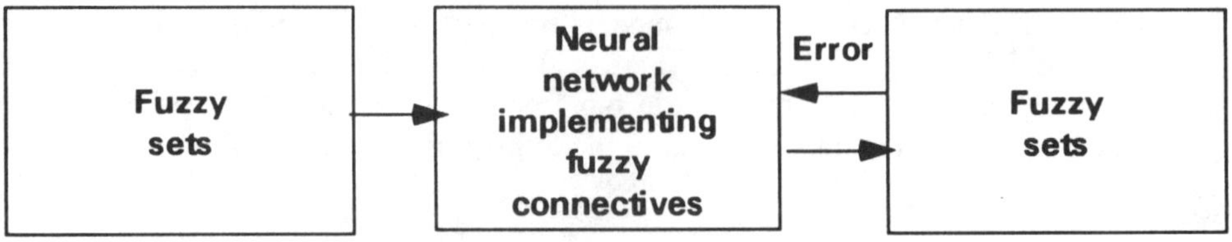

Figure 4: Neural network implementing fuzzy connectives

v: Modeling the error or instability or energy function of a neural network based systems using measures of fuzziness/uncertainty of a set.

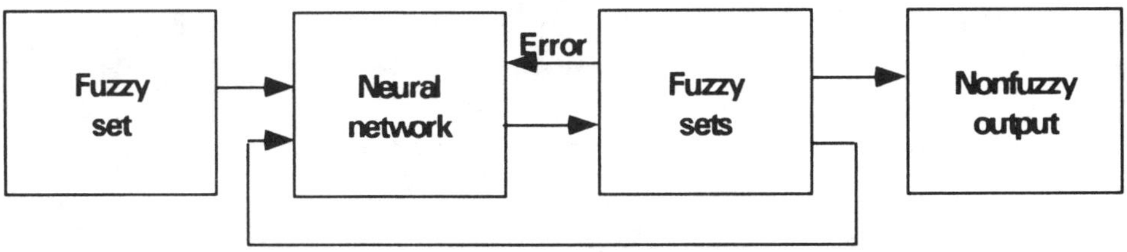

Figure 5. A layered network implementing self-organization

The first way of integration is to incorporate the concept of fuzziness into a neural network framework, i.e., to build *fuzzy neural networks.* For example, the target output of the neurons in the output layer during training can be fuzzy label vectors. In this case the network itself is functioning as a fuzzy classifier. Keller and Hunt[97] first suggested the incorporation of the concept of fuzzy pattern recognition into perceptron (single layer). They described a method for fuzzifying the labeled target data used for training the perceptron. Instead of giving hard labels to the training samples, membership functions denoting their degrees of belonging to the classes were used as labels. Instead of using the weight updation as

$$W \leftarrow W + cX_k \tag{2}$$

(c is a constant and X_k is the input data) they used

$$W \leftarrow W + \left|u_{1k} - u_{2k}\right|^m cX_k \tag{3}$$

where m is a constant and u_{ik} denotes the degree of belonging of x_k to the ith class. Incorporation of a membership function in the label vectors also acted as a good stopping criterion for linearly non-separable classes (where the classical perceptron oscillates). Later this idea has been exploited by Pal and Mitra[63] for designing a fuzzy version of the *MLP* for pattern classification. Pal and Mitra attempted to incorporate fuzziness in the input layer of the *MLP* by giving inputs in linguistic form (e.g., small, medium and high) so that it can handle imprecise data. Similar expressions have also been made by them for Kohonen's self-organizing network. It has been shown that these modified versions provide better performance for certain non-convex decision regions compared to the conventional ones. The concept of fuzzy labels has also been utilized in[98] for learning a network.

Sanchez[65] has pointed out the noticeable similarities (like training by example, dynamic adjustment of changes in the environment, ability to generalize, tolerance to noise, graceful degradation at the border of the domain of expertise, and ability to discover new relations between variables) between neural networks and expert systems. He developed a fuzzy version of the connectionist expert system of Gallant.[64] In such an expert classification system knowledge base is generated from a training set of examples and is stored as connection strengths. The weights (w_{ij}) between the input and the hidden layers are linguistic labels of fuzzy sets (identified by membership function) characterizing the variation of the input neurons, and are assumed to be known. The weights between the hidden layer and the output layers (b_{ij}) are determined by training. The output of the neurons are computed by combining the weights for inferencing as

$$v_i(t) = \min\{w_{ij}, b_{ij}\} \tag{4}$$

A possible utility of such an expert system in biomedical domain is also stated.

The second way to incorporate fuzziness into the standard neural network is by making the individual neurons fuzzy.[99,100] The idea is originally introduced by Lee and Lee[99] in 1975. The classical model of a neuron assumes its activity as an 'all-or-none' process. It fails to model a type of imprecision which is associated with the lack of sharp transition from the occurrence to nonoccurrence of an event. Some of the concepts of fuzzy set theory are employed by Lee and Lee for defining a fuzzy neuron, which is a generalization of the classical neuron. The activity of a fuzzy neuron is a fuzzy process.[99] The input to such a neuron is a fuzzy set and the outputs are equal to some positive numbers, μ_j's ($0 < \mu_j \leq 1$), if it is firing and zero if it is quiet, μ_j denoting the degree to which the jth output is fired. The output is, therefore, also a fuzzy set. Unlike conventional neurons, such a neuron has multiple outputs. A fuzzy neural network is defined as a collection of interconnected fuzzy neurons. The utility of fuzzy neural networks to synthesize fuzzy automata is also investigated by them.

The third fusion methodology is to use neural networks for a variety of computational tasks within the framework of a pre-existing fuzzy model (i.e., implementation of a fuzzy logic formalism using neural networks). The use of multi-layer feed forward neural network for implementing fuzzy logic rules (**if-then** rules) is introduced by Keller et al..[101-103] It has been shown that the networks designed for implementing fuzzy rules can learn and extrapolate complex relationships between antecedents and consequent clauses for rules containing single, conjunctive and disjunctive antecedent clauses. For rules having conjunctive clauses, the architecture has a fixed number of neurons in the input layer for each antecedent clause, a set of neurons in the hidden layer connected only to the neurons in the input layer associated with each antecedent clause.

The neurons in the output layer are connected to all the neurons in the hidden layer. For implementing rules with disjunctive antecedent clauses, one more hidden layer was necessary. In[101] attempts are made to embed apriori knowledge of each rule directly into the weights of the network. In other networks the standard back-propagation learning algorithm is applied for learning weights. An attempt is also made by Takagi et al.[104] to design structured neural networks to perform **if-then** fuzzy inference rules.

Attempts are also made to have membership function representation by neural networks.[105,106] A method has been suggested by Ishibuchi and Tanaka[105] to identify real-valued and interval-valued membership functions from a set of given input-output data using a feed-forward layered neural network and back-propagation of error. Suggestions are also given to design membership functions of fuzzy neurons in.[107]

A great deal of effort has also been given for designing neural network driven optimal decision rules for fuzzy controllers.[108-110] A system for implementing fuzzy logic controllers using a neural network is designed in.[109] The linguistic values associated with the fuzzy control rules are realized by separate neural network blocks. The network is crafted depending on the inference structure provided by fuzzy logic involving intersection operations of fuzzy sets. The importance of different rules in the system is learned by operating the whole system and employing a rule which is of the form of the generalized delta rule. Suggestions are also given[111] for learning and tuning of fuzzy logic controllers based on re-enforcement learning. Emphasizing is mainly used to adjust membership functions of the linguistic labels used in control rules.

Fuzzy version of Kohonen's self-organizing feature map algorithm is developed by Huntsberger and Ajjimarangsee[112] in order to generate continuous valued outputs (representing the degree of belonging) by adding one layer to the original Kohonen-network. Fuzziness is also incorporated in the learning rate by replacing the learning rate usually found in Kohonen-type update rules for the weight vectors with fuzzy membership of the nodes in each class. The proposed update rule is

$$W_j(t+1) = W_j(t) + \mu_{ik}\{x_k - W_j(t)\} \tag{5}$$

where μ_{ik} is the fuzzy membership of input x_k in class i instead of the standard updating rule

$$W_j(t+1) = W_j(t) + \alpha\{x_k - W_j(t)\} \tag{6}$$

with α as a constant. They also have shown that the results produced by this fuzzy version of Kohonen's algorithm are similar to those obtained by fuzzy c-means algorithms.[20] Parallel implementations of this technique are also suggested. Further modification on the rate of learning is done by Bezdek et al. in[113] and a relationship between the fuzzy version of Kohonen's algorithm and the fuzzy c-means algorithm is established.

Another way of fusion is to change the integration/transformation operation performed at each node so that they perform some sort of fuzzy aggregation (i.e., fuzzy union, intersection, aggregation). In[114] fuzzy set connectives are used in multi-layer network structures suitable for pattern recognition and other decision making systems. Various union, intersection, generalized mean and multiplicative hybrid operators (which are used by fuzzy sets to aggregate imprecise information in order to arrive at a decision in uncertain environments) are implemented by layered networks. A generalized mean operator was introduced by Yager.[115] Its form is

$$\overline{w} = \left(\frac{w_1^2}{\sum w_i^2} x_1^p + \frac{w_2^2}{\sum w_i^2} x_2^p + \cdots + \frac{w_n^2}{\sum w_i^2} x_n^p \right)^{\frac{1}{p}} \tag{7}$$

where w_is are the relative importance given to corresponding input x_is and $\sum_{i=1}^{n} w_i = 1$.

The hybrid (compensatory) model used was the γ-model of Zimmermann and Zysno[116] and is expressed as

$$y = \prod_i \left(x_i^{\delta^i} \right)^{1-\gamma} \left(1 - \prod_i (1 - x_i)^{\delta^i} \right)^{\gamma} \tag{8}$$

where $\sum_{i=1}^{m} \delta_i = m$ and $0 \le \gamma \le 1$; $x_i \in [0,1]$ are the inputs, δ_i is the weight associated with x_i and γ controls the degree of compensation. The hybrid operator can behave as union, intersection or mean operators with different sets of parameters which can be learned through training procedure. An iterative algorithm to determine the type of aggregation functions and its parameters at each node in the network is also provided thereby making the network more flexible. The learning procedure involved is the same as that of the *MLP*. The training procedure of the multiplicative γ-model is slow. To achieve faster convergence, the additive γ-model is studied, under the above framework, by Keller and Chen[117] as an alternative connective in such networks.

Gupta[118] suggested the use of generalized **AND** (which can be expressed using the notion of triangular norms) and **OR** (represented by triangular conorm) operations for fuzzy signals (signals bound by the graded membership function over the unit interval [0, 1]) instead of multiplication and summation operations as used in standard neural networks. Thus for fuzzy inputs, $\boldsymbol{x}(t) \in [0,1]^n$ and synaptic strengths $w(t) \in [0,1]^n$ the weighted synaptic signal $z(t) \in [0, 1]^n$ is defined as

$$z_i(t) = w_i(t) \textbf{ AND } x_i(t), \ i = 1,2, \ldots ,n \tag{9}$$

and the aggregated input to a neuron is

$$u_i(t) = \underset{i}{\textbf{OR}} z_i(t) \tag{10}$$

The non-liner mapping with threshold $w_o \in [0,1]$ is then defined as

$$v_i(t) = \left[u_i(t) \text{OR } w_o(t)\right]^\alpha \tag{11}$$

where α is a positive quantity. For $0 < \alpha \leq 1$ the operation corresponds to dilation operation of a fuzzy set and for $\alpha > 1$ it corresponds to concentration operation.

Pedrycz[119] tried to introduce fuzziness in neural networks in a different way. He pointed out the analogies between structures involving composite operators and a certain class of neural networks. Links are established between neural network architectures and relational systems in terms of fuzzy relational equations. The proposed architecture is based exclusively on set theoretical operations. The individual neurons perform logical operations (like *max, min*) which are mainly used in set theory instead of arithmetic operations. The problem of learning of connection strengths or weights was also studied and relevant learning rules were proposed. A performance index, called *equality index*, is also introduced for keeping track of these logical operations. He has also suggested[120] to design neural networks to implement logic operations used in fuzzy set theory.

The fifth way to integrate the concepts of fuzzy sets and neural networks is to use the fuzziness measures/uncertainty measures of a fuzzy set to model the error in neural networks. An attempt is made in this context in[83] by Ghosh et al. to incorporate various fuzziness measures in a multi-layer network for performing (unsupervised) self-organizing task in image processing, in general, and object extraction in particular. The network architecture is basically a feed forward one with back propagation of error, but unlike conventional *MLP* it does not require any supervised learning. Each layer has $M \times N$ neurons for an $M \times N$ image (each neuron correspond to an image pixel). Each neuron is connected to the corresponding neuron in the previous layer and its neighbors. Another structural difference from the standard *MLP* is that there exists a feed-back path from the output to the input layer. The status of neurons in the output layer is described as a *fuzzy set*. A fuzziness measure (e.g., index of fuzziness and entropy) of this set is used as a measure of error in the system (instability of the network) and back-propagated to correct weights.

The input value (U_i) to the ith neuron in any layer [except the input layer] is calculated using the formula

$$U_i = \sum_j W_{ij} o_j \tag{12}$$

where W_{ij} is the connection strength between the ith neuron of one layer and jth neuron of the previous layer and o_j is the output status of the jth neuron of the previous layer. j can either belong to the neighborhood of i or $j = i$ of the previous layer. The output is the obtained as

$$o_i = \frac{1}{1 + e^{-U_i}} \tag{13}$$

Starting from the input layer, this way the input pattern is passed on to the output layer and the corresponding output states are calculated. The output value of each neuron lies in [0,1].

After the weights have been adjusted by back-propagating the fuzziness measure of the output status of the neurons (which is treated as a fuzzy set) properly, the output of the neurons in the output layer is fed back to the corresponding neurons in the input layer. The second pass is then continued with this as input. The iteration (updating of weights) is continued as in the previous case until the network stabilizes, i.e., the error value (measure of fuzziness) becomes negligible. For quadratic index of fuzziness[29] the expression for weight updating is

$$\Delta W_{ij} = \begin{cases} \eta(-o_j)f'(I_j)o_i & \text{if } 0 \le o_j \le 0.5 \\ \eta(1-o_j)f'(I_j)o_i & 0.5 \le o_j \le 1.0 \end{cases} \tag{14}$$

for the output layer and

$$\Delta W_{ji} = \eta\left(\sum_k \delta_k W_{kj}\right)f'(I_j)o_i \tag{15}$$

for hidden layers; where $\delta_k = -\frac{\partial E}{\partial I_k}$ and η is a proportionality constant. In the converged state the ON neurons constitute one class, and the OFF neurons another. In[121] an approach is provided to design optimal network architecture by optimization of fuzziness of a set.

Note that these attempts of integration are mainly in the field of pattern recognition and to some extent in fuzzy logic control. Literature on neuro-fuzzy image processing is not adequate at this moment. For further references on this issue one can refer.[8,12,122-130]

5. FURTHER SCOPE OF WORK

Fusion of these two technologies are tried out in various labels (viz. input label, neuron characteristic, output label) mainly in pattern recognition problems. Literature is poor on such attempts for performing image processing tasks. Besides, most of the attempts are only preliminary. Their usefulness and validity should be studied rigorously by choosing problems from different aspects of image processing and pattern recognition. Recently attempts are also made to establish links between the existing fuzzy set theoretical algorithms and neural network based approaches.[112,113,131]

Attempts are made mainly to integrate the concepts of layered networks and fuzzy sets. Incorporation of fuzzy set theoretical concepts in various stages of self-organizing or auto-associative network extraction algorithms in order to handle ill-defined input data more efficiently may be tried out. In addition to these, fuzzy set theoretical algorithms can also be altered so that they can be implemented on a particular type of neural network architecture.

Attempts are made to design fuzzy neural networks by assigning fuzzy labels to training samples and using linguistic inputs. In this context the effect of fuzzification of input data (by using variable membership functions) on the recognition score can also be investigated. The variation on the amount of learning (i.e., variation of learning time) due to fuzzification of input data and output labels should be studied. Investigation can also be made to check the utility of fuzziness measures as an error measure of such neural network systems. Some of the fuzzy geometrical properties can also be used as features while training an artificial neural network. Designing of fuzzy labels in noisy environments may be an interesting problem of study. Since the learning in such networks is a strenuous process, development of methodologies based on genetic algorithms[132,133] for determining the global optimal set of parameters (weights, biases and network topologies) for neural network based systems should constitute another part of further investigation.

Preliminary attempt are made to design neural networks governed by fuzzy logic formalism based on **if-then** rules. A set of neurons is assigned for every conjunctive antecedent clause in the input layer. Attempts should be made to find out ways for implementing rules with different sets of antecedent clauses by the same network. Disjunctive antecedent clauses are implemented using hidden neurons. Use of input neurons to implement disjunctive clauses needs to be investigated. Designing optimal networks for implementing a given set of rules should constitute another part of further research. Fuzzy logic rules can also be altered so that they become mappable on a particular type of neural network architecture.

Attempts are made to identify membership functions by back-propagation learning. Feed forward layered networks can approximate continuous functions. This property can also be used to design membership functions. The concepts can also be used to design appropriate membership functions for image processing and pattern recognition tasks with suitable modification.

Only a few fuzzy set theoretical operators (like union, intersection and aggregation) are implemented by neural networks. Other standard operators like sharpening, complementing, bounded sum, bounded difference etc. may also be implemented. Implementation of fuzzy image processing systems involving these types of operations can also be made.

Various fuzziness measures have been used to implement the multi-layer perceptron as an unsupervised system. The intermediate outputs are fuzzy, but the final output is nonfuzzy. Incorporation of fuzzy hedges like *small, medium, high* in the input layer, and fuzzy membership in the output status needs to be done for handling the uncertainty in the input efficiently. The utility of some other ambiguity measure (like fuzzy geometrical properties[14,17] and the fuzzy correlation measure[21,56] as objective functions of neural network based systems can be investigated. The effectiveness of fuzziness measures as error measure in designing neural network based classifiers should also be investigated.

6. CONCLUSIONS

Relevance of fuzzy set theoretical approaches to image analysis and recognition is described. Characteristics of neural network and their applications to various image processing operations along with the relevance are provided. Utility and feasibility of the fusion of these two technologies is described. From the discussions it appears that this area will continue to be in the frontline of research for the next five to ten years, if not more.

REFERENCES

1. D. E. Rumelhart, J. McClelland, et al., *Parallel Distributed Processing: Explorations in the Microstructure of Cognition*. Vol. 1, Cambridge, MA: MIT Press, 1986.
2. T. Kohonen, *Self-organization and Associative Memory*. Berlin: Springer Verlag, 1989.
3. J. J. Hopfield, "Neurons with graded response have collective computational properties like those of two state neurons," in *Proceedings National Academy of Science*, (USA), pp. 3088-3092, 1984.
4. S. Grossberg (ed.), *Neural Networks and Natural Intelligence*. Reading, MA: MIT Press, 1988.
5. K. Fukushima, "Neocognitron: a self-organizing multilayer neural network model for a mechanism of pattern recognition unaffected by shift in position," *Biological Cybernetics*, vol. 36, pp. 193-202, 1980.
6. Y. H. Pao, *Adaptive Pattern Recognition and Neural Networks*. New York: Addison-Wesley, 1989.
7. P. D. Wassermann, *Neural Computing: Theory and Practice*. New York: Van Nostrand Reinhold, 1990.
8. B. Kosko, *Neural networks and Fuzzy Systems*. Englewood Cliffs, New Jersey: Prentice-Hall, 1992.
9. L. O. Chua and L. Yang, "Cellular neural network: theory," *IEEE Transactions on Circuits and Systems*, vol. 35, pp. 1257-1272, 1988.
10. A. Ghosh, N. R. Pal, and S. K. Pal, "Neural computing: an introduction and some applications," *Students' Journal of the Institute of Electronics and Telecommunication Engineers*, 1992 (accepted).
11. E. Gelenbe, "Theory of random neural network model," in *Neural Networks: Advances and Applications*, (E. Gelenbe, ed.), North Holland: Elsevier Publishers B. V., 1991.
12. J. C. Bezdek and S. K. Pal (eds.), *Fuzzy Models for Pattern Recognition: Methods that Search for Structures in Data*. New York: IEEE Press, 1992.
13. J. M. S. Prewitt, "Object enhancement and extraction," in *Picture Processing and Psycho-Pictorics*, (B. S. Lipkin and A. Rosenfeld, eds.), New York: Academic Press, 1970.
14. A. Rosenfeld, "Fuzzy geometry of image subsets," *Pattern Recognition Letters*, vol. 2, pp. 311-317, 1984.
15. S. K. Pal, "Fuzzy sets in image processing and recognition," in *Proceedings First IEEE International Conference on Fuzzy Systems*, (San Diego, USA), pp. 119-126, 1992.

16. R. Krishnapuram and J. M. Keller, "Fuzzy set theoretical approach to computer vision: an overview," in *Proceedings First IEEE International Conference on Fuzzy Systems*, (San Diego, USA), pp. 135-142, 1992.
17. S. K. Pal and A. Ghosh, "Fuzzy geometry in image analysis," *Fuzzy Sets and Systems*, vol. 48, no. 2, pp. 23-40, 1992.
18. L. A. Zadeh, "Fuzzy sets," *Information and Control*, vol. 8, pp. 338-353, 1965.
19. D. Dubois and H. Prade, *Fuzzy Sets and Systems: Theory and Applications*. New York: Academic Press, 1980.
20. J. C. Bezdek, *Pattern Recognition with Fuzzy Objective Function Algorithms*. New York: Plenum Press, 1981.
21. S. K. Pal and D. D. Majumder, *Fuzzy Mathematical Approach to Pattern Recognition*. New York: John Wiley (Halsted Press), 1986.
22. A. Kandel, *Fuzzy Mathematical Techniques with Applications*. New York: Addison-Wesley, 1986.
23. G. J. Klir and T. Folger, *Fuzzy Sets, Uncertainty and Information*. Reading, MA: Addison-Wesley, 1989.
24. T. L. Huntsberger, "Representation of uncertainty in low level vision," *IEEE Transactions on Computer*, vol. 23, no. 2, p. 145, 1986.
25. S. K. Pal and R. A. King, "Image enhancement using smoothing with fuzzy sets," *IEEE Transactions on Systems, Man, and Cybernetics*, vol. 11, pp. 494-501, 1981.
26. W. Pedrycz, "Fuzzy sets in pattern recognition: methodology and methods," *Pattern Recognition*, vol. 23, no. 1/2, pp. 121-146, 1990.
27. J. M. Keller and C. L. Carpenter, "Image segmentation in the presence of uncertainty," *International Journal of Intelligent Systems*, vol. 5, pp. 193-208, 1990.
28. R. R. Yager and L. A Zadeh (eds.), *An Introduction to fuzzy logic applications in intelligent systems*. Boston: Kluwer Academic Press, 1992.
29. A. Kaufmann, *Fuzzy Subsets - Fundamental Theoretical Elements*. New York: Academic Press, 1980.
30. A. Deluca and S. Termini, "A definition of a non probabilistic entropy in the setting of fuzzy set theory," *Information and Control*, vol. 20, pp. 301-312, 1972.
31. W. X. Xie and S. D. Bedrosian, "An information measure for fuzzy sets," *IEEE Transactions on Systems, Man, and Cybernetics*, vol. 14, pp. 151-156, 1984.
32. B. Kosko, "Fuzzy entropy and conditioning," *Information Sciences*, vol. 40, pp. 165-174, 1986.
33. N. R. Pal and S. K. Pal, "Object background segmentation using a new definition of entropy," *IEEE Proceedings, Part E*, pp. 284-295, 1989.
34. S. K. Pal and A. Rosenfeld, "Image enhancement and thresholding by optimization of fuzzy compactness," *Pattern Recognition Letters*, vol. 7, pp. 77-86, 1989.
35. S. K. Pal and A. Rosenfeld, "A fuzzy medial axis transformation based on fuzzy disk," *Pattern Recognition Letters*, vol. 12, no. 10, pp. 585-590, 1991.
36. S. K. Pal and A. Ghosh, "Index of area coverage of fuzzy image subsets and object extraction," *Pattern Recognition Letters*, vol. 11, no. 12, pp. 831-841, 1990.

37. D. Dubois and M. C. Jaulent, "A generalized approach to parameter evaluation in fuzzy digital pictures," *Pattern Recognition Letters*, vol. 6, pp. 251-259, 1987.
38. A. Rosenfeld and A. C. Kak, *Digital Picture Processing*. New York: Academic Press, 1982.
39. D. Marr, *Vision*. New York: W. H. Freeman and Company, 1982.
40. A. Rosenfeld, "The diameter of a fuzzy set," *Fuzzy Sets and Systems,* vol. 13, pp. 241-246, 1984.
41. A. Rosenfeld and R. Klette, "Degree of adjacency or surroundedness," *Pattern Recognition*, vol. 18, no. 2, pp. 169-177, 1985.
42. S. K. Pal, R. A. King, and A. A. Hashim, "Automatic gray level thresholding through index of fuzziness," *Pattern Recognition Letters*, vol. 1, pp. 141-146, 1983.
43. C. A. Murthy and S. K. Pal, "Fuzzy thresholding: mathematical framework, bound functions and weighted moving average," *Pattern Recognition Letters*, vol. 11, pp. 197-206, 1990.
44. N. R. Pal and S. K. Pal, "Higher order fuzzy entropy and hybrid entropy of a set," *Information Sciences*, vol. 61, no. 3, pp. 211-231, 1992.
45. S. K. Pal and A. Dasgupta, "Spectral fuzzy sets and soft thresholding," *Information Sciences*, 1992 (in press).
46. S. K. Pal and A. Ghosh, "Image segmentation using fuzzy correlation," *Information Sciences*, vol. 62, no. 3, pp. 223-250, 1992.
47. T. L. Huntsberger, C. L. Jacobs, and R. L. Cannon, "Iterative fuzzy image segmentation," *Pattern Recognition*, vol. 18, no. 2, pp. 131-138, 1985.
48. M. Trivedi and J. C. Bezdek, "Low-level segmentation of aerial images with fuzzy clustering," *IEEE Transactions on Systems, Man, and Cybernetics*, vol. 16, no. 4, pp. 589-598, 1986.
49. R. L. Cannon, R. Dave, and J. C. Bezdek, "Efficient implementation of fuzzy c-means clustering algorithms," *IEEE Transactions on Pattern Analysis and Machine Intelligence*, vol. 8, no. 2, 1986.
50. J. M. Keller, H. Qui, and H. Tahani, "The fuzzy integral in image segmentation," in *Proceedings NAFIPS*, (New Orleans, LA), pp. 324-338, 1986.
51. W. X. Xie, "An information measure for a color space," *Fuzzy Sets and Systems*, vol. 36, pp. 157-165, 1990.
52. K. Tanaka and M. Sugeno, "A study on subjective evaluations of printed color images," *International Journal Approximate Reasoning,* vol. 5, no. 3, pp. 213-222, 1991.
53. V. Goetcherian, "From binary to gray tone image processing using fuzzy logic concepts," *Pattern Recognition*, vol. 12, pp. 7-15, 1980.
54. S. K. Pal and R. A. King, "On edge detection of X-ray images using fuzzy sets," *IEEE Transactions on Pattern Analysis and Machine Intelligence*, vol. 5, pp. 69-77, 1983.
55. S. K. Pal, "Fuzzy skeletonization of images," *Pattern Recognition Letters,* vol. 10, pp. 17-23, 1990.
56. C. A. Murthy, S. K. Pal, and D. D. Majumder, "Correlation between two fuzzy membership functions," *Fuzzy Sets and Systems,* vol. 7, pp. 23-38, 1985.

57. J. M. Keller and L. Sztandera, "Spatial relation among fuzzy subsets of an image," in *Proceedings ISUMA*, (University of Maryland, USA), pp. 207-211, 1990.
58. S. K. Pal and L. Wang, "Fuzzy medial axis transformation (FMAT): practical feasibility," *Fuzzy Sets and Systems,* vol. 50, pp. 15-34, 1992.
59. "Special issue on neural network hardware design," *IEEE Transactions on Neural Networks,* vol. 1992.
60. U. Ramacher and U. Ruckert (eds.), *VLSI Design of Neural Networks*. Boston: Kluwer Academic Press, 1991.
61. C. Mead, *Analog VLSI and Neural Systems*. Reading, MA: Addison-Wesley, 1989.
62. R. P. Lippmann, "Pattern classification using neural networks," *IEEE Communications Magazine*, pp. 47-64, 1989.
63. S. K. Pal and S. Mitra, "Multilayer perceptron, fuzzy sets and classification," *IEEE Transactions on Neural Networks,* vol. 3, no. 5, pp. 683-697, 1992
64. S. I. Gallant, "Connectionist expert system," *Communication of the Association for Computing Machinery*, vol. 31, pp. 152-169, 1988.
65. E. Sanchez, "Fuzzy connectionist expert systems," in *Proceedings First International Conference on Fuzzy Logic and Neural Networks*, (Iizuka, Japan), pp. 31-35, 1990.
66. B. K. Parsi and B. K. Parsi, "On problem solving with Hopfield neural networks," *Biological Cybernetics*, vol. 62, pp. 415-423, 1990.
67. S. C. Newton and S. Mitra, "An adaptive fuzzy system for control and clustering of arbitrary data patterns," in *Proceedings First IEEE International Conference on Fuzzy Systems*, (San Diego, USA), pp. 363-370, 1992.
68. S. C. Newton, S. Pemmaraju, and S. Mitra, "Adaptive fuzzy leader clustering of complex data sets in pattern recognition," *IEEE Transactions on Neural Networks*, vol. 3, pp. 974-800, 1992.
69. D. J. Burr, "Experiments on neural net recognition of spoken and written text," *IEEE Transactions on Acoustics, Speech, and Signal Processing*, vol. 36, no. 7, pp. 1162-1168, 1988.
70. J. Lee, R. C. Weger, S. K. Sengupta, and R. M. Welch, "A neural network approach to cloud classification," *IEEE Transactions on Geoscience and Remote Sensing*, vol. 28, no. 5, pp. 846-855, 1990.
71. W. E. Blanz and S. L. Gish, "A connectionist classifier architecture applied to image segmentation," *International Journal of Pattern Recognition and Artificial Intelligence*, vol. 5, no. 6, pp. 603-617, 1991.
72. S. S. Yu and W. H. Tsai, "Relaxation by Hopfield neural network," *Pattern Recognition*, vol. 25, pp. 197-209, 1992.
73. N. Babaguchi, K. Yamada, K. Kise, and Y. Tezuku, "Connectionist model binarization," *International Journal of Pattern Recognition and Artificial Intelligence,* vol. 5, no. 6, pp. 629-644, 1991.
74. G. W. Cottrel, P. Munro, and D. Zipser, "Image compression by back propagation: an exemplar of extensional programming," Tech. Rep. ICS-8702, University of California, San Diego, 1987.
75. S. P. Luttrell, "Image compression using a multilayer neural network," *Pattern Recognition Letters*, vol. 10, pp. 1-7, 1989.

76. G. L. Bilbro, M. White, and W. Synder, "Image segmentation with neuro-computers," in *Neural Computers*, (R. Eckmiller and C. V. D. Malsberg, eds.), Springer-Verlag, 1988.
77. C. T. Chen, E. C. Tsao, and W. C. Lin, "Medical image segmentation by a constraint satisfaction neural network," *IEEE Transactions on Nuclear Science*, vol. 38, no. 2, pp. 678-686, 1991.
78. B. S. Manjunath, T. Simchony, and R. Chellappa, "Stochastic and deterministic networks for texture segmentation," *IEEE Transactions on Acoustics, Speech, and Signal Processing*, vol. 38, no. 6, pp. 1039-1049, 1990.
79. R. H. Silverman, "Segmentation of ultrasonic images with neural networks," *International Journal of Pattern Recognition and Artificial Intelligence*, vol. 5, no. 6, pp. 619-628, 1991.
80. A. Ghosh, N. R. Pal, and S. K. Pal, "Image segmentation using a neural network," *Biological Cybernetics*, vol. 66, no. 2, pp. 151-158, 1991.
81. A. Ghosh and S. K. Pal, "Neural network, self-organization and object extraction," *Pattern Recognition Letters*, vol. 13, no. 5, pp. 387-397, 1992.
82. A. Ghosh, N. R. Pal, and S. K. Pal, "Object background classification using Hopfield type neural network," *International Journal of Pattern Recognition and Artificial Intelligence*, vol. 6, no. 5, 1992 (in press).
83. A. Ghosh, N. R. Pal, and S. K. Pal, "Self-organization for object extraction using multilayer neural network and fuzziness measures," *IEEE Transactions on Fuzzy Systems*, vol. 1, no. 1, 1993 (in press).
84. A. Ghosh, N. R. Pal, and S. K. Pal, "On robustness of neural network based systems under component failure," *IEEE Transactions on Neural Networks*, 1992 (revised).
85. J. Shah, "Parameter estimation, multiscale representation and algorithms for energy-minimizing segmentation," in *Proceedings International Conference on Pattern Recognition*, pp. 815-819, 1990.
86. C. Cortes and J. A. Hertz, "A network system for image segmentation," in *Proceedings International Conference on Neural Networks*, pp. 121-125, 1989.
87. L. Bedini and A. Tonazzini, "Neural network use in maximum entropy image restoration," *Image and Vision Computing*, vol. 8, pp. 108-114, 1990.
88. Y. T. Zhou et al., "Image restoration using a neural network," *IEEE Transactions on Acoustics, Speech, and Signal Processing*, vol. 36, pp. 940-943, 1988.
89. S. C. Newton, B. S. Nutter, and S. Mitra, "Statistical and neural network classifiers in model-based 3-D object recognition," in *Proceedings of the SPIE: Intelligent Robotics and Computer Vision*, (Boston, MA), pp. 209-218, 1990.
90. N. M. Nasrabadi and W. Li, "Object recognition by a Hopfield neural network," *IEEE Transactions on Systems, Man, and Cybernetics*, vol. 21, no. 6, 1991.
91. T. A. Jamison and R. J. Schalkoff, "Image labeling: a neural network approach," *Image and Vision Computing*, vol. 6, no. 4, pp. 203-213, 1988.
92. Y. L. Cun *et al.*, "Back-propagation applied to handwritten zip-code recognition," *Neural Computing*, vol. 1, pp. 541-551, 1989.
93. I. Guyon, "Application of neural networks to character recognition," *International Journal of Pattern Recognition and Artificial Intelligence*, vol. 4, pp. 221-223, 1992.
94. K. Fukushima, "Character Recognition with neural networks," *Neurocomputing*, vol. 4, pp. 221-233, 1992.

95. S. Geman and D. Geman, "Stochastic relaxation, Gibbs distribution, and the Bayesian restoration of images," *IEEE Transactions on Pattern Analysis and Machine Intelligence*, vol. 6, pp. 721-741, 1984.
96. J. J. Hopfield and D. W. Tank, "Neural Computation of decisions in optimization of problems," *Biological Cybernetics,* vol. 52, pp. 141-152, 1985.
97. J. M. Keller and D. J. Hunt, "Incorporating fuzzy membership functions into the perceptron algorithm," *IEEE Transactions on Pattern Analysis and Machine Intelligence*, vol. 7, no. 6, pp. 693-699, 1985.
98. L. O. Hall, "Learning on fuzzy data with a backpropagation scheme," in *Proceedings NAFIPS*, (Missouri-Columbia), pp. 329-332, 1991.
99. S. C. Lee and E. T. Lee, "Fuzzy neural networks," *Mathematical Biosciences*, vol. 23, pp. 151-177, 1975.
100. T. Yamakawa and S. Tomada, "A fuzzy neuron and its applications to pattern recognition," in *Proceedings Third IFSA Congress*, (Seattle), pp. 30-38, 1989.
101. J. M. Keller, R. R. Yager, and H. Tahani, "Neural network implementation of fuzzy logic," *Fuzzy Sets and Systems*, vol. 45, no. 1, pp. 1-12, 1992.
102. J. M. Keller and H. Tahani, "Backpropagation neural networks for fuzzy logic," *Information Sciences*, vol. 62, pp. 205-221, 1992.
103. J. M. Keller and R. Krishnapuram, "Evidence aggregation networks for fuzzy logic inference," *IEEE Transactions on Neural Networks*, vol. 3, pp. 761-769, 1992.
104. H. Takagi *et al.*, "Neural networks designed on approximate reasoning architecture and their applications," *IEEE Transactions on Neural Networks*, vol. 3, pp. 752-760, 1992.
105. H. Ishibuchi and H. Tanaka, "Identification of real-valued membership functions," in *Proceedings First International Conference on Fuzzy Logic and Neural Networks*, (Iizuka, Japan), pp. 179-182, 1990.
106. H. Takagi and I. Hayashi, "Artificial neural network driven fuzzy reasoning," *International Journal of Approximate Reasoning*, vol. 5, pp. 191- 212, 1991.
107. T. Yamakawa and M. Furukawa, "A design algorithm of membership functions for a fuzzy neuron using example-based learning," in *Proceedings First IEEE International Conference on Fuzzy Systems*, (San Diego, USA), pp. 75-82, 1992.
108. M. M. Gupta, W. Pedrycz, and J. Kiszka, "Fuzzy control: from fuzzy controller to cognitive controller," in *Proceedings Third IFSA Congress*, (Seattle), pp. 258-261, 1989.
109. R. R. Yager, "Fuzzy logic controllers using neural network," *Fuzzy Sets and Systems*, vol. 48, pp. 53-64, 1992.
110. Y. Hayashi, E. Czogala, and J. J. Buckley, "Fuzzy neural controller," in *Proceedings First IEEE International Conference on Fuzzy Systems*, (San Diego, USA), pp. 197-202, 1992.
111. H. R. Berenji and P. Khedkar, "Learning and tuning of fuzzy logic controllers through reinforcement," *IEEE Transactions on Neural Networks*, vol. 3, pp. 724-740, 1992.
112. T. L. Huntsberger and P. Ajjimarangsee, "Parallel self-organizing feature maps for unsupervised pattern recognition," *International Journal of General Systems*, vol. 16, pp. 357-372, 1990.

113. J. C. Bezdek, E. C. K. Tsao, and N. R. Pal, "Fuzzy Kohonen clustering networks," in *Proceedings First IEEE International Conference on Fuzzy Systems*, (San Diego, USA), pp. 1035-1046, 1992.
114. R. Krishnapuram and J. Lee, "Fuzzy connective based hierarchical aggregation networks for decision making," *Fuzzy Sets and Systems*, vol. 46, pp. 11-27, 1992.
115. R. R. Yager, "Fuzzy decision making including unequal objectives," *Fuzzy Sets and Systems,* vol. 1, pp. 87-95, 1978.
116. H. J. Zimmermann and P. Zysno, "Decisions and evaluations by hierarchical aggregation of information," *Fuzzy Sets and Systems*, vol. 10, pp. 243-260, 1983.
117. J. M. Keller and Z. Chen, "Learning in neural networks utilizing adaptive hybrid operators," in *Proceedings Second International Conference on Fuzzy Logic and Neural Networks*, (Iizuka, Japan), pp. 85-87, 1992.
118. M. M. Gupta, "Fuzzy logic and neural networks," in *Proceedings Second International Conference on Fuzzy Logic and Neural Networks*, (Iizuka, Japan), pp. 157-160, 1992.
119. W. Pedrycz, "Neurocomputations in relational systems," *IEEE Transactions on Pattern Analysis and Machine Intelligence*, vol. 13, pp. 289-297, 1991.
120. W. Pedrycz, "Fuzzy neural networks with reference neurons as pattern classifiers," *IEEE Transactions on Neural Networks*, vol. 3, pp. 770-775, 1992.
121. K. J. Kios and N. Liu, "A machine learning method for generation of a neural network architecture: a continuous ID3 algorithm," *IEEE Transactions on Neural Networks*, vol. 3, pp. 280-291, 1992.
122. *Proceedings First IEEE International Conference on Fuzzy Systems*, (San Diego, USA), 1992.
123. *Proceedings Second IEEE International Conference on Fuzzy Logic and Neural Networks,* (Iizuka, Japan), 1992.
124. N. P. Archer and S. Wang, "Fuzzy set representation of neural network classification boundaries," *IEEE Transactions on Systems, Man, and Cybernetics*, vol. 21, pp. 735-742, 1991.
125. G. A. Carpenter, S. Grossberg, and D. B. Rosen, "Fuzzy ART: fast stable learning and categorization of analog patterns by an adaptive resonance system," *Neural Networks*, vol. 4, 1991.
126. "Special issue on fuzzy logic and neural networks," *International Journal of Approximate Reasoning*, vol. 6, no. 2, 1992.
127. M. E. Cohen and D. L. Hudson, "Approaches to the handling of fuzzy input data in neural networks," in *Proceedings First IEEE International Conference on Fuzzy Systems*, (San Diego, USA), pp. 93-100, 1992.
128. "Special issue on fuzzy logic and neural networks," *IEEE Transactions on Neural Networks*, vol. 3, no. 5, 1992.
129. P. J. Werbos, "Neurocontrol and fuzzy logic: connections and designs," *International Journal of Approximate Reasoning*, vol. 6, pp. 185-219, 1992.
130. "Special issue on neural networks," *International Journal of Pattern Recognition and Artificial Intelligence*, vol. 6, no. 1, 1992.
131. J. C. Bezdek, "On the relationship between neural networks, pattern recognition, and intelligence," *International Journal of Approximate Reasonings*, Vol. 6, no. 2, pp. 85-107, 1992.

132. L. Davis, *Handbook of Genetic Algorithms*. New York: Van Nostrand Reinhold, 1991.
133. D. E. Goldberg, *Genetic Algorithms: Search, Optimization and Machine Learning*. New York: Addison-Wesley, 1989.

Adaptive Pattern Recognition by Self-organizing Neural Networks.

Sunanda Mitra, Yong Soo Kim, and Surya Pemmaraju

Computer Vision and Image Analysis Laboratory
Department of Electrical Engineering
M.S. 3102
Texas Tech University
Lubbock, Texas 79409-3102

ABSTRACT

Nonlinear decision boundaries separating similar regions in unlabeled data sets are encountered in many real applications but are extremely difficult to generate by statistical decision models. Supervised multilayered neural networks are capable of generating some nonlinear decision boundaries but require extensive training procedures leading to computational burden as well. Unsupervised self-organizing neural nets are capable of forming crisp clusters from unlabeled data sets. A recent trend has been to integrate the concept of fuzzy sets with adaptive learning inherent to neural nets. Such integration yields not only better clustering of similar data groups, it may also provide a method for generating nonlinear boundaries among clusters in close proximity. The weaknesses and strengths of such integrated self-organizing neuro-fuzzy models for adaptive pattern recognition are described. Successful classification of standard data sets and generation of nonlinear decision boundaries among neighboring clusters from computer generated data are demonstrated with a recently developed integrated adaptive fuzzy clustering (IAFC) model.

1. INTRODUCTION

Supervised learning and known data distributions form the basis of designing most traditional pattern recognition systems [1,2]. In many practical applications such as medical diagnosis, radar signatures, etc., such training sets as well as knowledge of data structures are not always available. Unsupervised classification or clustering of unlabeled data sets has been widely used to select similar representatives in such cases and thus obtain decision boundaries separating similar data groups [3].

Crisp clusters in unlabeled datasets can be discovered by a number of iterative clustering approaches [4,5]. However complex decision boundaries of overlapping or closely located clusters are extremely difficult to generate. Nonlinear decision models based on neural networks and fuzzy logic are some approaches considered lately [3]. The consideration of fuzzy membership function of the data samples has led to the well known fuzzy c-means (FCM) clustering algorithm [6]. Among some of the other fuzzy clustering algorithms developed, the use of a generalized covariance metric for shape identification and performance measures for cluster validity have been addressed [7]. The fuzzy objective function used in FCM has been recently reformulated by relaxing the probabilistic constraint of membership grades [8].

A quite recent trend is to combine the strength of adaptive learning in neural networks with the fuzzy membership values of the data samples with respect to similar groups or clusters for avoiding misclassifications. Self-organizing neural networks are suitable for discovering crisp clusters in unlabeled datasets [9]. In general, learning rates of neural networks decrease with time to guarantee convergence of the weights. Such learning rates, however, do not consider how similar the input pattern is to the prototype of the corresponding cluster. The fuzzy membership value accounts for the degree of correspondence between the input pattern and the cluster centroid. Both Kohonen and ART-type self-organizing networks have been used to develop neuro-fuzzy clustering algorithms [10]. The fuzzification of Kohonen network and its modification for accepting continuous valued inputs have been achieved [11]. However the problems inherent in Kohonen clustering network (KCN) such as forced termination, dependence on data sequence, and lack of an optimization model for the learning strategy have been addressed recently and modified fuzzy KCN (FKCN) and fuzzy LVQ (FLVQ) algorithms have been developed [12]. The equivalence of FKCN and FLVQ to FCM under certain criteria has also been demonstrated. Models for fuzzy clustering based on the ART-type self-organizing neural networks have also been developed in recent years [13-15]. The original ART-1 [16] was capable of accepting binary inputs only. ART-1 has been modified to ART-2 [17] which is capable of accepting continuous-valued inputs. However fuzzification of ART-1 algorithm also allows continuous-valued inputs and avoids the complexity of ART-2. Among the fuzzy ART-type clustering algorithms, those developed by Simpson [13,18] and Carpenter et al. [14] are very similar and can be categorized as fuzzy min-max clustering algorithms. The adaptive fuzzy leader clustering (AFLC) [19] algorithm developed by Newton et al. differs from the fuzzy min-max clustering algorithms in a number of ways, namely the representation of the clusters, the fuzzy membership function, and the updating rule of the cluster centroids.

However, all of the neuro-fuzzy clustering algorithms developed so far suffer from restrictions in identifying the exact decision boundaries of the clusters in close proximity. The inability to obtain exact decision boundaries stems from the choice of the similarity measure and the representation of the clusters. The integrated adaptive fuzzy clustering (IAFC) [20] has been developed recently to obtain better decision boundaries among neighboring clusters by introducing a new similarity measure. This new algorithm also eliminates some of the inherent structural problems present in other fuzzy ART-type algorithms by developing a new learning rule. This learning rule is based on fuzzifying the weight updating rule in Kohonen clustering network by incorporating the fuzzy membership function derived in the FCM model in the updating rule. Therefore, the new learning rule integrates the advantages of FCM and KCN, into an ART-type neural network architecture.

This chapter presents a brief review of the existing self-organizing neuro-fuzzy clustering algorithms. The clustering results obtained with a standard data set namely Anderson IRIS data set as well as experimental evidence of convergence of the IAFC algorithm are presented and compared with the results of other clustering algorithms. The effect of the new similarity measure developed for the IAFC algorithm on generating nonlinear decision boundaries among overlapping and neighboring clusters is also demonstrated.

2. FUZZY SELF-ORGANIZING NETWORKS

2.1 Fuzzy Kohonen Networks

The incorporation of fuzzy logic into Kohonen self-organizing feature map has been achieved by replacing the learning rate with the fuzzy membership value as follows [11]:

$$\mathbf{v}_i(t+1) = \mathbf{v}_i(t) + u_{ij} * N_i(t) * (\mathbf{x}_j - \mathbf{v}_i(t)), \qquad (1)$$

where $\mathbf{v}_i$ is the centroid of the ith cluster , u_{ij} is the fuzzy membership value of the input pattern $\mathbf{x}_j$ in the ith cluster, and $N_i(t)$ is the size of the neighborhood. The fuzzy membership value of the input pattern in a cluster depends on the inverse of squared distance between the input pattern and a cluster centroid, i.e.,

$$u_{ij} = \left(\sum_{k=1}^{c} \left(\frac{d_{ij}}{d_{ik}} \right) \right)^{-1}, \tag{2}$$

where c is the number of clusters and $d_{ij} = (\mathbf{x}_j - \mathbf{v}_i)^T (\mathbf{x}_j - \mathbf{v}_i)$. Thus the algorithm is forced to terminate by just reducing the size of the neighborhood to zero. Such a modification does not guarantee convergence of the weights.

Bezdek et al. fuzzified the Kohonen clustering network, by integrating the FCM model into the learning rule of the Kohonen clustering network [12]. In Kohonen clustering network, the learning rates and the size of the neighborhoods are heuristically decreased for the convergence of weights. But, in FKCN, the distribution of learning rates and the size of neighborhoods are controlled by changing the value of weight exponents with time. The FKCN guarantees convergence of the weights by minimizing the objective function as in the FCM model.

FKCN controls learning rates and size of the neighborhoods in KCN by changing the value of weight exponents with time as follows:

$$m_t = m_0 - t^* \Delta m = m_0 - t^*(m_0 - 1)/t_{max}, \tag{3}$$

where m_0 is the initial value of weight exponent and is greater than one and t_{max} is the iteration limit. The fuzzified learning rate is defined as:

$$\alpha_{ik,t} = (u_{ik,t})^{m_t}, \qquad 1 \le i \le c \tag{4}$$

where $u_{ik,t}$ is the fuzzy membership value of the input pattern $\mathbf{x}_k$ with respect to the ith cluster. When m_t approaches one, $u_{ik,t}$ can be zero or one as in the FCM model. In this limit, the learning rule of the FKCN becomes the competitive learning rule with the learning rate in the winning cluster being one and the learning rates in other clusters being zero. The summary of this algorithm is as follows:

1. Fix the number of clusters c, and the distance measure $\| \cdot \|_A$.
Select $\varepsilon > 0$.

2. Select t_{max} and $m_0>1$. Initialize cluster centroids, $\mathbf{v}_1(0)$, $\mathbf{v}_2(0)$, ..., $\mathbf{v}_c(0)$.
3. For t = 1, 2, ..., t_{max}.
   3. 1 For k = 1, 2, ..., n (the number of input patterns).
      (a) Calculate the learning rates from equations (3) and (4), where

$$u_{ik,t}=\left[\sum_{j=1}^{c}\left(\frac{\left\|\mathbf{x}_k-\mathbf{v}_{i,t-1}\right\|_A}{\left\|\mathbf{x}_k-\mathbf{v}_{j,t-1}\right\|_A}\right)^{2/m_t-1}\right]^{-1}, for \quad 1\leq i\leq c, \tag{5}$$

      (b) If k = n stop; Otherwise, next k.
   3. 2 Update all weights:

$$\mathbf{v}_{i,t}=\mathbf{v}_{i,t-1}+\frac{\left[\sum_{k=1}^{n}\alpha_{ik,t}\left(\mathbf{x}_k-\mathbf{v}_{i,t-1}\right)\right]}{\sum_{k=1}^{n}\alpha_{ik,t}}. \tag{6}$$

   3. 3 Compute

$$E_t=\left\|\mathbf{v}_t-\mathbf{v}_{t-1}\right\|^2. \tag{7}$$

   3. 4 If $E_t\leq\varepsilon$ stop; Otherwise, next t.

In the fuzzified LVQ (FLVQ), the weight exponent may have either increasing or decreasing value. The problem with the FKCN and FLVQ algorithms is that they both require large memory capacity due to non-sequential learning.

2.2 Fuzzy Adaptive Resonance Theory Neural Nets

The similarities between adaptive resonance theory and leader clustering algorithm are well-known [9,21,22]. A number of researchers have recently attempted fuzzification of the ART neural networks to obtain better structures in unlabeled data sets. In the fuzzy adaptive resonance theory neural network developed by Simpson, he fuzzified the ART-1 neural network by using fuzzy min-max operator [13]. This fuzzy neural network can cluster analog data because of the replacement of the logical AND operator in ART-1 by fuzzy min operator and other modifications.

The fuzzy adaptive resonance theory neural network has been modified recently to the fuzzy min-max clustering neural network [18].

Fuzzy ART by Carpenter et al. [14] is similar to the fuzzy adaptive resonance theory neural network by Simpson. Fuzzy ART can also cluster analog data in addition to binary data. This generalization is accomplished by replacing the logical AND operator in ART-1 by the fuzzy min operator. Such a replacement can produce category proliferation problem where weights approach to zero because of the unidirectional learning rule. This problem was prevented by using complement coding. Complement coding has the same emphasis on off-cells as on on-cells to represent the input pattern. Each category is represented by a hyper-box determined by on-cells/off-cells responses that are similar to min-max points in the fuzzy adaptive resonance theory neural network [16]. This hyper-box can only grow because the weights can only decrease. But the hyper-box can grow only within the bound determined by the vigilance parameter. If the value of the vigilance parameter is low, the size of the hyper-box is large and vice versa. The similarity measure used for vigilance criterion is fuzzy hamming distance from the edges of the hyper-box. We describe the fuzzy ART first and then the fuzzy min-max clustering neural network.

2.2.1 Fuzzy ART [14]

In Fuzzy ART proposed by Carpenter et al., the M-dimensional input pattern $\mathbf{a}$ is preprocessed into 2M-dimensional complement coding form:

$$\mathbf{I} = (\mathbf{a}, \mathbf{a}^c) = (a_1, a_2, \ldots, a_M, 1\text{-}a_1, 1\text{-}a_2, \ldots, 1\text{-}a_M). \tag{8}$$

Each component of the input pattern $\mathbf{a}$ should be in the interval [0, 1]. With the complement coding, a weight $\mathbf{w}_j = (w_{j1}, \ldots, w_{jM})$ of the LTM traces is changed to complement coding form:

$$\mathbf{w}_j = (\mathbf{u}_j, \mathbf{v}_j^c), \tag{9}$$

where $\mathbf{u}_j = (\wedge_j \mathbf{a}) = \min\{\mathbf{a}: \mathbf{a}$ has been categorized in category j$\}$, $\mathbf{v}_j = (\vee_j \mathbf{a}) = \max\{\mathbf{a}: \mathbf{a}$ has been categorized in category j$\}$. $\mathbf{u}_j$ and $\mathbf{v}_j$ define corners of the hyper-box R_j. The size of hyper-box R_j is defined as

$$|R_j| = |\mathbf{v}_j - \mathbf{u}_j| = |\vee_j \mathbf{a} - \wedge_j \mathbf{a}|, \tag{10}$$

which is equal to the sum of the height and the width of R_j.

In Fuzzy ART, it uses the choice function $T_j(\mathbf{a})$ to decide the order of search, rather than following the procedure of finding the category that has the maximum of the inner product of the input pattern and the weights. The choice function $T_j(\mathbf{a})$ for category j is defined as:

$$T_j(\mathbf{a})=\frac{|\mathbf{a}\wedge\mathbf{u}_j|}{\alpha+|\mathbf{u}_j|}, \tag{11}$$

where α is a choice parameter that is larger than 0, $(\mathbf{a},\mathbf{u})_i=\min(a_i,u_i)$, and

$|\mathbf{x}|=\sum_{i=1}^{M}|x_i|$. The $\alpha \to 0$ case is called the conservative limit because small values of α tend to minimize the change in weights during learning. The Jth category becomes a winner where

$$T_J(\mathbf{a}) = \max\{T_j(\mathbf{a}): j = 1,\dots,N\}. \tag{12}$$

After a winner is decided, it performs the matching procedure. If a winner satisfies the vigilance criterion:

$$\frac{|\mathbf{I}\wedge\mathbf{w}_J|}{|\mathbf{I}|}\geq\rho, \tag{13}$$

where ρ is the vigilance parameter, the weight is updated. With the complement coding, the vigilance criterion is

$$\left|R_J^{(new)}\right|=\left|R_J^{(old)}\oplus\mathbf{a}\right|=\left|(\mathbf{a}\vee\mathbf{v}_J)-(\mathbf{a}\wedge\mathbf{u}_J)\right|\leq M(1-\rho). \tag{14}$$

If the expanded hyper-box $R_j^{(new)}$, the smallest hyper-box that includes $R_j^{(old)}$ and **a**, satisfy the vigilance criterion with complement coding, the weight of a winner neuron is updated.

If the chosen category J satisfies the vigilance criterion, the weight $\mathbf{w}_J$ is updated according to the learning rule

$$w_J^{(new)}=\beta\left(I\wedge w_J^{(old)}\right)+(1-\beta)w_J^{(old)}, \tag{15}$$

where β is the learning rate in the interval [0, 1]. Under the assumption of fast learning (β = 1), the learning rule in the form of complement coding is

$$\mathbf{w}_J^{(new)} = \mathbf{I} \wedge \mathbf{w}_J^{(old)} = \left(\mathbf{u}_J^{(new)}, \left(\mathbf{v}_J^{(new)} \right)^c \right). \tag{16}$$

The weights with complement coding are initialized to

$$w_{j,1} = \ldots = w_{j,M} = \ldots = w_{j,2M} = 1. \tag{17}$$

Fuzzy ART does not use the fuzzy membership function, which is the basis of fuzzy clustering model, but uses fuzzy min-max operator. Therefore, the input pattern is normalized to the range [0, 1] which can yield poor performance in real applications.

2.2.2 Fuzzy Min Max Clustering Neural Network [18]

Simpson improved his earlier fuzzy adaptive resonance theory neural network [13] by incorporating the hyper-box membership function and adding the contraction process. The fuzzy min-max clustering neural network adjusts hyper-boxes if there are overlapping hyper-boxes to which the input pattern completely belongs. This fuzzy neural network also clusters analog data and does not assume the number of clusters a priori. It updates the number of clusters in the same manner as in ART-1. To find a winner, it does not use the maximum dot product between the input pattern but uses the hyper-box membership function. The category in which the input pattern has the highest degree of membership becomes a winner. Two parameters are used to tune the performance of this fuzzy neural network. One is the threshold of hyper-box θ, $0 \leq \theta \leq 1$, that is similar to the vigilance parameter in ART-1 and controls the maximum size of the hyper-box. The other is the sensitivity parameter γ, that controls the fuzziness of the cluster membership and regulates how fast the hyper-box membership value decrease when the input pattern is located outside of the hyper-box. The larger the value of γ, the more crisp the value of the hyper-box membership. There are two weights from each input neuron to the output neuron. One weight, $\mathbf{v}_j$, represents the min value for the jth hyper-box and the other weight, $\mathbf{w}_j$, represents its max value. The summary of this algorithm is as follows:

1. Select the threshold of hyper-box θ, $0 \leq \theta \leq 1$, and the sensitivity parameter γ, $0 \leq \gamma \leq 1$. Initialize all the min points ($\mathbf{v}_j$) and the max points ($\mathbf{w}_j$):

   $\mathbf{v}_j = (1,\ldots,1)$,

   $\mathbf{w}_j = (0,\ldots,0)$.
2. Apply the n-dimensional input pattern **a**.

3. If there is no committed neuron, assign the input pattern **a** as the first category,

$$\mathbf{v_j} = \mathbf{w_j} = \mathbf{a},$$

and go to step 2; otherwise, go to step 4.

4. Calculate the hyper-box membership value:

$$b_j(\mathbf{a},\mathbf{v}_j,\mathbf{w}_j)=\frac{1}{n}\sum_{i=1}^{n}\left[1-f(a_i-w_{ij},\gamma)-f(v_{ij}-a_i,\gamma)\right], \tag{18}$$

where

$$f(x,\gamma)=\begin{cases} 1, & if\ x\gamma \succ 1, \\ x\gamma, & if\ 0 \le x\gamma \le 1, \\ 0, & if\ x\gamma \prec 0, \end{cases} \tag{19}$$

for all committed neurons j.

5. Select a winner.

$$b_J(\mathbf{a},\mathbf{v}_J,\mathbf{w}_J)=\max\left\{b_{j'}(\mathbf{a},\mathbf{v}_{j'},\mathbf{w}_{j'})\right\}, \tag{20}$$

where j' consists of the output neurons which are not temporarily reset.

6. Vigilance Test.
If

$$\sum_{i=1}^{n}\left[\max(w_{ij},a_i)-\min(v_{ij},a_i)\right] \le \theta, \tag{21}$$

go to step 8; otherwise, go to step 7.

7. Reset temporarily the winning neuron. If all committed neurons are reset, activate the first uncommitted neuron, form a new category, and go to step 8; otherwise, go to step 5.

8. Update weights of the winning neuron.

$$\begin{aligned} v_{iJ}^{new} &= \min\left(v_{iJ}^{old},a_i\right), \quad \forall i=1,\ldots,n, \\ w_{iJ}^{new} &= \max\left(w_{iJ}^{old},a_i\right) \quad \forall i=1,\ldots,n. \end{aligned} \tag{22}$$

9. Overlap test: Determine whether the recent expansion caused any overlap between hyper-boxes.

10. Contraction: if there is any overlapping hyper-boxes, contract the hyper-boxes to eliminate the overlap.

11. Enable reset neurons. Go to step 2.

Although fuzzy min-max clustering neural network improved the original fuzzy adaptive resonance theory neural network by incorporating the hyper-box membership function and adding the contracting procedure, it still has restrictions in the shape of clusters formed and the problem associated with normalization of the input data.

2.2.3 Modified Adaptive Fuzzy Leader Clustering [23]

Another fuzzy ART-1 type neural network is the adaptive fuzzy leader clustering (AFLC) algorithm developed by Newton et al. [19]. AFLC incorporates an incremental updating rule into the ART-1 architecture. This incremental updating rule is based on the FCM equations. It does not update all cluster centroids as in the FCM model but updates only the centroid of the winner cluster. This algorithm does not assume the number of clusters a priori but updates the number of clusters during processing. Each category is represented by a cluster centroid. This algorithm uses a ratio of the Euclidean distance of the current input sample from the winner cluster centroid to the average Euclidean distance of all input patterns belonging to that category as the similarity measure for the vigilance criterion. AFLC requires a large memory capacity because of the learning rule used and the similarity measure used for the vigilance criterion. The clusters generated by this algorithm are hyper-spherical in shape.

In the modified AFLC, updating process is followed by a verification procedure whose function is to check if the previous classification is still valid. The location of the samples which join the cluster at a later stage can often cause the prototype (centroid) of the cluster to shift in a particular direction. The extent and the direction of the shift depend on the order in which the data is fed to the algorithm. As a result, the distance of some of the data samples from the new centroid of the cluster to which they originally belonged, will increase drastically and hence the vigilance condition might not be satisfied any more. This could result in misclassifications if these samples are found to be closer to another neighboring cluster and satisfy the vigilance condition with respect to it. The modified AFLC avoids this problem by means of a verification procedure which tests if all the samples in the updated cluster conform to the original classification. A sample that does not satisfy the original classification condition is reclassified by selecting the cluster prototype with maximum membership value to be the winner.

However classification of noisy data sets often poses a difficult problem as the noise sample points can severely bias the clustering procedure. Isolation of random noise samples has been achieved recently by the modified AFLC by introducing a noise cluster that contains all the outliers which can not be matched to form a separate data cluster. To avoid the danger of treating a couple of neighboring noise samples as a separate valid cluster, the user can define the minimum number of samples required to form a proper cluster. Besides, if the noise points are in small groups, these groups can be merged and separated based on an additional feature like pixel intensity, in a typical digital image processing application.

The modified AFLC algorithm can be summarized as follows:
1. Start with no cluster prototype, C = 0.
2. The first input data sample is stored as a noise sample in the zeroth cluster or the noise cluster.
3. The next input data sample [$\mathbf{X}_j = \{ X_{j1}, X_{j2}, ..., ..., X_{jp}\}$ for $2 \le j \le N$] is compared with all the samples in the zeroth cluster. If any one of the samples in the noise cluster satisfies the following criteria,

$$Y_i = \max_i \left\{ \sum_{k=1}^{p} x_{jk} b_{ki} \right\}; \quad 1 \le j \le N \tag{23}$$

$$R = \frac{\sqrt{d^2(\mathbf{X}_j, \mathbf{v}_i)}}{\frac{1}{N_i} \sum_{k=1}^{N_i} \sqrt{d^2(\mathbf{X}_k, \mathbf{v}_i)}} \le \tau. \tag{24}$$

where
N is the total number of samples,

N_i is the number of samples in the class i and $d^2(\mathbf{X}_j, \mathbf{v}_i) = \|\mathbf{X}_j - \mathbf{v}_i\|^2$ is the Euclidean distance,
x_{jk} is the kth component of the normalized X_j,
b_{ki} is the bottom-up weight connecting the kth component of the input vector to the ith cluster prototype, and
τ is the vigilance parameter,

then, it becomes the winner and forms a new cluster of its own. The current input sample is added to that cluster. Using the hybrid equations (25) and (26), based on partial FCM, the centroid and membership values are updated as follows.

$$\mathbf{v}_i = \frac{1}{\sum_{j=1}^{N_i} (\mu_{ij})^m} \sum_{j=1}^{N_i} (\mu_{ij})^m \mathbf{X}_j, \tag{25}$$

$$\mu_{ij} = \frac{\left(\frac{1}{\left\| \mathbf{X}_j - \mathbf{v}_i \right\|^2} \right)^{\frac{1}{m-1}}}{\sum_{k=1}^{C} \left(\frac{1}{\left\| \mathbf{X}_j - \mathbf{v}_k \right\|^2} \right)^{\frac{1}{m-1}}}, \tag{26}$$

where
v_i is the centroid of the ith cluster and
μ_{ij} is the fuzzy membership value of the jth sample with respect to the ith cluster.
Output the index of the new cluster. Go to Step 7.
4. If none of the samples in the zeroth cluster satisfies the above specified conditions, the input sample is compared with the prototypes of all the other existing data clusters to find the winner cluster.
5. If the input sample belongs to one of the clusters, it means that it is a sample from that pre-existing cluster and is appended to that cluster. The centroid and membership values are updated. Go to Step 7.
6. If none of the cluster centroids or the samples in the zeroth cluster satisfy the similarity condition in equation (24), the input sample is stored in the zeroth cluster along with the outliers. This sample is considered to be an outlier as long as it is present in the zeroth cluster, i.e. till other samples of its category come and join it to form a new class. Go to Step 3.
7. After updating the centroid, a verification test is performed to check if the previous classifications are still valid. This is done by verifying if all

the samples in the updated cluster satisfy the vigilance condition. If any sample violates this condition, it is reclassified by selecting the cluster prototype with maximum membership value to be the winning cluster. Go to Step 3.

2.2.4 Integrated Adaptive Fuzzy Clustering (IAFC) Model [20]

Existing neuro-fuzzy clustering algorithms suffer from restrictions in the shape of the decision boundaries of closely located clusters formed. FKCN and AFLC also require large memory capacity due to non-sequential learning. IAFC solves the above-mentioned problems by incorporating a new similarity measure for the vigilance criterion and a new learning rule into the neural network architecture similar to ART-1. IAFC does not assume the number of clusters in the data set a priori, but updates it following the process in ART-1.

The new similarity measure for the vigilance criterion, proposed in IAFC, incorporates a fuzzy membership value to Euclidean distance. Such an incorporation yields more flexibility to the clusters formed since it considers not only the Euclidean distance between the input pattern and the centroid of a winner but also the relative proximity of the input pattern to the currently existing cluster centroids as the degree of similarity.

The new learning rule, developed in IAFC, incorporates a fuzzy membership value, an intra-cluster membership value, and a function of the number of iterations into Kohonen-type learning rule. The fuzzy membership value used in IAFC is based on the membership formula of the FCM model. But, the number of clusters in IAFC is updated dynamically. Because of the effect of the new similarity measure, cluster centroids can drift around. Such a phenomenon can prevent weights from converging fast. The use of an intra-cluster membership value eliminates this problem. An intra-cluster membership value is decided by the distance between the input pattern and the centroid of the chosen category. The combination of a function of the number of iterations and an intra-cluster membership value guarantees weights or cluster centroids to converge.

The new similarity measure for the vigilance criterion, used in IAFC, is defined as follows:

$$e^{-\gamma\mu_i}\|\mathbf{X}-\mathbf{v}_i\|, \qquad (27)$$

where μ_i is the fuzzy membership value of the input data pattern in the ith winning cluster, γ is a multiplicative factor that controls the shape of clusters,

X is the input data pattern,

$\mathbf{v}_i$ is the centroid (top-down weights) of the ith winning cluster,

$\| . \|$ is the Euclidean distance.

This new similarity measure may also yield better classification in specific applications since an input data point can be assigned to a winning cluster due to its large fuzzy membership value in that cluster whereas the same data point may be considered as an outlier when an Euclidean measure is used. However, the extent of the outlier location is determined by the vigilance parameter that a user can decide. This measure allows more flexibility to the shape of a cluster than current similarity measures such as the Euclidean distance and the Mahalanobis distance. When the input data point is located in the overlapping regions of clusters, the effect of the new similarity measure is outweighed by the competition among the clusters. Figures 1(a), 1(b), and 1(c) demonstrate the effect of the new similarity measure and multiplicative factor γ in forming clusters with a deviation of the cluster boundary from circular shape.

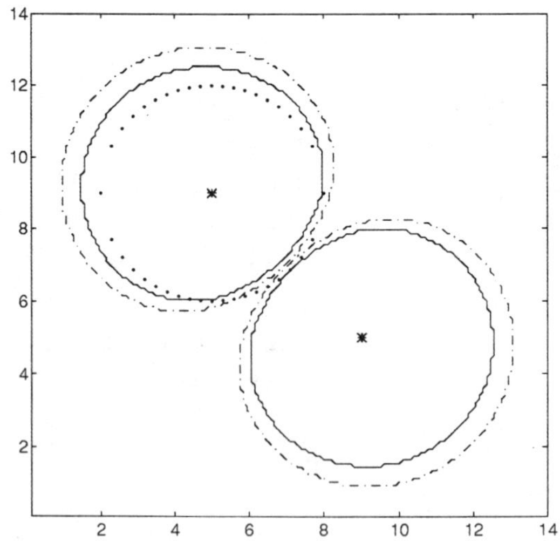

-------- $\gamma = 1.0$

-.-.-.-.-. $\gamma = 1.2$

.... circle

Figure 1(a). Cluster boundaries for different values of γ.

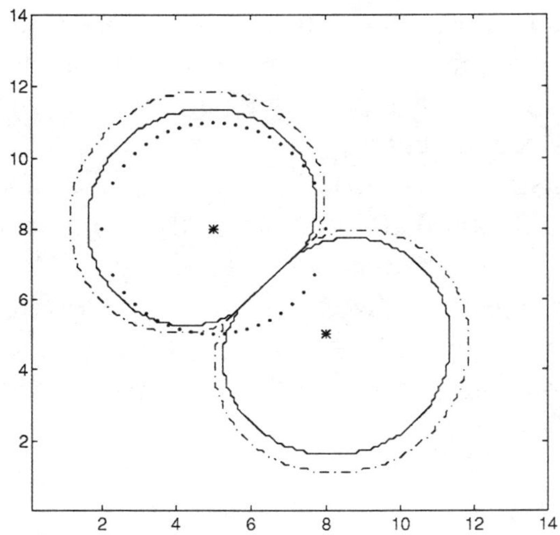

------- $\gamma = 1.0$

-.-.-.-.- $\gamma = 1.2$

.... circle

Figure 1(b). Cluster boundaries for different values of γ when cluster centroids are located too close.

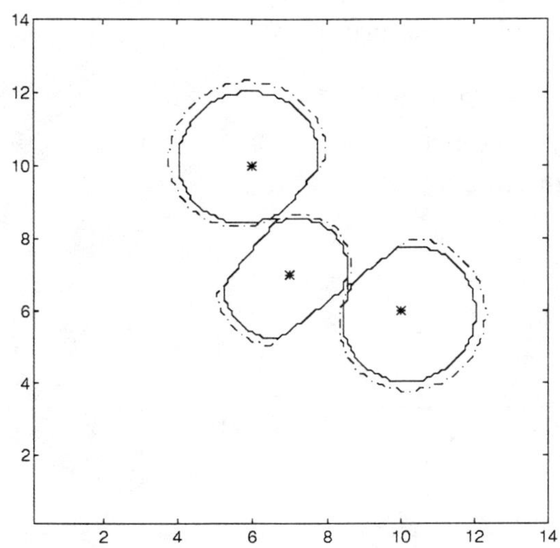

------- $\gamma = 1.0$

-.-.-.-.- $\gamma = 1.2$

Figure 1(c). Cluster boundaries for different values of γ when the number of clusters is three.

The IAFC algorithm consists of three major procedures: deciding a winning cluster, performing the vigilance test, and updating the centroid of a winning cluster. The input data pattern is normalized prior to presentation to the neural network and this normalized input data pattern (**I**) is fed to the neural network in parallel with the input data pattern. The normalized input data pattern is used only to find a winning cluster. After the normalized input data pattern is applied to the bottom-up weights, which are the normalized version of the centroids of clusters, competition among neurons in an output layer occurs in a winner-take-all fashion. The neuron that receives the largest input activation Y_i wins the competition:

$$Y_i = \sum_{k=1}^{p} I_k b_{ki}, \tag{28}$$

where

I_k is the kth component of the normalized input data pattern, and

b_{ki} is the bottom-up weight from the kth input node to the ith output node(cluster).

In order to justify the validity of the ith winning cluster, a new vigilance criterion is proposed using the unnormalized input data pattern(**X**). After selecting a winning cluster, the IAFC algorithm performs the vigilance test according to the vigilance criterion:

$$e^{-\gamma\mu_i} \|\mathbf{X} - \mathbf{v}_i\| \le \tau, \tag{29}$$

where τ is the vigilance parameter. The value of γ is normally chosen to be one. The fuzzy membership value is decided as follows [6, 24]:

$$\mu_i = \frac{\left(\frac{1}{\|\mathbf{X} - \mathbf{v}_i\|^2}\right)^{\frac{1}{m-1}}}{\sum_{j=1}^{n} \left(\frac{1}{\|\mathbf{X} - \mathbf{v}_j\|^2}\right)^{\frac{1}{m-1}}}, \tag{30}$$

where

n is the number of currently existing clusters, and

$m \in [1, \infty)$ is a weight exponent.

The value of m is experimentally set to 2. The number of currently existing clusters, n, is updated during clustering. In FCM [6], all c centroids of clusters are updated regardless of winning or losing.

If the winning cluster satisfies the vigilance test, the centroid is updated as follows:

$$\mathbf{v}_i^{(new)} = \left(1-\lambda_{fuzzy}\right)\mathbf{v}_i^{(old)} + \lambda_{fuzzy}\mathbf{X}, \tag{31}$$

where

λ_{fuzzy} is $f(l)\cdot\pi\left(\mathbf{X};\mathbf{v}_i^{(old)},\tau\right)\cdot\mu_i^2$,

l is the number of iterations, and

$\pi(\mathbf{X};\ \mathbf{v}_i^{(old)},\ t)$ decides the intra-cluster membership value of the input data pattern in the ith winning cluster.

The π-function [25] is defined as

$$\pi(\mathbf{X};\mathbf{v}_i^{(old)},\tau) = \begin{cases} 1-2(\|\mathbf{X}-\mathbf{v}_i^{(old)}\|/\tau)^2, & \text{when} \quad 0 \le \|\mathbf{X}-\mathbf{v}_i^{(old)}\| \le \tau/2, \\ 2(1-\|\mathbf{X}-\mathbf{v}_i^{(old)}\|/\tau)^2, & \text{when} \quad \tau/2 \le \|\mathbf{X}-\mathbf{v}_i^{(old)}\| \le \tau, \\ 0, & \text{when} \quad \|\mathbf{X}-\mathbf{v}_i^{(old)}\| \ge \tau. \end{cases} \tag{32}$$

We use

$$f(l) = 1/\{k(l-1)+1\}, \tag{33}$$

where k is a constant. The centroid of each cluster is used as the top-down weight related with each cluster.

The IAFC algorithm initially starts with the number of clusters (n) set to 0. When the first input pattern is presented, it becomes the first cluster centroid. For the next input pattern, the membership value will obviously be 1, since only one cluster is present. In this case, the vigilance criterion used is just $\|\mathbf{X} - \mathbf{v}_i\| \le \tau$, as the fuzzy membership value has no physical meaning. If it satisfies the vigilance test, the centroid is updated by the statistical average between two points. After more than one clusters are formed, the IAFC algorithm uses the fuzzy membership value for the vigilance test and the intra-cluster membership value for updating the centroid of a winning cluster. While determining a winning cluster, the winner is decided by the direction between the input data and the centroids of clusters. It can cause misclassifications because a cluster of which the centroid has the smallest difference of angle wins the competition even though it is located farther from the input data than other cluster centroids. To avoid misclassifications, the IAFC algorithm calculates the membership value after selecting the winning cluster. If the fuzzy membership value is less than the THRESHOLD, decided by a user, the IAFC algorithm activates the procedure to find the cluster that is closest to the input pattern. The closest cluster becomes a new winner. The IAFC algorithm can be summarized by the following steps:

1. Initialize parameters (n = 0, τ, THRESHOLD), output neurons, and input neurons.
2. Apply the next input pattern **x**.
3. Find a winning cluster (output neuron).
4. Calculate the fuzzy membership value of the input pattern in the winning cluster.
5. If the fuzzy membership value is less than the THRESHOLD, find the closest cluster to the input pattern as a new winning cluster.
6. If the winning cluster satisfies the vigilance test, go to step 8; otherwise, go to step 7.
7. Temporarily reset the winning output neuron. If all committed output neurons are reset, activate the first uncommitted output neuron, form a new cluster, and go to step 8; otherwise, go to step 3.
8. Update the centroid of the winning cluster and both bottom-up and top-down weights. Enable reset output neurons and go to step 2.

3. PERFORMANCE OF FUZZY SELF-ORGANIZING NEURAL NETWORK MODELS FOR PATTERN RECOGNITION

Since the misclassifications in the overlapping clusters of Anderson's IRIS data set has been used by researchers as a performance measure for many clustering or pattern recognition algorithms, we have also used it to compare the performance of two fuzzy self-organizing neural network models namely FKCN, and IAFC with that of FCM.

Figure 2 shows the convergence rate of the above algorithms and the number of misclassifications obtained by each algorithm. It is noteworthy that IAFC yields minimum number of misclassifications only after 6 iterations although the desired termination limit ($\in$) of 0.00001 is achieved after 31 iterations. From this graph, it is experimentally evident that the IAFC algorithm does not result in oscillations and can be said to be more stable as compared to the other algorithms. Figures 3(a) - 3(d) present the three dimensional graphic display of the Anderson's Iris data as four different combinations of three out of four features.

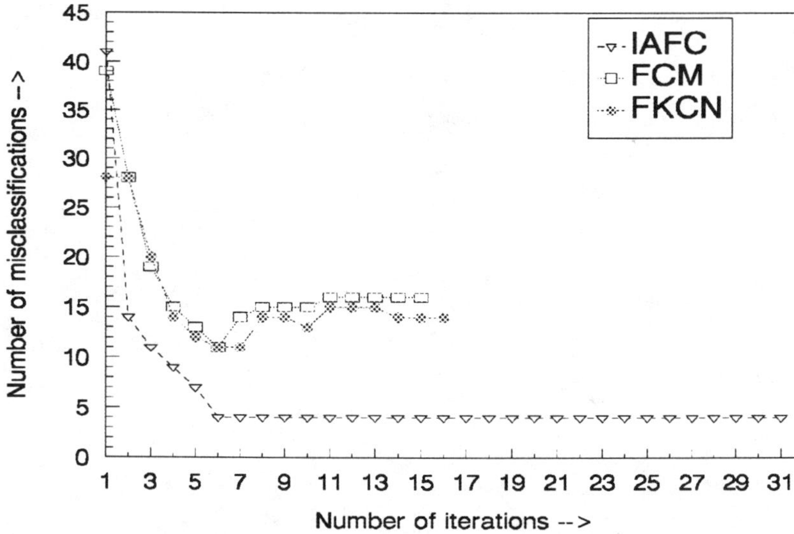

Figure 2. Number of iterations vs. misclassifications for FCM, FKCN and IAFC for a termination limit of 0.00001.

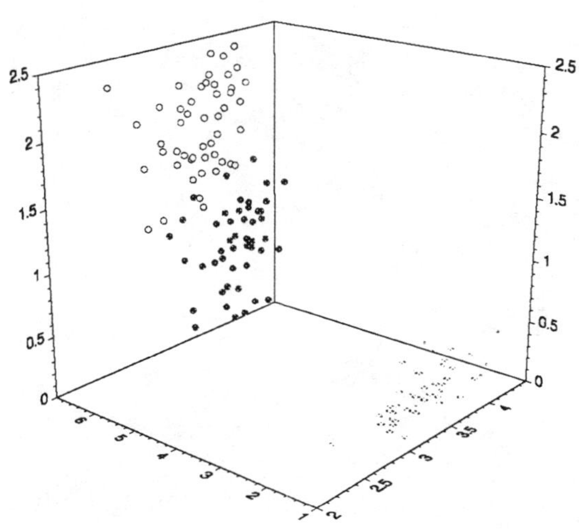

Figure 3(a). Actual Iris data clusters in 2nd, 3rd and 4th dimensions.

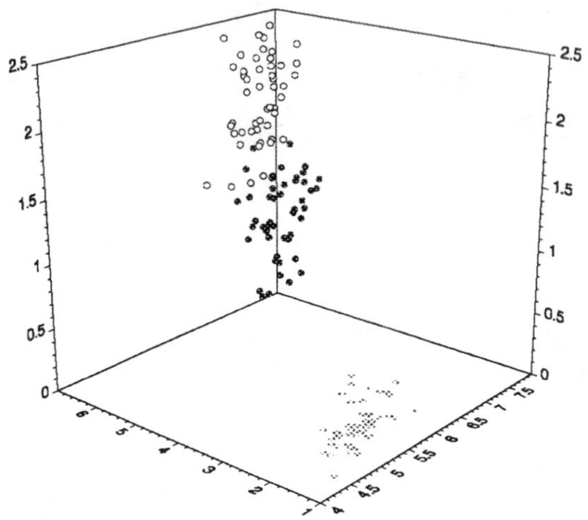

Figure 3(b). Actual Iris data clusters in 1st, 3rd and 4th dimensions.

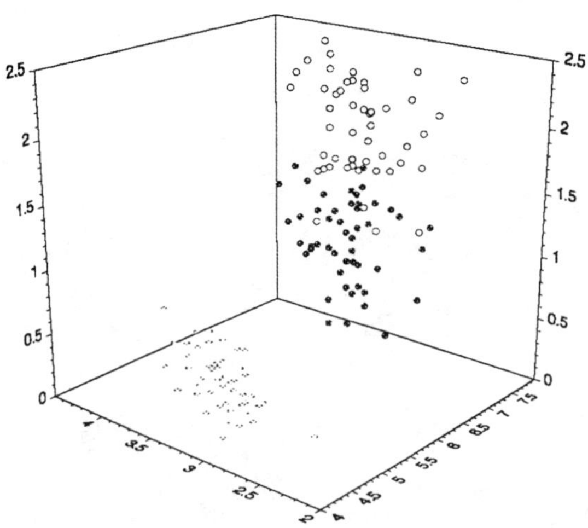

Figure 3(c). Actual Iris data clusters in 1st, 2nd and 4th dimensions.

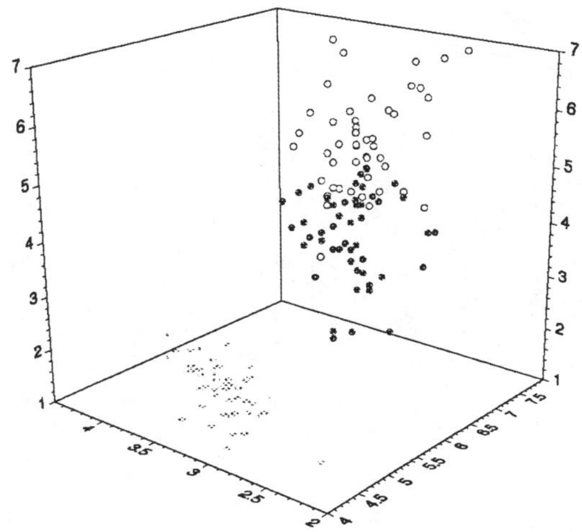

Figure 3(d). Actual Iris data clusters in 1st, 2nd and 3rd dimensions.

Figures 4 (a) - 4 (d) show the output display using K-means, FCM, AFLC, and IAFC algorithms for the 1st, 2nd, and 3rd dimensions. The number of clusters has been set to be equal to three for the K-means and FCM algorithms. The centroids of these clusters have been initialized with three different random samples belonging to the three clusters. Figure 5 shows the misclassifications yielded by these four algorithms in the form of a confusion matrix.

Table 1 shows the actual centroid locations from the statistical average of the input data set and the updated centroids computed from the four algorithms.

A new set of results shown in Figure 6 include comparison of the FCM and FKCN algorithms for the case when the centroid values are initialized with randomly generated numbers $\in [0,8]$. A termination limit of 0.00001 yields 4 misclassifications in the case of IAFC algorithm and 14 and 16 misclassifications for FKCN and FCM algorithms respectively. Since AFLC is not an iterative algorithm, changes in termination parameter or initial conditions does not affect its classification results.

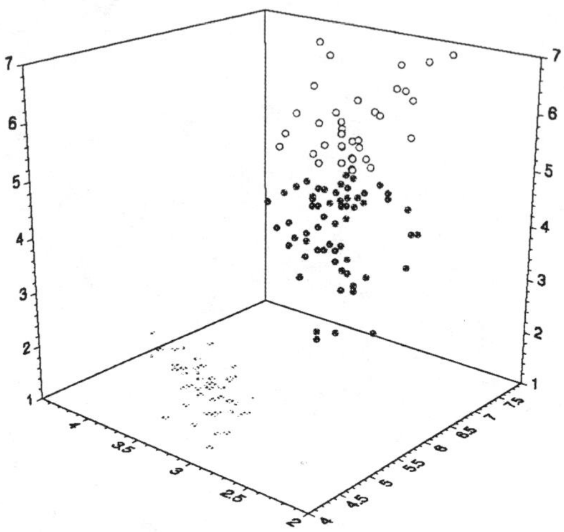

Figure 4(a). Display of 1st, 2nd, and 3rd features of Iris data output from K-means algorithm.

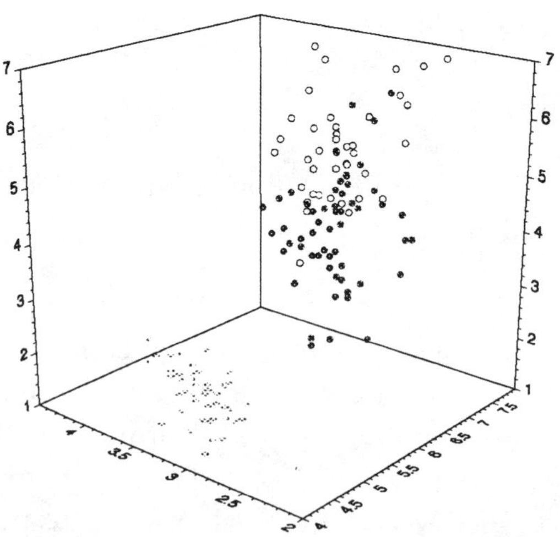

Figure 4(b). Display of 1st, 2nd, and 3rd features of Iris data output obtained from AFLC.

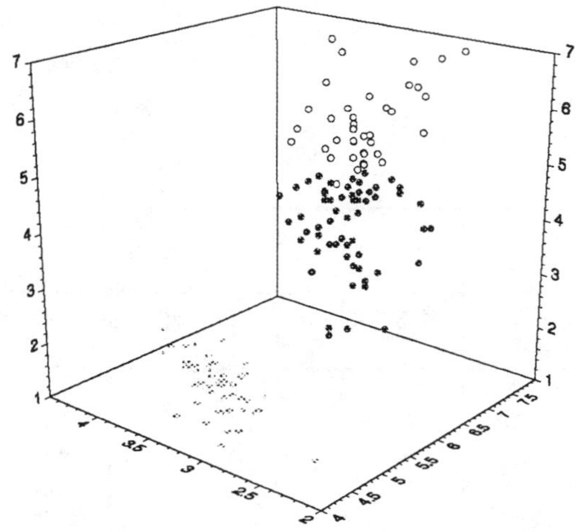

Figure 4(c). Display of 1st, 2nd, and 3rd features of Iris data output using Fuzzy C-means.

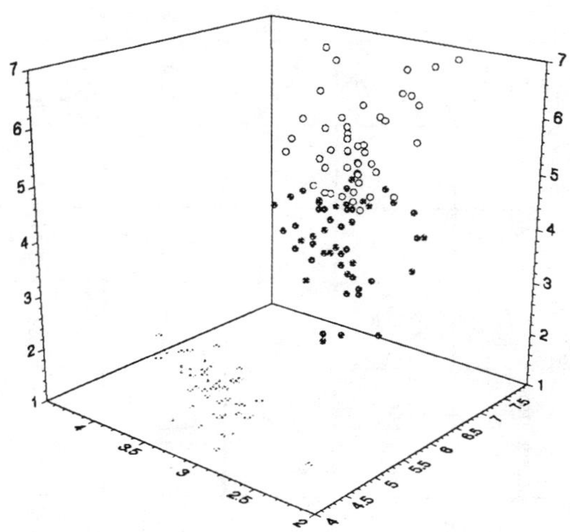

Figure 4(d). Display of 1st, 2nd, and 3rd features of Iris data output obtained from IAFC.

Table 1. Comparison of Computed Centroids of Iris Data Clusters

Algorithm	Cluster 1				Cluster 2				Cluster 3			
Statistical Average	5.01	3.43	1.46	0.25	5.94	2.77	4.26	1.33	6.59	2.97	5.55	2.03
K-means	5.00	3.43	1.46	0.25	5.90	2.75	4.39	1.43	6.85	3.07	5.74	2.07
Fuzzy C-means	5.00	3.41	1.48	0.25	5.89	2.76	4.37	1.40	6.78	3.05	5.65	2.05
AFLC	5.00	3.42	1.46	0.24	5.95	2.76	4.33	1.34	6.72	3.05	5.66	2.14
IAFC	4.99	3.41	1.47	0.25	5.77	2.76	4.17	1.28	6.57	3.05	5.43	2.08

K-means

Output \ Actual	1	2	3
1	50		
2		48	14
3		2	36

Fuzzy C-means

Output \ Actual	1	2	3
1	50		
2		47	13
3		3	37

AFLC

Output \ Actual	1	2	3
1	50		
2		50	7
3			43

IAFC

Output \ Actual	1	2	3
1	50		
2		47	4
3		3	46

Figure 5. Comparison of Iris data classification results.

Figure 7 shows the effect of τ on the number of clusters formed. The optimal range of τ when the IRIS data set is classified into three clusters is between 1. 5 and 2. 2 when other parameters i.e. k in the expression for f(l) and membership threshold value are chosen to be 0. 5 and 0. 4 respectively. While clustering unlabeled data, one can use a validity measure to find optimal range of τ. The number of clusters vs. τ graph of Figure 7 is not affected by changes in the value of k, so long the value of τ is not too low.

Fuzzy C-means

Output \ Actual	1	2	3
1	50		
2		47	13
3		3	37

FKCN

Output \ Actual	1	2	3
1	50		
2		47	11
3		3	39

AFLC

Output \ Actual	1	2	3
1	50		
2		50	7
3			43

IAFC

Output \ Actual	1	2	3
1	50		
2		49	3
3		1	47

Figure 6. Iris data classification results for termination limit = 0.00001.

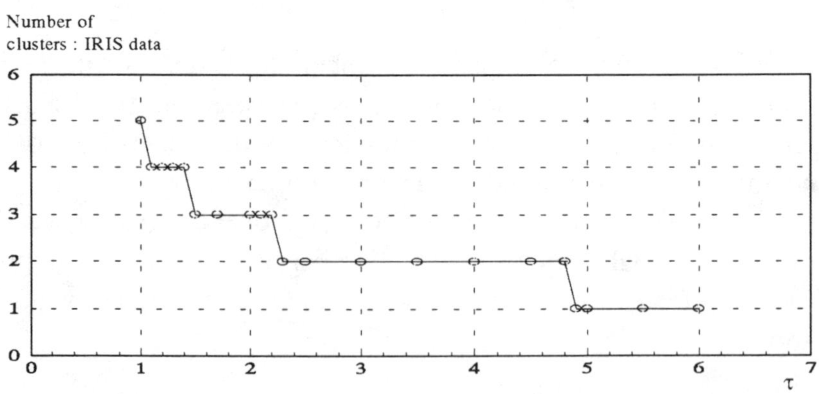

Figure 7. Number of clusters vs. the vigilance parameter, τ, graph for a range of τ when $f(l) = 1/\{0.5(l-1)+1\}$ and THRESHOLD = 0.4.

4. CONTROL RULES BY FUZZY SELF-ORGANIZING CLUSTERING ALGORITHMS :

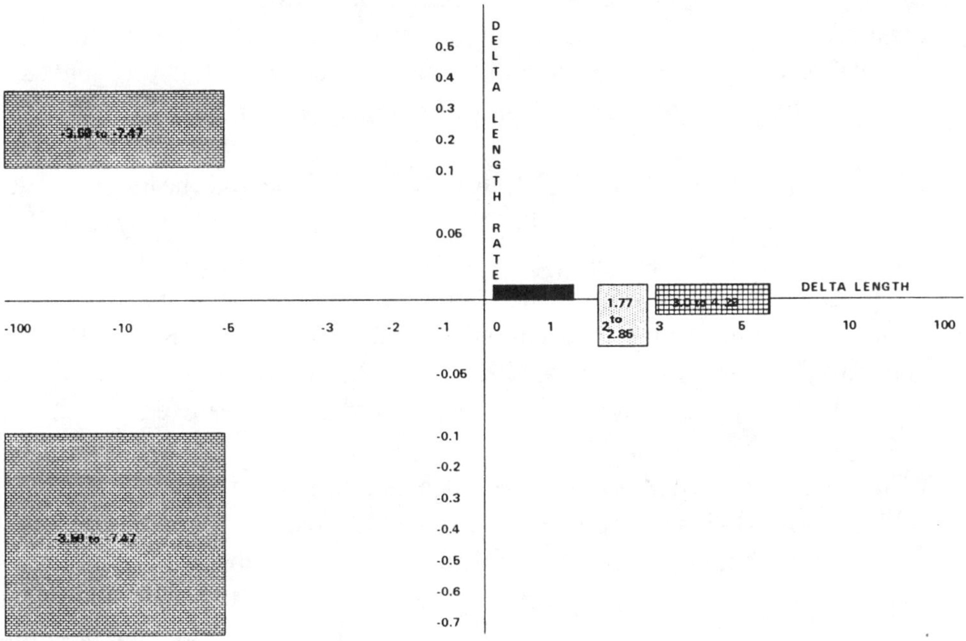

Figure 8. Control Delta Voltages Corresponding to Length Error and Length Rate Error

Although this chapter is mainly concerned with application of fuzzy self-organizing clustering algorithms to pattern recognition, such clustering can be used to find control rules in many nonlinear dynamical systems [26-28]. We have used both AFLC and IAFC algorithms to find the control actions for a tethered satellite system (TSS) from the simulation data [29]. The input parameters to the system are the length error and the length rate error and the output parameter is the delta voltage. This data set consisting of 1765 samples was collected at intervals of 50 seconds. Figure 8 shows a plot displaying the clusters obtained from the AFLC algorithm in a two-dimensional feature space. Each cluster represents a specific range of input and output parameters. Hence, these clustering results help us in predicting the output of the system for a particular range of the input parameters. Incorporation of these clustered delta voltages into the simulator should provide smoother operational characteristics of the TSS system.

5. CONCLUSION

The adaptivity of neural networks and the concept of fuzzy membership functions of data samples have been integrated in the recent self-organizing neuro-fuzzy models for better classification of similar regions in unlabeled data sets. Such models also have potential for generating control rules for nonlinear dynamical systems[26-28].

Although the well-known FCM algorithm [6] achieves the integration of fuzzy membership function into crisp C-means clustering by modifying the within-cluster sum-of-squared-error criterion and more recent integration of FCM model into Kohonen clustering network [12] to eliminate some of the inherent structural problems in KCN, the non-sequential algorithms such as FKCN, FCM as well as AFLC [19,23] require large memory capacity due to the updating rules used. Both fuzzy ART [14] and fuzzy adaptive resonance theory [13] require normalization of the input data yielding poor performance in some real applications. In AFLC and IAFC, the input data is normalized for finding the centroids, however the vigilance test is performed using the unnormalized data thus avoiding some misclassifications. The IAFC [20] model demonstrates fewer misclassifications than those obtained by FCM and FKCN [Figure 2] for the standard data set used when the termination limit $\in= 0.00001$, and also shows more stability. In terms of reaching the termination limit, IAFC converges slower than FCM and FKCN, however it yields minimum misclassifications (only 4) just after 6 iterations, as compared to FCM and FKCN which result in more misclassifications (14 and 16 respectively) after around 16 iterations. At this time a mathematical stability analysis has not been carried out to predict the general behavior of the IAFC model.

One specific aspect of the IAFC model is the introduction of a new similarity measure that allows the model to generate nonlinear decision boundaries among data structures having overlapping regions or regions in close proximity. Further development in this area may yield a method of generating nonlinear decision boundaries more accurately.

6. ACKNOWLEDGMENT

This work was supported in part by a grant from NASA-JSC under contract #NAG-9-673 and a grant from Northrop Co. under contract # HHH 880856K.

7. REFERENCES

1. K. Fukunaga, Introduction to Statistical Pattern Recognition (Academic Press, San Diego, 1990)
2. R.O. Duda and P.E. Hart, Pattern Classification and Scene Analysis (New York: Wiley, 1973)
3. Y. H. Pao, Adaptive Pattern Recognition and Neural Networks (Addison-Wesley, Reading, 1989).
4. A. K. Jain and R. C. Dubes, Algorithms for Clustering Data (Englewood Cliffs, N.J., Prentice Hall, 1988).
5. G. H. Ball and D. J. Hall, ISODATA, An Iterative Method of Multivariate Data Analysis and Pattern Classification, Proc. IEEE Intl. Comm. Conf. Philadelphia, PA (June 1966).
6. J. C. Bezdek, Pattern Recognition with Fuzzy Objective Function Algorithms (Wiley, New York, 1973).
7. J. C. Bezdek and S. K. Pal, Eds. Fuzzy Models for Pattern Recognition (IEEE Press, New York, 1982).
8. R. Krishnapuram and J. M. Keller, A possibilistic approach to clustering, vol. 1, no. 2 (1993), 98-110.
9. R. Lippman, An introduction to computing with neural networks, IEEE ASSP Magazine, no. 4 (1987), 4-21 .
10. S. Mitra and Y. S. Kim, Neuro-fuzzy models in pattern recognition, SPIE Proc. 2061, Boston, MA (Sept. 8 -10, 1993).
11. T. L . Huntsberger and P . Ajjimarangsee, Parallel self-organizing feature maps for unsupervised pattern recognition, Intl. J. General Systems, vol. 16, no. 4 (1990), 357-372.
12. J. C. Bezdek, E. C. Tsao, and N. R. Pal, Fuzzy Kohonen clustering networks, First Fuzz-IEEE Proc. (1992), 1035-1043.
13. P. Simpson, Fuzzy adaptive resonance theory, Proceedings of Neuro-engineering Workshop (1990) .
14. G. A. Carpenter, S. Grossberg, and D. Rosen, Fuzzy ART: Fast stable learning and categorization of analog patterns by an adaptive resonance system, Neural Networks, vol. 4 (1991), 759-771.

15. S. C. Newton and S. Mitra , "Self-organizing leader clustering in a neural network using a fuzzy learning rule," SPIE Proc., vol. 1565 (1991), 331-337.
16. G. A. Carpenter and S. Grossberg, A massively parallel architecture for a self-organizing neural pattern recognition machine, Computer Vision, Graphic & Image Processing, vol. 37 (1987), 54-115.
17. G. A. Carpenter and S. Grossberg, ART 2: Self-organization of stable category recognition codes for analog patterns, Applied Optics, vol. 2 (1987), 4919-4938.
18. P. K. Simpson, Fuzzy min-max neural networks - Part 2: clustering, IEEE Transactions on Fuzzy systems, vol. 1, no. 1 (1993), 32-45.
19. S. C. Newton, S. Pemmaraju and S. Mitra, Adaptive fuzzy leader clustering of complex data sets in pattern recognition, IEEE Transactions on Neural Networks, vol. 3, no. 5 (1992), 794-800.
20. Y. S. Kim and S. Mitra, Integrated adaptive fuzzy clustering (IAFC) algorithm, Second Fuzz-IEEE Proc. (1993).
21. B. Moore, ART 1 and pattern clustering, Proc. 1988 Connectionist Models Summer School (1988).
22. L. I. Burke, Clustering characterization of adaptive resonance, Neural Networks, vol. 4 (1991), 485-491
23. S. Pemmaraju and S. Mitra, Identification of noise outliers in clustering by a fuzzy neural network, Second Fuzz-IEEE Proc., San Francisco (1993).
24. J. M. Keller, M. R. Gray, and J. A. Givens, Jr., A fuzzy K-nearest neighbor algorithm, IEEE Trans. on Syst., Man, Cybern., vol. SMC-15, no. 4, (1985), 580-585.
25. S. K. Pal and S. Mitra, Fuzzy dynamic clustering algorithm, Pattern Recognition Letters, vol. II (1990), 525-535.
26. L. Wang, Training of fuzzy logic systems using nearest neighborhood clustering, Second Fuzz-IEEE Proc., San Francisco (1993), 13-17.
27. S. K. Sin and Rui J. P. deFigueiredo, Fuzzy system design through fuzzy clustering and optimal predefuzzification, Second Fuzz-IEEE Proc., San Francisco (1993), 190-195.
28. S. Mitra and S. Pemmaraju, Clustering of tethered satellite system simulation data by an adaptive neuro-fuzzy algorithm, NAFIPS 1992 Workshop Proc., (1992).
29. R. N. Lea, J. Villareal, Y. Jani and C. Copeland, Fuzzy logic based tether control, NAFIPS 1991 Workshop Proc., (1991), 398-402.

Fuzzy Morphological Algorithms

Edward R. Dougherty, Center for Imaging Science,
Rochester Institute of Technology, Rochester, New York 14623

Divyendu Sinha, Department of Computer Science,
College of Staten Island and Graduate Center
City University of New York, Staten Island, New York 10301

Purnendu Sinha, Department of Electrical Engineering,
Stevens Institute of Technology, Hoboken, New Jersey 07030

Abstract. *Processing gray-scale realizations of images that are ideally binary (such as gray-scale realizations of printed characters) is problematic due to the fact that gray-scale processing should be consistent with the binary nature of the ideal image. Essentially, any final decision (such as the recognition of a specific character at a specific location) should reflect the content of the ideal image, which is generally unknown. Too often, a gray-scale realization of an ideal binary image is processed using methods appropriate for gray-scale realizations of ideal gray-scale images. These should not be expected to lead, ipso facto, to decision procedures appropriate to binary images. Fuzzy morphological algorithms do not assume probabilistic knowledge of the degradation process; however, they mirror the processing that one would have performed were the ideal binary image known. Thus, they lead to decision procedures consistent with those that would have been taken following processing of the ideal binary image.*

The theory of fuzzy mathematical morphology has been fully characterized in terms of a set of axioms for subset fuzzification. Within the general framework there exists an infinite number of realizations of fuzzy mathematical morphology and the determination of which theory is relevant is application dependent; nevertheless, there does exist a general paradigm for the construction of fuzzy algorithms from corresponding binary algorithms. In this chapter we discuss this general paradigm for algorithm construction, and in particular, we discuss morphological algorithms for smoothing, geometric filtering, edge detection, object detection, and peak detection. The efficacy of these algorithms are underlined by the various simulations presented herein.

1 Introduction

Although mathematical morphology began in the context of random processes [21], until recently the great majority of research has concentrated on deterministic analysis, especially upon the relationship of mathematical morphology to the theory of lattices [2,15-17,24] and the characterization of deterministic shapes [3,31,32,36]. Within the past couple of years there has been a significant effort to study the statistical effects of morphological transformations [1], find statistically optimal morphological filters [5-7,20,22], and find statistically efficient shape detectors [10]. These efforts have lead to the development of efficient morphological filters, especially for restoration; however, they are hampered by the mathematical difficulties inherent in the study of nonlinear operations upon random processes. For the most part, optimization has involved filter representations, and practical design has been accomplished by placing various constraints upon filters, the result being the derivation of suboptimal filters. Utilization of fuzzy morphological algorithms can circumvent some of these difficulties, while at the same time not requiring the kind of probabilistic process knowledge necessary for statistical optimization. The two approaches are not at odds; rather, they are complementary.

Fuzzy morphologies [25-29] (as developed by Sinha and Dougherty) represent an extension of the original binary mathematical morphology that is distinct from gray-scale morphology [13,23,34,35]. The latter extends binary morphology by treating the space beneath a gray-scale image as a binary set called the umbra [34,35]. The customary binary operations are then applied to the umbrae of both the image and structuring element and the output is reconstructed by taking the "surface" of the resulting binary output. In fact, the umbra is an artifice and more recent formulations [15-17,24] employ a strictly supremum-infimum lattice theoretic approach. The lattice-based extension encompasses both binary and gray-scale morphology because both the subsets of the Euclidean plane and the extended-real-valued functions form complete lattices. Even in the abstract lattice-theory setting, gray-scale morphology remains in the spirit of the umbra methodology. In none of the generalizations is there an algorithmic paradigm in which all binary morphological algorithms naturally lift to gray-scale algorithms that remain faithful to their binary counterparts. For example, gray-scale erosion does not act like a marker of fits and consequently the gray-scale version of Serra's hit-or-miss transform [23] does not act as a pattern detector in the way the binary transform does. Fuzzy morphology, on the other hand, represents an extension of binary mathematical morphology that is different in essence from the umbra-motivated extension and in which there is a natural lifting of algorithms.

In binary morphology, the primary operation is erosion and erosion is based on the ordinary concept of subsethood (or "fitting" in the terminology of morphology); in fuzzy morphology, the morphological operations are grounded on a fuzzification of subsethood [26,30], the result being an intrinsic fuzzification of morphology. The axioms upon which fuzzifications of morphology depend are actually the axioms governing subset fuzzification. This should not be too surprising since the key relation in morphology is fitting, which is defined via the notion of subsethood. This is quite distinct from previous fuzzy-set-theory-based generalizations, which did not introduce any new intrinsic fuzziness in the system. In the previous efforts in defining fuzzy morphology [11,12,14,18], fuzziness was introduced only in the modeling of gray-scale images as fuzzy binary images, and not in the operations.

Recent efforts have concentrated on a characterization of all fuzzy morphologies [27-29]. The need for such a characterization became apparent upon completion of the first attempt at an intrinsically fuzzy mathematical morphology [25]. It was apparent that the main body of fuzzy Minkowski algebra, as there formulated, was grounded on a small number of basic properties of fuzzy erosion and that, in turn, these properties were based on the notion of fuzzy subsethood employed in the algebra. The program was clear: first characterize subset fuzzification in terms of an axiom set governing an indicator function for subsethood and then characterize mathematical morphology relative to the subsethood axioms.

1.1 Crisp-set morphology

We begin with a brief outline of the basic morphological operations that are relevant to the chapter. For more complete accounts, see [4], [13] or [23]. In the present section, all sets will be crisp.

The basic idea of mathematical morphology is to probe an image with a template (usually called a **structuring element**) and to quantify the manner in which the structuring element fits (or does not fit) within the image. By marking the locations at which the structuring element fits within the image, we derive structural information concerning the image. This information depends on both the size and shape of the structuring element. In general, the structuring element has a simple shape and is very small (compared to the image being investigated).

The translation of a set A by a point x is defined by

$$\mathcal{T}(A; x) = \{a + x : a \in A\}. \tag{1}$$

The fundamental operation of mathematical morphology is **erosion** of a set A by another set (structuring element) B. It is defined by

$$A \ominus B = \{x : \mathcal{T}(B; x) \subseteq A\} \tag{2}$$

$A \ominus B$ consists of all points x for which the translation of B by x fits inside of A. Fig. 1 illustrates erosion. Dual to erosion is **dilation** (or, Minkowski addition), defined by

$$A \oplus B = \bigcup_{a \in A} \mathcal{T}(B; a) = \{x : \mathcal{T}(B; x) \cap A^c \neq \emptyset\}. \tag{3}$$

Dilation is dual to erosion, in that it can be obtained from erosion by means of set complementation: $A \oplus B = [A^c \ominus (-B)]^c$ where $-B = \{-b : b \in B\}$ denotes **reflection through the origin**.

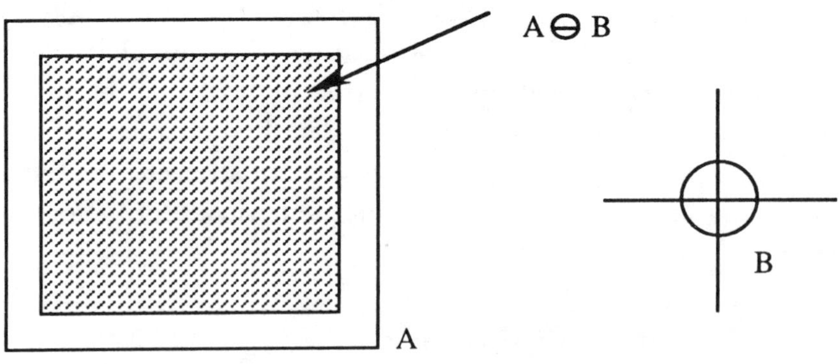

Figure 1: Illustration of binary erosion.

A basic secondary operation of mathematical morphology is **opening**. The opening of set A by structuring element B is defined by

$$A \circ B = (A \ominus B) \oplus B \tag{4}$$

A more conceptually useful characterization of opening is as the union of all translates of the structuring element that fit inside the input image:

$$A \circ B = \bigcup_{\mathcal{T}(B;x) \subseteq A} \mathcal{T}(B; x). \tag{5}$$

Fig. 2 illustrates opening. Notice how the contour (boundary) of the image has been smoothed by the round structuring element. This smoothing represents a filtering of the image from the inside and it is due to the shape of the structuring element. Different shaped structuring elements produce different filtering effects. Rather than view the opened image itself as the final output of the processing, we can take a different view. We can consider the set-theoretic subtraction $A \setminus (A \circ B)$, called the **top-hat transform**. In Fig. 2, the top-hat output consists of input-image corners that protrude into the background. The top-hat is employed for extracting features.

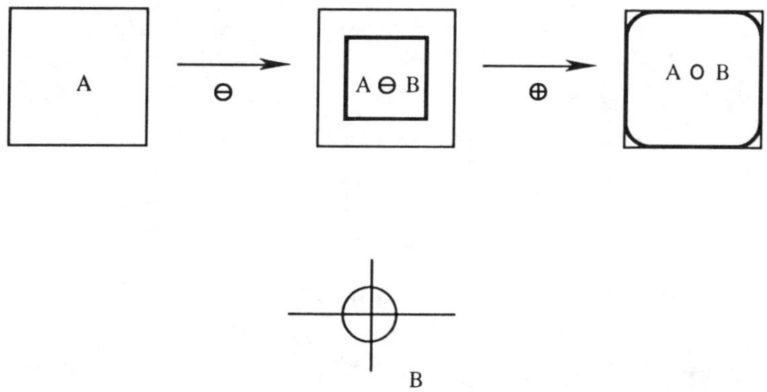

Figure 2: Illustration of binary opening and top-hat transform.

The dual operation to opening is **closing**, which is defined as dilation followed by erosion:

$$A \bullet B = [A \oplus (-B)] \ominus (-B) \tag{6}$$

Closing is dual to opening because $A \bullet B = (A^c \circ B)^c$. Whereas opening filters from the inside, closing filters from the outside.

A key application of opening is for filtering a binary image S corrupted by union clutter noise N, so that the image to be filtered is $S \cup N$. If the structuring element does not fit within the noise-clutter components not contiguous with S, then these are eliminated by opening. This restoration effect is illustrated in Fig. 3, where it is seen that a side effect of the opening is contour smoothing of the original image. Again, this contour smoothing depends on the choice of structuring element.

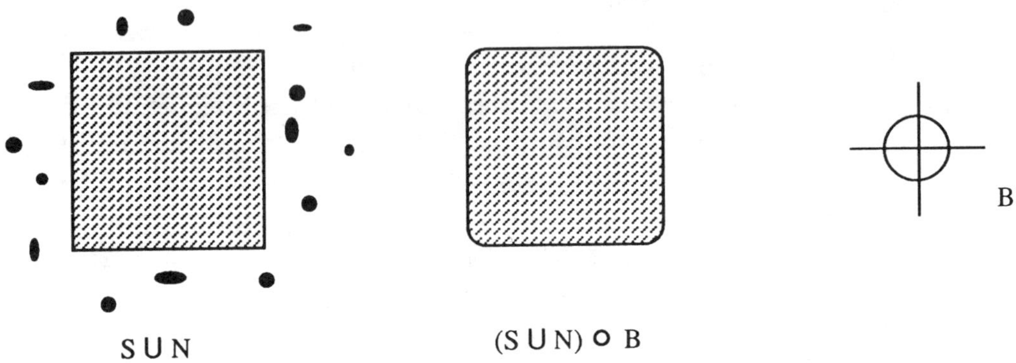

Figure 3: Noise removal via binary opening.

It is customary in morphology to study the effects of an operation for a fixed structuring element. For example, to study the effects of opening by a fixed structuring element B, we define a unary operator $\psi(A) = A \circ B$. Such unary operators are called **filters**. A few algebraic properties of these filters are of considerable interest:

Translation invariant if $\psi[\mathcal{T}(A; x)] = \mathcal{T}[\psi(A); x]$.
Increasing if $A \subseteq C$ implies $\psi(A) \subseteq \psi(C)$.
Antiextensive if $\psi(A) \subseteq A$.
Extensive if $\psi(A) \supseteq A$.
Idempotent if $\psi(\psi(A)) = \psi(A)$.

As filters, erosion, dilation, opening, and closing are translation invariant and increasing. Opening and closing are idempotent. Furthermore, opening [closing] is antiextensive [extensive].

Erosion marks pixels according to the manner in which a structuring element fits inside an image; similar marking can be applied to the manner in which a structuring element fits into the complement of an image, which means that the structuring element "misses" the image. The **hit–or–miss transform** [23] is defined via fitting inside and outside simultaneously with a "hit" structuring element and a "miss" structuring element, respectively. It requires two structuring elements, H and M, which are treated as a structuring-element pair $\langle H, M\rangle$, and is defined by

$$A \otimes \langle H, M\rangle = (A \ominus H) \cap (A^c \ominus M) \tag{7}$$

A point is in the hit–or–miss output if and only if H translated to the point fits inside A and M translated to the point fits outside A. It is assumed H and M are disjoint, for otherwise it would be impossible for both fits to occur simultaneously. In reference [3] it is assumed that $M = H^c \cap W$ for some *a priori* fixed set W (called a **window**) but not in [10,36]. The hit–or–miss transform is a powerful image processing tool. In particular, it is used for shape detection [3,10,36]. Fig. 4 illustrates the manner in which the hit–or–miss transform is employed to locate a shape (triangle) among other shapes. As a shape detector, the hit–or–miss transform is very sensitive to noise and a key motivation for the development of fuzzy morphology has been the desire for a fuzzy hit–or–miss transform [26,28,33]. The hit–or–miss transform is a translation-invariant non-increasing mapping.

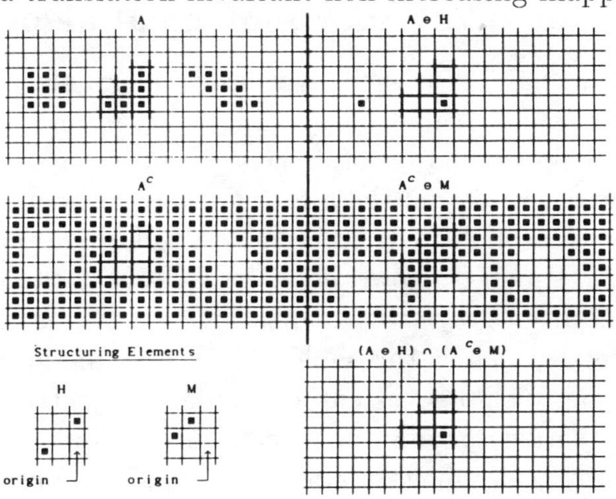

Figure 4: Shape detection via binary hit–or–miss transform.

Edge detection is often performed by the **morphological gradient**, which for an *isotropic* structuring element B, usually the 3×3 square in digital processing, is defined by

$$GRAD(A, B) = (A \oplus B) \setminus (A \ominus B) \tag{8}$$

In the binary case, the gradient locates the edge by expanding the image (via dilation), shrinking it (via erosion), and then taking the set-theoretic difference of the two. Operation of the gradient is illustrated in Fig. 5. The morphological gradient is also a translation-invariant non-increasing mapping.

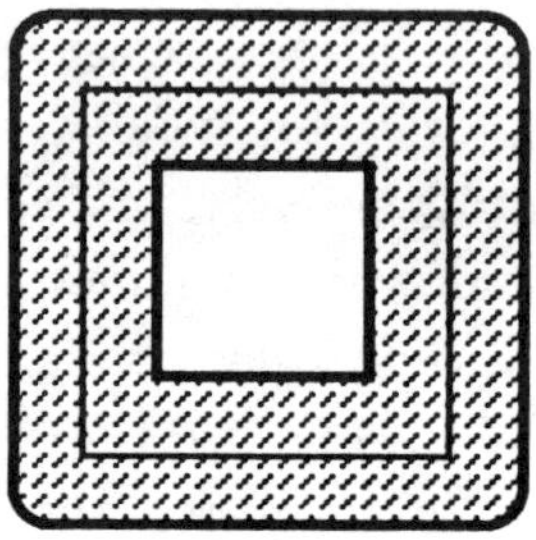

$(A \oplus B) - (A \ominus B)$

Figure 5: Edge detection via binary morphological gradient.

1.2 Need for fuzzy morphological algorithms

To this point we have discussed five basic binary morphological algorithms: contour smoothing by openings, feature extraction by the top-hat transform, restoration of images corrupted by union noise by opening, shape detection via the hit–or–miss transform, and edge detection via the morphological gradient. All of these operations are defined on binary images; however, it is likely that these binary images have come from gray-scale realizations by means of some binarization procedure, such as thresholding. We can model the situation by assuming there is some essentially binary object captured by a gray-scale acquisition system and because of various effects, termed "noise," the gray-scale image is not two-valued. Hence the need for binarization. If we take the simple case of global thresholding, we are confronted with operating on a binarization of a realization of an ideal image, with the hope that operations on the binarized realization will somehow mirror what we would like to accomplish on the ideal image. Subsequently, we will formally characterize the algorithmic problem, but first we provide an example relevant to restoration of an image corrupted by union noise to illustrate the problem to be addressed by fuzzy algorithms (see [26] for an example relating to shape recognition by the hit–or–miss transform). For visualization purposes the example will be in the context of one-dimensional signals.

Consider the ideal gray-scale shape of Fig. 6(a). The shape is essentially binary, its only gray-scale characteristic being its single gray value y, this value depending on the manner of image capture. In fact, from a phenomenological perspective it can be considered to be binary. Thresholding it at any value $t \leq y$ provides a perfect $\{0, 1\}$-binarization.

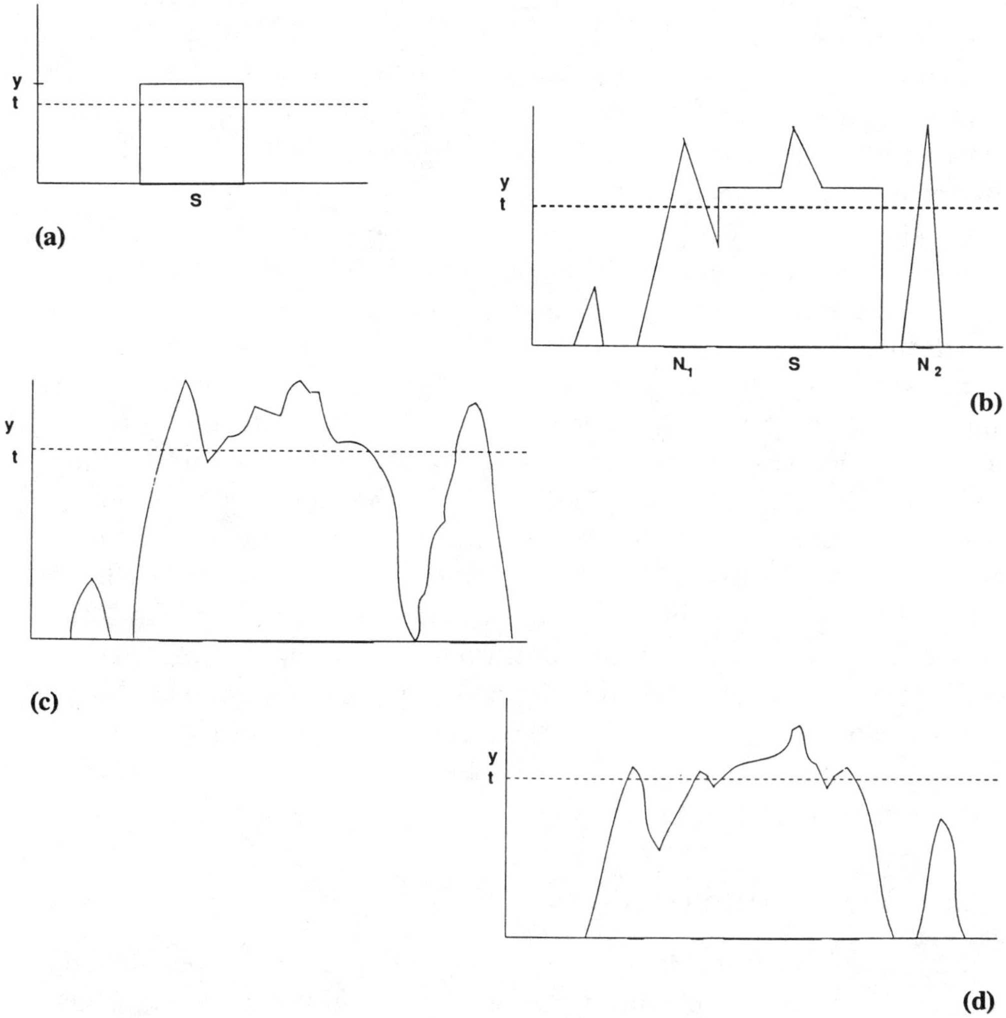

Figure 6: Illustration of the problems caused by clutter and random noise.

Suppose that the ideal shape is viewed as existing among clutter in some ideal gray-scale world. Then, ideally, it might be of the form depicted in Fig. 6(b). If we denote the ideal shape by f and ideal clutter by g, then the ideal corrupted shape is $\max[f, g]$. When binarized by thresholding at t, the ideal shape yields the binary image S and the ideally cluttered shape yields the binary image $S \cup N$, where the ideal noise is given by $N = N_1 \cup N_2$. A binary opening on $S \cup N$ with a structuring segment B with length midway between that of S and N_1 provides perfect restoration of S in the ideal setting.

In a more realistic setting, neither the ideal shape nor the ideal clutter appear as in Fig. 6(b); rather, a realization of the image-union-clutter might well appear as in Fig. 6(c). This realization is quite close to the ideal, so that thresholding at t yields a binarization quite close to the ideal binarization. Thus, opening by the structuring element B employed in the ideal case of Fig. 6(b) will eliminate the noise components and leave a single shape component very close to S. Rather than having an image-clutter realization close to the ideal, suppose the captured realization is very light (light grays corresponding to small gray values). Then thresholding at t might result in a situation such as the one depicted in Fig. 6(d), where thresholding yields an S-binarization with holes in it, so that opening by B leaves a greatly diminished filtered binary image. Even if we were able to find a statistically optimal threshold value t, unless there is very little variability among the different realizations, the situation of Fig. 6(d) would likely still occur at times, and when it does the filtering process will provide poor $\{0, 1\}$ results. Fuzzy opening will mitigate the problem by allowing the opening to take place on a $[0, 1]$-normalization of the realization, and it will do so in a manner fully consistent with binary opening.

1.3 Fuzzy morphology

Fuzzy morphology involves (fuzzy) morphological operations defined on fuzzy realizations of binary subsets of the universal set $\mathcal{U}$, which is usually taken to be two-dimensional discrete (or, Euclidean) space. The membership values are interpreted as indicators of spatial certainty (and, therefore, uncertainty) present in the image. In the usual manner, $\mu_A(x)$, $x \in \mathcal{U}$, denotes our faith in the proposition "pixel x belongs to set A." The erosion of a fuzzy set A by a fuzzy structuring element (set) B is written $\mathcal{E}(A, B)$ and is defined via its membership function $\mu_{\mathcal{E}(A,B)}(x)$, the latter valuation being interpreted as our belief in the proposition, "the translated fuzzy set $\mathcal{T}[B; x]$ is a subset of fuzzy set A." Thus, fuzzy erosion introduces fuzziness into the system and, as with the original crisp-set erosion, it serves as a marker of "fits," albeit "fuzzy fits."

Thus the study of a fuzzy-set-theory-based morphological framework must revolve around the notion of "degree of subsethood." In particular, an **indicator for fuzzified set inclusion** $\mathcal{I}$ is defined in such a way that $\mathcal{I}(X, Y)$ measures belief in the proposition "X is a subset of Y" with X and Y as fuzzy sets. As mentioned earlier, we define erosion in terms of $\mathcal{I}$ via its membership function:

$$\mu_{\mathcal{E}(A,B)}(x) = \mathcal{I}[\mathcal{T}(B; x), A]. \tag{9}$$

Dilation ($\mathcal{D}$) is then defined by duality

$$\mathcal{D}(A, B) = \mathcal{E}(A^c, -B)^c. \tag{10}$$

where $-B$ denotes the reflection of image B. Opening and closing are defined in the usual iterative fashions in terms of erosion and dilation:

$$\mathcal{O}(A, B) = \mathcal{D}[\mathcal{E}(A, B), B] \tag{11}$$

$$\mathcal{C}(A, B) = \mathcal{E}[\mathcal{D}(A, -B), -B] \tag{12}$$

In order for the **inclusion grade operator**, as $\mathcal{I}$ is usually called in the fuzzy literature, to model our intuitive notions about approximate subsethood, we must impose certain restrictions on it. A key aspect regarding the original development of fuzzy morphology [25] is the articulation of nine (non-independant) properties satisfied by ordinary set inclusion that must also be satisfied by the indicator defining erosion. Therein it was proposed that the indicator function satisfy the following seven independant axioms:

Axiom 1: $\mathcal{I}(A, B) = 1 \iff A \subseteq B$.
Axiom 2: $\mathcal{I}(A, B) = 0 \iff \{x : \mu_A(x) = 1 \wedge \mu_B(x) = 0\} \neq \emptyset$.
Axiom 3: $B \subseteq C \Longrightarrow \mathcal{I}(A, B) \leq \mathcal{I}(A, C)$.
Axiom 4: $\mathcal{I}(-A, -B) = \mathcal{I}(A, B)$ and for all $x \in \mathcal{U}$,
$\mathcal{I}(A, B) = \mathcal{I}[\mathcal{T}(A; x), \mathcal{T}(B; x)]$
Axiom 5: $\mathcal{I}(A, B) = \mathcal{I}(B^c, A^c)$.

Axiom 6: $\mathcal{I}(A, B \cap C) = \min[\mathcal{I}(A, B), \mathcal{I}(A, C)]$.
Axiom 7: $\mathcal{I}(A, B \cup C) \geq \max[\mathcal{I}(A, B), \mathcal{I}(A, C)]$.

The conditions stated in the axioms A1 through A7 have been found by investigating the proofs of various binary morphological results in order to find the properties of (crisp) set inclusion needed to make the proofs go through. It should be mentioned that these axioms can be justified solely from the point of view of "soft fitting" [26,30]. For example, Axioms 1 and 2 merely state that if the sets are cirsp, then we obtain the usual set inclusion operator; Axiom 5 is a counterpart of the tautology (for crisp set inclusion) $A \subseteq B \iff B^c \subseteq A^c$; Axiom 6 states that the degree to which A is subset of $B \cap C$ is determined solely by the degrees to which A fits inside the sets B and C seperately.

One such indicator, the one studied in [25], is given by

$$\begin{aligned} \mathcal{I}(X,Y) &= \inf_{\alpha \in \mathcal{U}} \min\left[1, 1 - \mu_X(\alpha) + \mu_Y(\alpha)\right] \\ &= \inf_{\alpha \in \mathcal{U}} \mu_{X^c \triangle Y}(\alpha) \end{aligned} \tag{13}$$

where $\triangle$ denotes the **bold union** operator of Giles:

$$\mu_{A \triangle B}(x) = \min[1,\ \mu_A(x) + \mu_B(x)]. \tag{14}$$

Using this definition of indicator in the Eqs. (9) through (12), the fuzzy Minkowski algebra of the original study results:

M1. Erosion, dilation, opening, and closing are monotonically increasing.

M2. Erosion, dilation, opening, and closing are spatially translation invariant.

M3. Dilation distributes over union: $\mathcal{D}(A_1 \cup A_2,\ B) = \mathcal{D}(A_1, B) \cup \mathcal{D}(A_2, B)$.

M4. Erosion is right-distributive over intersection: $\mathcal{E}(A, B \cap C) = \mathcal{E}(A, B) \cap \mathcal{E}(A, C)$, and left-antidistributive over union: $\mathcal{E}(A \cup B, C) = \mathcal{E}(A, C) \cap \mathcal{E}(B, C)$.

M5. Opening and closing are dual: $\mathcal{O}(A, B) = \mathcal{C}(A^c, B)^c$.

M6. Opening is antiextensive: $\mathcal{O}(A, B) \subseteq A$. Closing is extensive: $\mathcal{C}(A, B) \supseteq A$.

M7. Opening and closing are idempotent: $\mathcal{O}[\mathcal{O}(A,B),B] = \mathcal{O}(A,B)$, $\mathcal{C}[\mathcal{C}(A,B),B] = \mathcal{C}(A,B)$.

In addition, erosion, dilation, opening and closing possess fitting characterizations that result from the fitting characterization of the bold-union-based indicator function.

All the salient algebraic features of the usual binary Minkowski algebra are preserved in the fuzzy Minkowski algebra. In addition, erosion, dilation, opening and closing possess fitting characterizations that result from the fitting characterization of the indicator function given by Eq. (13). In particular, it was shown that for Minkowski algebra to generalize to fuzzy settings, set inclusion need not be a two-valued operator.

As has been demonstrated in the general study of fuzzy morphology [28,29], except for those properties depending upon fitting characterizations, the basic Minkowski properties follow from the indicator axioms themselves, whereas the fitting characterizations and their consequences depend on the particular form of the indicator that is adopted. A fundamental point of [28,29] is that, so long as the axioms are satisfied, different realizations of the indicator can yield different Minkowski algebras, and, therefore, different realizations of fuzzy mathematical morphology.

The upshot of the foregoing considerations is this: if we assume that the preceding seven axioms must be satisfied by an indicator, then the meaning of fuzzy erosion (fuzzy marking) depends upon the analytic formulation of indicators that satisfy the axioms, which, in reality, means a careful study of the fuzzification of set inclusion. This has been recently accomplished by the authors [26,30], where based on the seven axioms and a rather general postulation of a mathematical form for indicators, classes of suitable indicators have been developed.

Let us assume the indicator is of the form

$$\mathcal{I}(A,B) = \theta\left\{\psi\Big[\lambda\left(\mu_A(x)\right),\ \ \phi\left(\mu_B(x)\right)\Big]:\ \ x \in \mathcal{U}\right\}. \tag{15}$$

where λ, $\phi : [0,1] \to [0,1]$, $\psi : [0,1] \times [0,1] \to [0,1]$, $\theta : \mathcal{M}([0,1]) \to [0,1]$, with $\mathcal{M}([0,1])$ denoting the collection of all multi-subsets of the set $[0,1]$. Then necessary and sufficient conditions on the functions θ, ψ, λ, and ϕ can be given to determine whether or not the proposed indicator satisfies the seven axioms [26]. Going further, a more specific form of the indicator is proposed, namely,

$$\begin{aligned} \mathcal{I}(X,Y) &= \inf_{\alpha \in \mathcal{U}} \min\left[1, \lambda(\mu_X(\alpha)) + \lambda(1-\mu_Y(\alpha))\right] \\ &= \inf_{\alpha \in \mathcal{U}} \mu_{\gamma(X)\Delta\gamma(Y^c)}(\alpha) \end{aligned} \tag{16}$$

where γ is a unary operation on fuzzy sets that is range induced by λ:

$$\forall\, \alpha \in \mathcal{U}, \quad \mu_{\gamma(X)}(\alpha) = \lambda\left[\mu_X(\alpha)\right]. \tag{17}$$

Thus the development of suitable indicators for fuzzified set inclusion depends on the form of the λ function. Now, it has been shown (see [26]) that necessary and sufficient conditions for indicator $\mathcal{I}$ defined via Eq. (16) to satisfy the seven axioms are given by the following five properties for λ:

P1. λ is a non-increasing function.

P2. $\lambda(0) = 1$.

P3. The equation $\lambda(x) = 0$ has a single solution.

P4. For all $\alpha \in [0.5, 1]$, the equation $\lambda(x) = \alpha$ has a single solution.

P5. For all x, $\lambda(x) + \lambda(1-x) \geq 1$.

As an immediate consequence, γ behaves like a generalized complement operation on fuzzy sets: it satisfies all algebraic properties of complement except for involution (see [26]). A few examples of acceptable λ are:

$$\lambda(x) = 1 - x^n, \qquad n \geq 1, \tag{18}$$

$$\lambda(x) = \frac{1-x}{1+nx}, \qquad 0 \geq n > -1, \tag{19}$$

$$\lambda(x) = \frac{1}{1+x^n} - \frac{x}{2}, \qquad n \geq \frac{Ln(3)}{Ln(2)} = 1.5849\cdots \tag{20}$$

To the extent that the seven axioms faithfully encapsulate our fundamental understanding of inclusion-ness, the set-inclusion fuzzification they engender determines the kind of fuzzy morphologies we can encounter; to the extent that the formulation of Eq. (16) describes the class of analytic expressions we wish to consider for the indicator, the preceding five properties P1 through P5 determine the class of fuzzy morphologies and the corresponding fuzzy Minkowski algebras.

To get an intuitive feel for the action of fuzzy erosion and opening, we consider the special case of Eq. (16) with $\lambda(x) = 1 - x$. This is the case we studied exclusively in [25]. To find the erosion of an image A by structuring element B at the point x, we find the smallest nonnegative number τ so that $\mathcal{T}(B;x)$ lowered by τ fits beneath image A. The eroded image at x has the value $1-\tau$. This "lowering down" of the membership function is accomplished via the **scalar addition** operation ($\bowtie$): the scalar addition of fuzzy set X and a constant $\tau \in [-1,1]$ is defined to be

$$\mu_{X \bowtie \tau}(x) = \min\left[1,\ \max(0,\ \tau + \mu_X(x))\right]. \tag{21}$$

In Fig. 7 we illustrate the computation of $\mathcal{I}(X,Y)$ for a general λ.

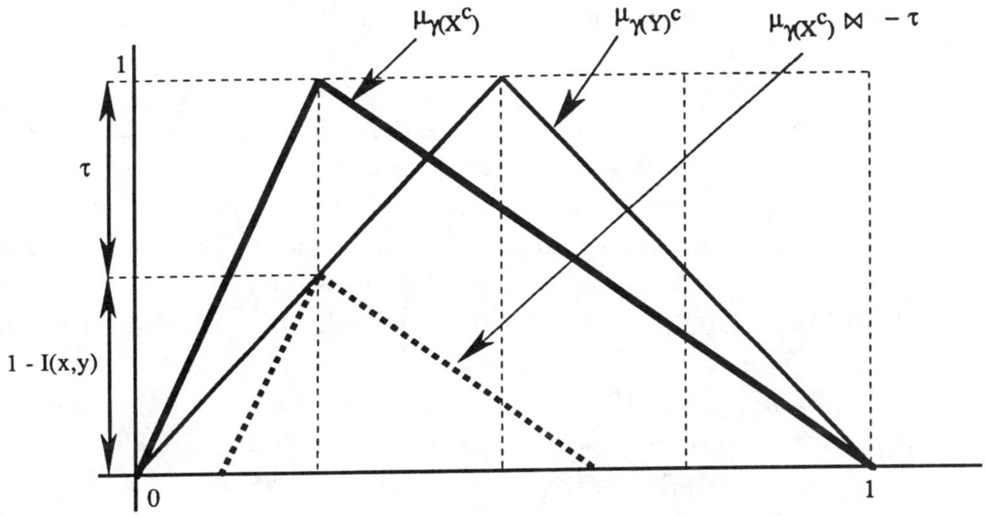

Figure 7: Computation of $\mathcal{I}(X,Y)$.

To find the opening when $\lambda(x) = 1 - x$, we form a (fuzzy) union of all lowered-down translates $\mathcal{T}(B;x)$ that fit beneath A. These two geometric descriptions of erosion and opening are the "fitting characterizations" for fuzzy morphology in the special case $\lambda(x) = 1 - x$. They derive from general fitting characterizations provided in

[27,29]. In general, we have the following result for fuzzy openings [29]:

$$\mathcal{O}(A,B) = \bigcup_{\substack{\mathcal{T}[\gamma(B)^c; z] \bowtie (\alpha - 1) \subseteq \gamma(A^c), \\ z \in \mathcal{U}, \text{ and } \alpha \in [0,1]}} \Big[\mathcal{T}[\gamma(B)^c; z] \bowtie [-\lambda(\alpha)]\Big]. \tag{22}$$

The following two families of structuring elements suffice for most of the tasks discussed in this chapter:

- $C(r,h)$ denotes a cylinderical structuring element of radius r, height h. This may be thought of as a fuzzy realization of a circle of radius r whose center is at $\langle 0,0 \rangle$.

- $S(l,h)$ denotes a rectangular-box structuring element of width l, length l, and height h. This may be thought of as a fuzzy realization of a square of size l and, if l is odd, then its center is assumed to be at $\langle 0,0 \rangle$.

Example 1 *As an illustrative example of shrinking via erosion, consider an image which contains an isoceless triangle of height* 20 *and base* 39, *a square of length* 20, *and a circle of radius* 15. *Fig. 8(a) shows a noisy gray-scale realization of this image. Figs. 8(b) through 8(d) show the outcome of eroding the image of Fig. 8(a) by images* $C(5,0.8)$, $C(9,0.7)$, *and* $C(15,0.8)$ *respectively. We have assumed that* $\lambda(x) = 1 - x$. *The "flat bed" in Figs. 8(b) and 8(d) has gray-value* 0.2; *the "flat bed" in Fig. 8(c) has gray-value* 0.3. *The "flat bed" represents the degree to which* $C(\bullet, h)$ *can be classified as a subset of* $\emptyset$, *that is, the value of* $\mathcal{I}\Big[C(\bullet,h),\emptyset\Big]$. *When* $\lambda(x) = 1 - x$, *this value is* $1 - h$. *As was proposed in [25,28], we should view the "flat bed" as the new background — the entire background has been painted with the gray-value* $1 - h$ *instead of* 0. *As* h *gets smaller we start loosing distinction between foreground and background, and eventually (when* $h = 0.5$ *in this case), the foreground and background will become one and the same.*

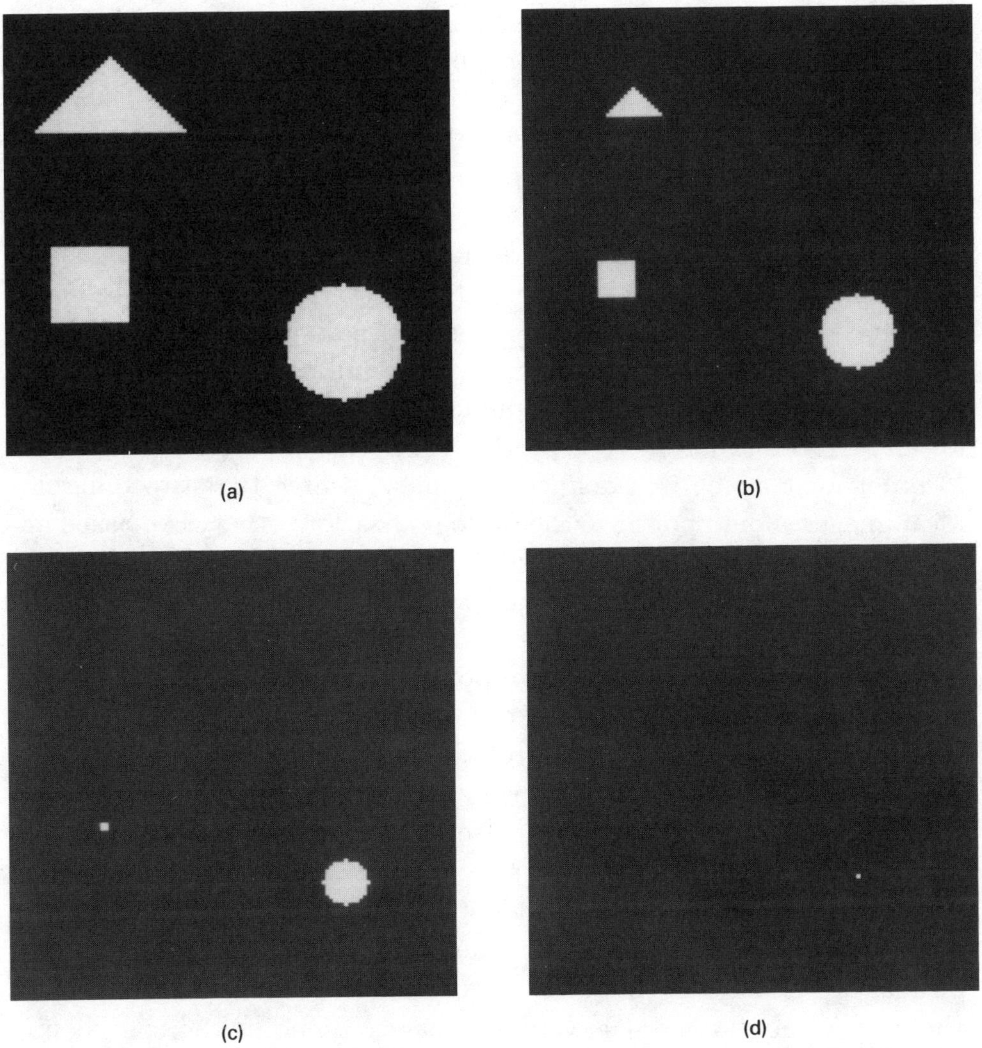

Figure 8: Illustration of contour shrinking via fuzzy erosion. See Example 1 for details.

2 Paradigms for algorithm development

The current study concerns the development of fuzzy morphological algorithms appropriate for processing gray-scale realizations of images that are ideally binary in nature. Given a collection of idealized binary images $\mathbf{B}$, we assume the existence of a randomization operation ρ that provides us with gray-scale realizations $\mathbf{R}$ of binary images. In particular, with each image $A \in \mathbf{B}$ we associate a unique realization $\rho(A) \in \mathbf{R}$. The simplest approach to solve the problem is to convert the realizations into binary images via some binarization operation β and then proceed with the usual binary processing of images. (See Fig. 9.) The binarization should be, ideally speaking, such that $\beta[\rho(A)] = A$ for all images $A \in \mathbf{B}$, that is, $\beta = \rho^{-1}$. But, in practice, ρ is ill-defined, and, in general, $\rho^{-1}(A)$ is multi-valued, that is, $\rho(A) \subseteq \mathbf{R}$. Thus, optimistically speaking, given an unknown but well-defined function $\rho : \mathbf{B} \rightarrow 2^{\mathbf{R}}$, we need to find a binarization $\beta : \mathbf{R} \rightarrow \mathbf{B}$ so that for all $A \in \mathbf{B}$ and for all $A' \in \rho(A)$, the relationship $\beta(A') = A$ holds. In order for β to be well-defined, we must assume that two different ideal images should not have the same realization; otherwise, based on the realizations, it will be impossible to tell them apart. Finding such a β, and then ensuring that β will be well-defined are two extremely non-trivial tasks.

A few paradigms have been proposed recently to handle this problem [9,28]. In these paradigms, one works directly with the realizations in $\mathbf{R}$ and the aim is to mirror the intent of the process that one would have performed on binary images. Suppose that one wishes to perform a binary morphological operation ψ on the ideal images, and then the output of this operation is to be subjected to a decision process Δ. In Fig. 9, we let $\mathbf{D}$ denote the set of all decisions.

We first normalize the gray-value range to [0,1] to obtain the set of images $\mathbf{F}$. The elements of $\mathbf{F}$ are treated as fuzzy (binary) images: the gray value $f(x, y)$ of any fuzzy image f denotes the spatial uncertainty at the pixel $\langle x, y \rangle$. Note that although we have employed a simple linear range normalization, a different image fuzzification operator ζ could be employed, perhaps one that is designed specifically to yield estimates of pixelwise uncertainty. In the general form of the paradigm, we carry the spatial uncertainty to the decision process. The intent here is to define a fuzzy morphological operation $\overline{\psi} : \mathbf{F} \rightarrow \mathbf{F}$ and a decision procedure $\overline{\Delta} : \mathbf{F} \rightarrow \mathbf{D}$ such that $\Delta[\psi(A)] = \overline{\Delta}[\overline{\psi}(A)]$.

In order for this approach to be successful, operation $\overline{\psi}$ must preserve the nuances of fuzzy set theory (in so far as modeling of spatial uncertainty is concerned) and that of the binary morphological operation ψ.

The paradigm of Fig. 9 provides us with considerable latitude in defining $\overline{\psi}$ and is in the true spirit of other artificial intelligence methodologies. For example, in granulometric analysis, one wishes to find the area of the opened image (that is, $\psi \equiv$ opening and $\Delta \equiv$ area). It is possible, though not necessarily desirable, to define $\overline{\Delta}$ and $\overline{\psi}$ such that $\psi(A)$ and $\overline{\psi}(\overline{A})$ are totally different images but the area of $\psi(A)$ is $\overline{\Delta}[\overline{\psi}(\overline{A})]$.

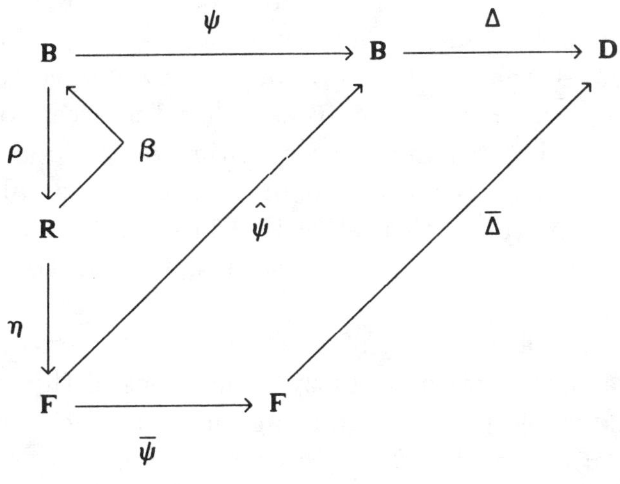

Figure 9: Various paradigms for morphological processing of noisy gray-scale realizations of ideal binary images.

There exist certain useful operations that do not fit into the preceding paradigm. For example, the various types of binarization techniques that produce a binary image as the output. In order to model such algorithms within the morphological framework, we have proposed a slightly different paradigm in [8]. Given the realization process ρ, the normalization operator η, and the ideal morphological transformation $\psi : \mathbf{B} \rightarrow \mathbf{B}$, the aim here is to define a different type of fuzzy morphological operation $\hat{\psi} : \mathbf{F} \rightarrow \mathbf{B}$ so that we obtain $\psi(A) = \hat{\psi}(\overline{A})$ for any ideal binary image A and its realizations $\overline{A}$.

The process of defining the operators $\hat{\psi}$ and $\overline{\psi}$ based on the knowledge of operator ψ is often called *lifting*. Among the various forms of lifting some are more natural than the others, and, of course, some are more appropriate than others. The appropriateness of a lifting technique is hard to generalize since it is application specific. Some successful efforts have already been made by the authors in defining fuzzy techniques for edge detection, object detection, peak detection, and smoothing [9,33]. We present a much more elaborate discussion of these in the sequel.

Assume algorithm ψ is composed of operations that are part of the binary Minkowski algebra: union, intersection, complement, spatial translation, reflection, erosion, and dilation. As shown in [25,28], each operation of the binary Minkowski algebra lifts directly to a fuzzy operation, so that ψ at once induces a fuzzy operator that can be employed for $\overline{\psi}$. This is not to say that there cannot be other candidates for $\overline{\psi}$, but certainly direct lifting of the operations is a natural choice, and we term the resulting $\overline{\psi}$ the **natural lifting** of ψ. Every binary morphological algorithm possesses a natural lifting; indeed, under suitable subset fuzzifications, the Minkowski algebra itself lifts [25,29].

In any event, the meaning of ψ as an operator on ideal objects is lifted to $\overline{\psi}$ as an operator on fuzzy realizations of the ideal image. And this is the whole point of requiring an axiomatic formulation of subset fuzzification: as a marker, binary erosion takes its meaning from fitting (subsethood), and naturally lifting it to multi-scaled realizations of binary images requires only that the meaning of fitting (subsethood) be preserved insofar that it is relevant to the Minkowski algebra.

Yet lifting of the operations is not the end of the story. Morphological operations require the specification of structuring elements. Lifting the binary erosion $A \ominus B$ to the fuzzy erosion $\mathcal{E}(\overline{A}, \overline{B})$ leaves one with the choice of the fuzzy structuring element $\overline{B}$. This choice is a design problem and must be made in accordance with the demands of subsequent processing.

Concerning the "lifting" of decision procedures, we need to take into consideration two forms of uncertainty: **vagueness** and **ambiguity**. Traditionally, vagueness is associated with the difficulty of making a choice between two or more alternatives, and ambiguity is associated with situations in which the choice between two or more alternatives is left unspecified [19]. In image processing vagueness arises because of the practical limitations imposed on the algorithms (like finite extent of the range for gray-values and truncation/roundoff errors), while ambiguity arises from the fact that two different ideal images can have the same gray-scale realization (and then no algorithm can resolve the ambiguity). A fuzzy–set–theory–based approach can not only handle both these types of uncertainty, but it also lends itself to practical implementations [19].

3 Fuzzy algorithms

We now provide some examples of binary morphological algorithms lifted to fuzzy algorithms. In each case we state the binary algorithm along with its fuzzy counterpart and then illustrate the fuzzy algorithm.

In what follows, we have assumed that the fuzzy set subtraction is defined by

$$\mu_{A\backslash B}(x) = \max[0,\ \mu_A(x) - \mu_B(x)]. \tag{23}$$

Zadeh calls this operation **bounded set difference**.

Rules of thumb

When using Eqs. (18) and (20) to define λ, we must use very small values of n if high noise-sensitivity is required, and slightly larger values of n (usually, more than 5) if very low noise-sensitivity is required. In particular, when n is large, the indicator $\mathcal{I}$ cannot distinguish between "signal" and "noise" — considering both of them to be one and the same. The converse holds for λ defined via Eq. (19): The closer n is to -1, the more noise-sensitive indicator $\mathcal{I}$ will be, and vice-versa.

Suppose that we are employing the λ function from the family given by Eq. (19). After extensive experimentation, we have found that moderate variations (from the optimal) in the value of n do not adversely affect the outcome of fuzzy algorithms. The exact opposite holds for the λ functions given by Eq. (18).

3.1 Shape detection via fuzzy hit-or-miss transform

The ordinary binary hit–or–miss transform detects objects in accordance with the paradigm illustrated in Fig. 4. The naturally lifted algorithm is given by

$$S \otimes \langle H, M\rangle = \mathcal{E}(S, H) \cap \mathcal{E}(S^c, M) \tag{24}$$

Suppose that in the ideal binary case (cf. Eq. (7)) we would choose the structuring element pair $\langle \mathcal{H}, \mathcal{M}\rangle$. Then, in Eq. (24), H and M are appropriately choosen fuzzy subsets of $\mathcal{H}$ and $\mathcal{M}$ respectively. The advantage of the fuzzy hit–or–miss is that shapes can be detected (with some degree of certainty) in light images because H can still fit in S to some degree and they can be detected in dark images because M can still fit within S^c to some degree. See [33] for more details.

Rules of thumb

As mentioned earlier, when $n \to \infty$ in Eqs. (18) and (20) (or, when $n \to -1$ in Eq. (19)), the indicator treats "signal" and "noise" as the same. The net effect of this, so far as the hit-or-miss transform is concerned, is that the transformed image becomes the universal set in the limiting situation. Hence, the value of n should be very small: $n \leq 2.25$ in Eq. (18), $n \geq -0.6$ in Eq. (19) and $n \leq 2.5$ in Eq. (20).

If the background is not constant, then the hit–or–miss transform may not show the spike(s) clearly at the center of object(s) and thereby require a more nontrivial peak detection algorithm. One such algorithm is: convolve the hit–or–miss transform output either by the mask M_1 or mask M_2, where

$$M_1 = \begin{pmatrix} -\frac{1}{4} & -\frac{1}{4} & -\frac{1}{4} \\ -\frac{1}{4} & \textcircled{1} & -\frac{1}{4} \\ -\frac{1}{4} & -\frac{1}{4} & -\frac{1}{4} \end{pmatrix}, \qquad M_2 = \begin{pmatrix} -\frac{1}{12} & -\frac{1}{12} & -\frac{1}{12} & -\frac{1}{12} & -\frac{1}{12} \\ -\frac{1}{12} & -\frac{1}{12} & -\frac{1}{12} & -\frac{1}{12} & -\frac{1}{12} \\ -\frac{1}{12} & -\frac{1}{12} & \textcircled{1} & -\frac{1}{12} & -\frac{1}{12} \\ -\frac{1}{12} & -\frac{1}{12} & -\frac{1}{12} & -\frac{1}{12} & -\frac{1}{12} \\ -\frac{1}{12} & -\frac{1}{12} & -\frac{1}{12} & -\frac{1}{12} & -\frac{1}{12} \end{pmatrix}$$

and then threshold at a small positive number.

Note that neither these masks nor the convolution can be justified from the perspective of fuzzy set theory and morphology. (In fact, the convolved image has $[-2, 1]$ as the range for gray-values.) However, it is not inconsistent with our methodology as this procedure is part of the decision process and not of the fuzzy morphology.

Example 2 *Consider a* 51×51 *image which has three* 5×5 *squares and whose background intensity varies linearly from one end to the other. An extremely noisy realization of this image is shown in Fig. 10(a). Suppose that we wish to identify the location of each of these three squares. The hit structuring element is* $S(5, h)$ *and the miss structuring element is* $S(l, h) \setminus S(5, h)$. *The values for* l *and* h *must be appropriately chosen. Because the background is very bright in the bottom-right portion of the image (at least as bright as the leftmost square), we expect the hit-or-miss transform to be slightly confused: the hit part will have a relatively high value but the miss part will have a relatively small value.*

Let $h = 0.9$ *and* $l = 9$. *Figs. 10(b) and 10(c) show the outcome of the fuzzy hit-or-miss transform when* $n = 1$ *in Eq. (18) and* $n = 1.6$ *in Eq. (20), respectively. In all these two figures, the three spikes, each of which is at the center of a* 5×5 *square, can be detected by convolving the image with mask* M_1 *and then thresholding the resultant image at approximately* $\frac{1}{40}$. *In Fig. 10(d) we show the outcome of this operation for the image of Fig. 10(c).*

Our next example illustrates the usage of "sparse" structuring elements.

Example 3 *Fig. 11(a) shows an out-of-focus camera image of the text "GENERAL". Suppose that we wish to identify the letter 'L.' The structuring elements* H *and* M *should be such that we can differentiate 'L' from the letters 'A', 'E', 'G', 'N', or 'R'. An easy, but inefficient way to accomplish this is to use the window approach: let* $\mathcal{H}$ *denote the ideal letter 'L' and* $\mathcal{M}$ *be a windowed-complement of* $\mathcal{H}$. *A more efficient approach is possible: Consider the image* $\mathcal{H}$ *shown in Fig. 11(b). Erosion by this structuring element can differentiate 'L' from 'A', 'N', and 'R'. Hence, the miss part should only concern itself with differentiating 'L' from 'E' and 'G'. The structuring element* $\mathcal{M}$ *could be defined as shown in Fig. 11(c). The fuzzy hit-or-miss transform with* $\lambda(x) = 1 - x$ *is shown in Fig. 11(d). The detector is slightly confused at letter 'G' — the confusion arising from the sparseness of the chosen miss structuring element.*

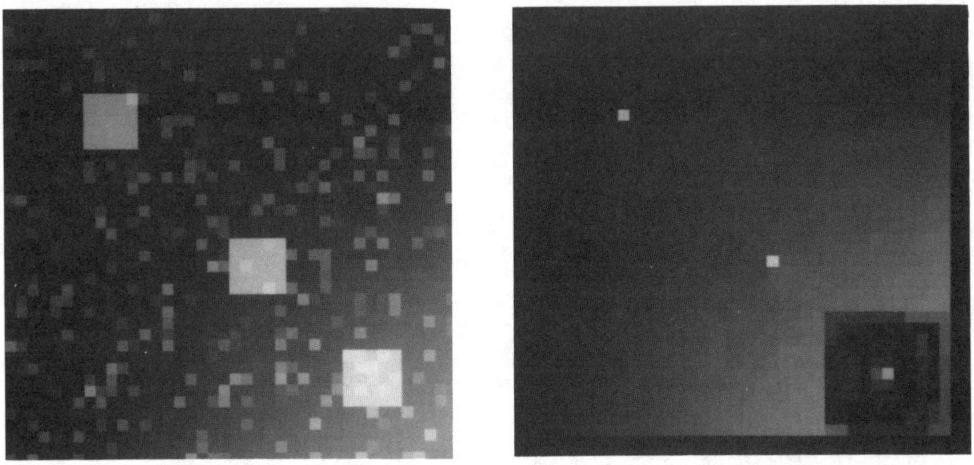

(a) (b)

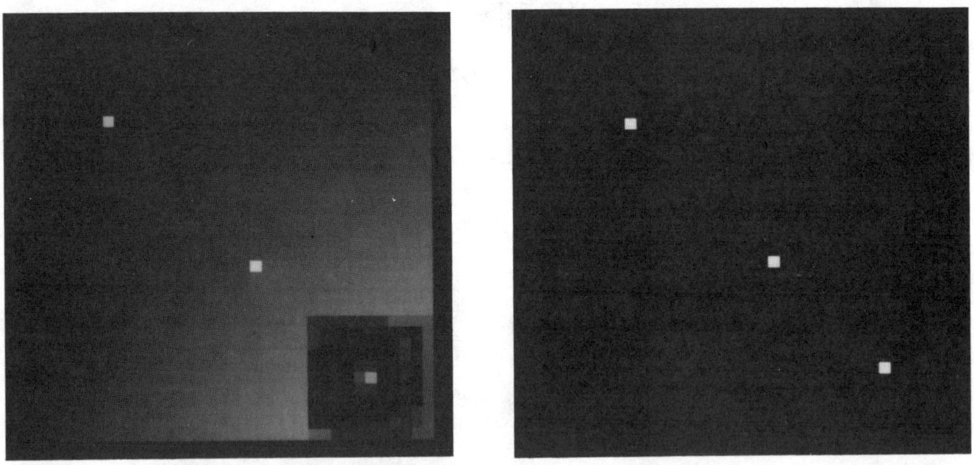

(c) (d)

Figure 10: Illustration of shape detection via fuzzy hit-or-miss transform. The aim is to detect the three squares in the noisy image (a). See Example 2 for details.

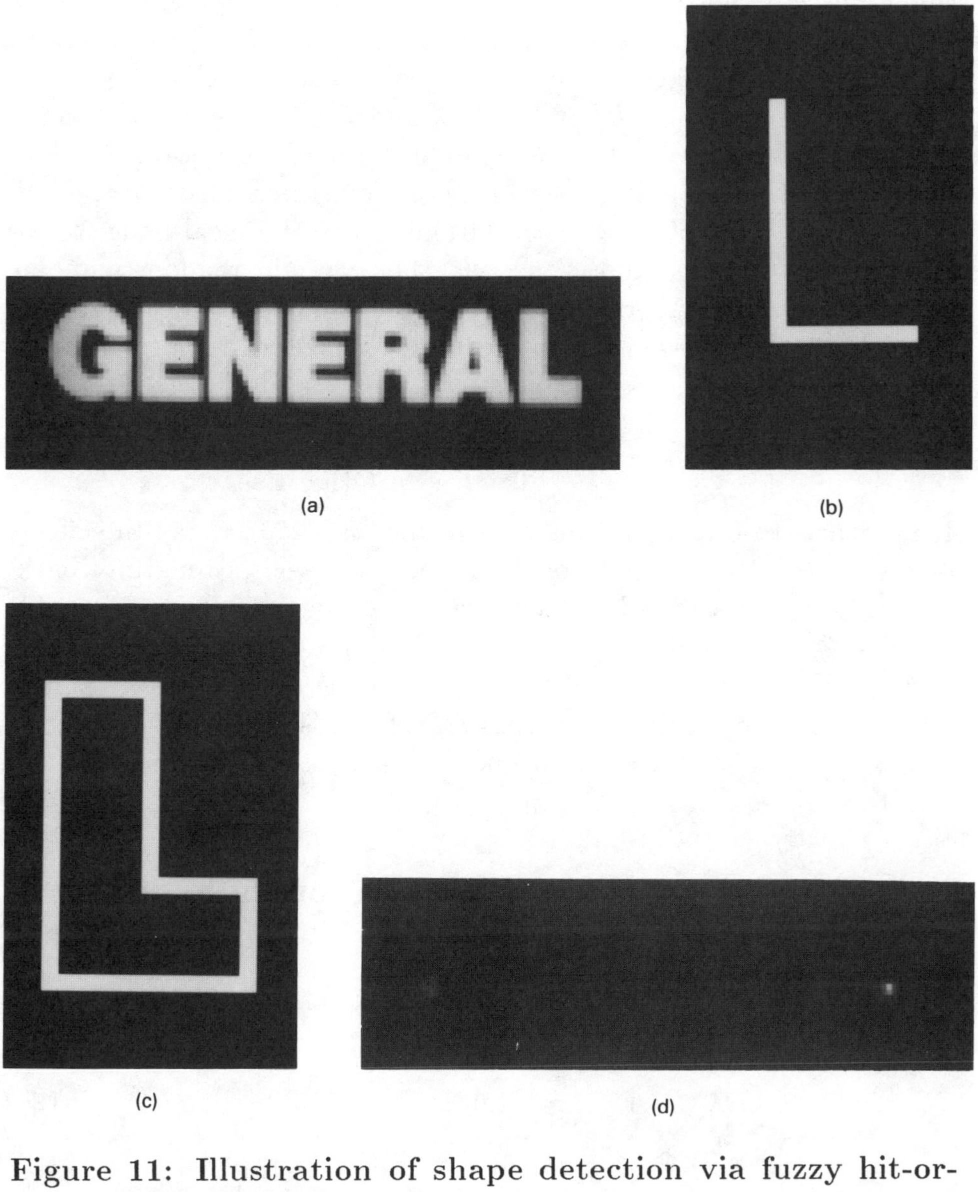

Figure 11: Illustration of shape detection via fuzzy hit-or-miss transform. The aim is to detect the letter 'L'. See Example 3 for details.

In [33] we have carried out a detailed noise-sensitivity analysis of the hit–or–miss transform based shape detector. As has been shown there, so long as we abide by the "rules of thumb" stated earlier, the shape detector is extremely insensitive to uniform random additive and/or subtractive noise. We present a small sample of that analysis.

Consider the image of Fig. 10(b) (cf. Example 2) and suppose that we wish to employ $h = 0.9$, $l = 9$, and $\lambda(x) = 1 - x$ in the shape detector. Let s denote the (percentage) strength of noise and let N denote the number of pixels to which noise must be added/subtracted. (There are a total of $51 \times 51 = 2601$ pixels in the ideal image.) For example, if $s = 10\%$ and $N = 100$, then we add random numbers between $[-0.1, 0.1]$ to randomly selected 100 pixels. Let α denote the maximum gray-value in the resultant image. The normalization process works as follows:

$$\mu_{NORM(X)}(v) = \begin{cases} \mu_X(v)/\alpha & \text{If } \mu_X(v) \geq 0 \\ 0 & \text{Otherwise} \end{cases}.$$

It is natural to investigate the transformed image to study the effects of s and N on the strength of three peaks vis-a-vis strength of background in the neighborhood of these three peaks. It is reasonable to assume the neighborhood as a 5×5 square-shaped region. Table 1 clearly shows that the detection of peaks is a considerably simple job because the fuzzy hit–or–miss transform is relatively insensitive to noise. In [33] we have also studied the affects of the values of h and l on the subsequent recognition.

3.2 Edge detection by fuzzy morphological gradient

The ordinary binary morphological gradient (cf. Eq. (8)) is naturally lifted by means of the fuzzy algorithm

$$GRAD(A, B) = \mathcal{D}(A, B) \setminus [A \cap \mathcal{E}(A, B)]. \tag{25}$$

Since $A \cap (A \ominus B) = A \ominus B$ holds in binary morphology when the structuring element contains the origin, we could rewrite Eq. (8) as $(A \oplus B) \setminus [A \cap (A \ominus B)]$. Now, the lifting is apparent. The most obvious lifting is extremely noise-sensitive and results in broken edges even for moderately noisy images. (As an aside, the relationship $\mathcal{E}(A, B) \subseteq A$ rarely holds in the fuzzy morphology.)

noise strength	number of noisy pixels	Leftmost square		Center square		Rightmost square	
		peak	maximum background	peak	maximum background	peak	maximum background
±5%	500	0.76	0.26	0.87	0.36	0.75	0.27
	1000	0.76	0.26	0.85	0.36	0.72	0.27
	1500	0.76	0.26	0.85	0.36	0.72	0.27
	2000	0.74	0.24	0.83	0.36	0.74	0.30
	2500	0.74	0.24	0.83	0.35	0.73	0.28
±10%	500	0.73	0.28	0.84	0.40	0.74	0.30
	1000	0.70	0.28	0.80	0.38	0.67	0.30
	1500	0.70	0.26	0.78	0.38	0.67	0.32
	2000	0.70	0.26	0.78	0.38	0.67	0.32
	2500	0.70	0.26	0.78	0.39	0.67	0.29
±15%	500	0.69	0.30	0.78	0.41	0.62	0.33
	1000	0.67	0.24	0.76	0.41	0.64	0.36
	1500	0.67	0.24	0.75	0.41	0.64	0.33
	2000	0.65	0.24	0.74	0.42	0.66	0.32
	2500	0.65	0.21	0.74	0.42	0.66	0.32
±20%	500	0.65	0.33	0.73	0.42	0.61	0.36
	1000	0.64	0.32	0.72	0.42	0.60	0.38
	1500	0.63	0.32	0.72	0.38	0.60	0.39
	2000	0.63	0.32	0.72	0.38	0.59	0.35
	2500	0.63	0.26	0.72	0.32	0.59	0.36

Table 1: Noise–insensitivity analysis of the fuzzy hit-or-miss transform based shape detector. This experiment was performed on the image of Figure 10(b).

Like the ordinary gradient, the fuzzy gradient gives a (fuzzy) band about the image marking its edge. Like the Sobel edge, we may need to perform thresholding on the morphological gradient to clearly "visualize" the edges. We, however, believe that such a thresholding defeats some of the purposes of modeling images as fuzzy sets. The membership values along the edges, if left alone, may be used in later decision making.

Rules of thumb

The structuring element should be $S(3, h)$ with $h \approx 1$. Even though the structuring element is more–or–less crisp, the corresponding fuzzy gradient is different than the gray-scale gradient. Choosing $h \leq 0.8$ will result in loss of edge information.

Consider the λ defined by Eq. (18). We have found that the edge intensity decreases very sharply with increase in the value of n. After a certain extent (around $n = 2.5$) we begin disconnecting the edge. While dealing with the families of Eqs. (19) and (20), we have observed that the rate of decrease in the edge intensity is lower, and that the family of Eq. (19) has the lowest rate of decrease. The value of n in Eqs. (18) through (20) should be such that $n \leq 1.8$, $n \geq -0.4$, and $n \leq 3.5$ respectively.

Example 4 *As the first example on edge detection consider the camera image of a joystick as shown in Fig. 12(a). Let the structuring element be* $S(3,1)$. *In Figs. 12(b) through 12(d) we show the fuzzy morphological gradient when* $n = 1$ *in Eq. (18),* $n = -0.1$ *in Eq. (19), and* $n = 1.6$ *in Eq. (20) respectively. These represent the best results we obtained for each of three different families of* λ *functions.*

Example 5 *For a second realistic example, consider the camera image of a 3M diskette shown in Fig. 13(a). We again let the structuring element be* $S(3,1)$. *Figs. 13(b) and 13(c) show the fuzzy morphological gradient when* $n = 1$ *in Eq. (18),* $n = -0.2$ *in Eq. (19) respectively. The unthresholded Sobel edge is shown in Fig. 13(d).*

Example 6 *Consider the camera image of a rectangular block sitting on top of another rectangular box as shown in Fig. 14(a). Figs. 14(b) and 14(c) show the fuzzy morphological gradient when* $n = 1$ *in Eq. (18) and* $n = 1.6$ *in Eq. (20) respectively. For comparison purposes, in Fig. 14(d) we show the fuzzy morphological gradient when* $n = 1$ *in Eq. (18) and the structuring element is* $S(3, 0.8)$.

Example 7 *As our final example on edge detection, consider the image of Fig. 15(a) in which we show a noiseless gray-scale realization of a circle which has a square-shaped dark patch in its center and there is an even darker triangular patch inside the square. In Fig. 15(b) we show the same image corrupted by noise. The noise in this image is considerably high and consequently the choice of threshold value in the Sobel edge is very critical. We show the (best) thresholded Sobel edge in Fig. 15(c). Assume that the structuring element*

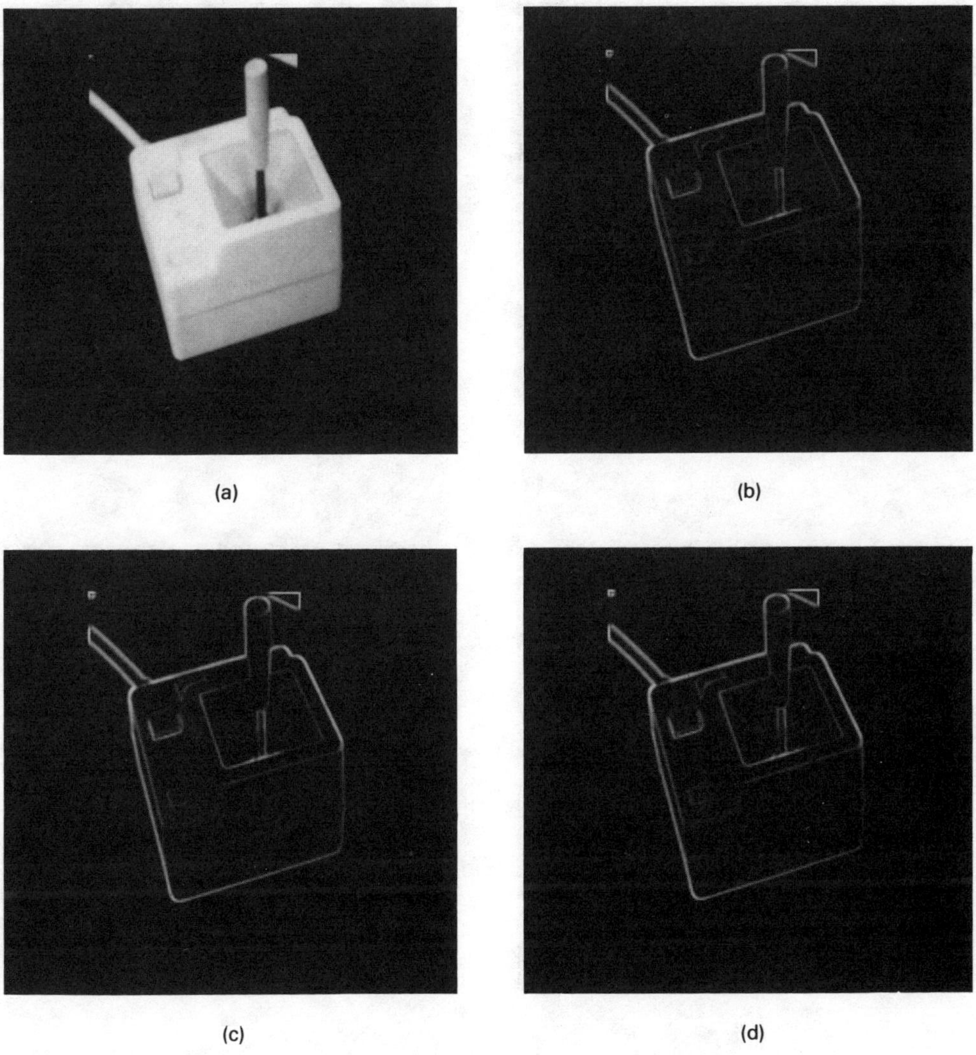

Figure 12: Illustration of edge detection via fuzzy morphological gradient. See Example 4 for details.

(a) (b) (c) (d)

Figure 13: Illustration of edge detection via fuzzy morphological gradient. See Example 5 for details.

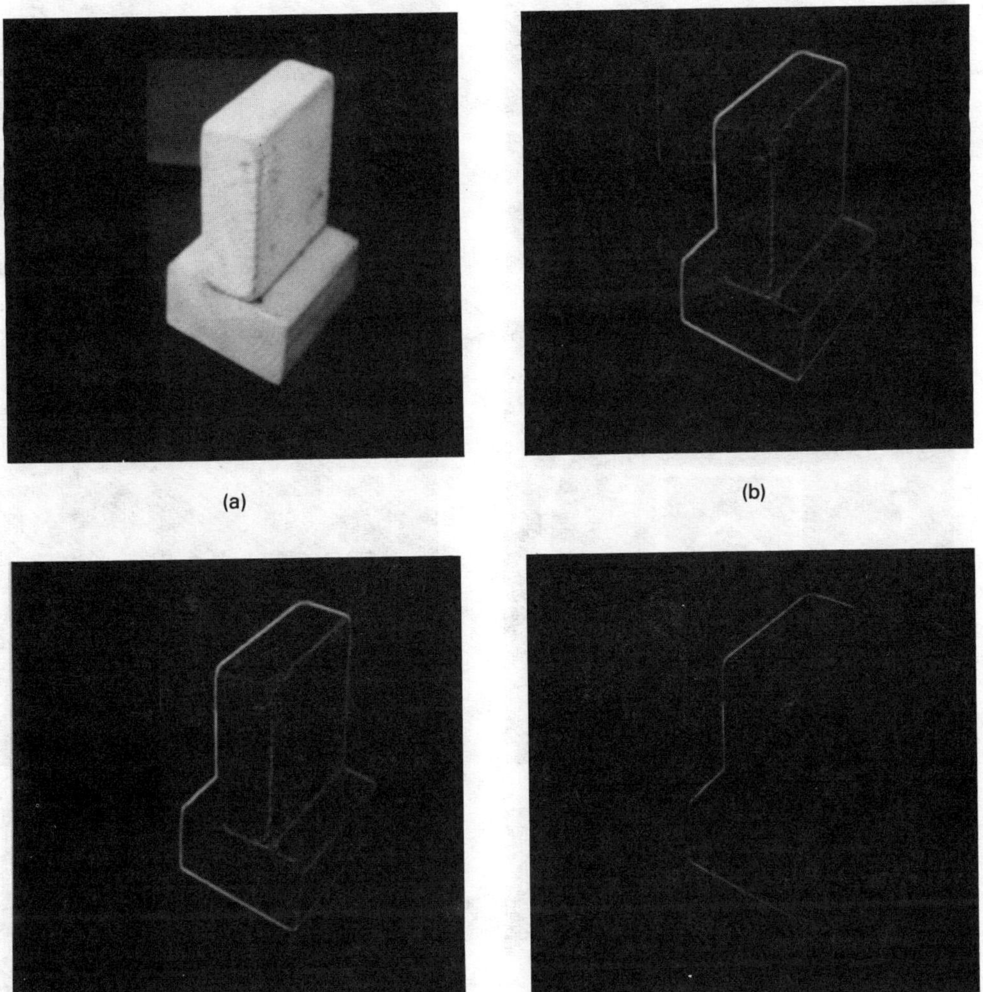

Figure 14: Illustration of edge detection via fuzzy morphological gradient. See Example 6 for details.

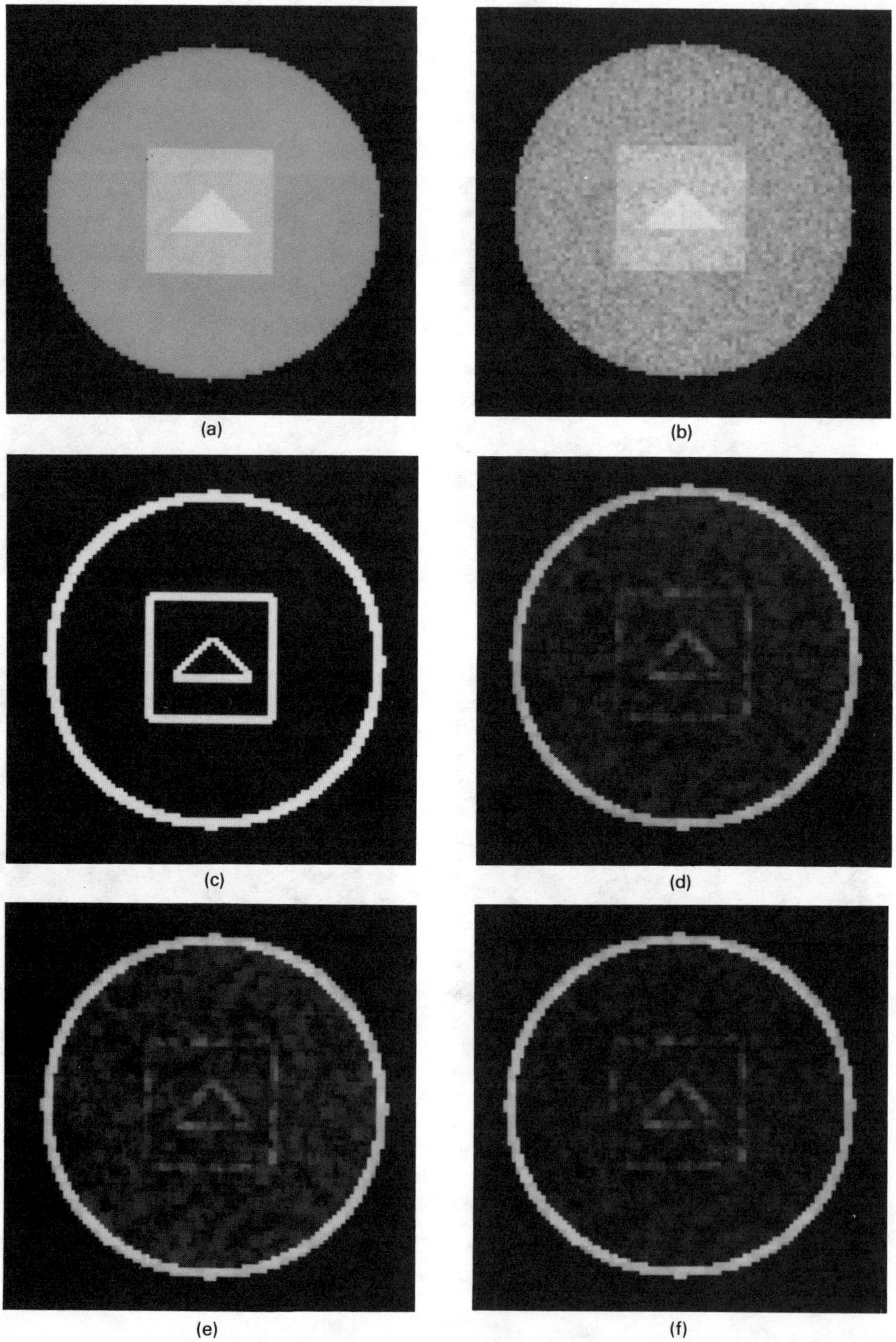

Figure 15: Illustration of edge detection via fuzzy morphological gradient. See Example 7 for details.

is $S(3,1)$. In Figs. 15(d) through 15(f) we show the fuzzy morphological gradient when $n = 1$ in Eq. (18), $n = 1.6$ in Eq. (20), and $n = -0.1$ in Eq. (19) respectively.

3.3 Peak detection by fuzzy top-hat transformation

In our discussion of binary operators we noted how the top-hat transform can be used to locate contour features by subtracting the opening of an image from the original image. Lifting the top-hat yields the fuzzy algorithm $\psi(A) = A \setminus \mathcal{O}(A, B)$, with subtraction being defined via Eq. (23). It turns out that a very interesting application of this operator is very much like the typical application done in ordinary gray-scale morphology using the gray-scale top-hat transform (see [4]), that being the location of peaks in the topography of the image intepreted as a gray-scale graph. In the fuzzy setting, the fuzzy top-hat can be used to locate regions of relatively high certainty, thereby finding regions in which belief in the presence of the object is relatively high (in comparison to pixels in other regions).

Rules of thumb

In general, a fuzzy top-hat can simultaneously detect wide as well as narrow peaks with the same structuring element. The value of n in Eqs. (18) through (20) should be close to 1, 0, and 1.6 respectively.

If the image is noisy, then the output of fuzzy top-hat will also contain noise. In such cases, it is advisable to compute the fuzzy top-hat of image $\mathcal{O}(A, B)$, where B is a suitably choosen structuring element (to perform smoothing). When performing the tophat transform directly on a noisy image, the structuring element should be quite small. For excessively noisy realizations, a single pixel structuring element may be needed to obtain decent results. The λ function given by $\lambda(x) = 1 - x$ works very well for noisy images.

Example 8 *Consider the image of Fig. 15(a) again. Suppose that we wish to remove the cylindrical bed (whose height is 0.52). We can perform the fuzzy opening by $C(r, 0.52)$ and then subtract the opened image from the original. r can be any number less than the radius of the circle in the image but, for efficiency, we perhaps will choose a* *small number (say 2). Fig. 16(a) shows the outcome for* $r = 2$ *and* $\lambda(x) = 1 - x$.

Figs. 16(b) through 16(d) show the outcomes of fuzzy top-hat transform on the noisy image of Fig. 15(b) when $n = 1$ *in Eq. (18),* $n = -0.1$ *in Eq. (19), and* $n = 1.6$ *in Eq. (20) respectively. The structuring element was* $C(0, 0.52)$ *(this image has only one pixel). As stated earlier, we should choose a very small radius so as to negate the effect of excessive random point noise.*

3.4 Contour Smoothing by Fuzzy Opening

As illustrated in Fig. 2, performing a binary opening with a round, say circular, structuring element can have a smoothing, or "lowpass" effect from the inside. Letting $\mathcal{B}$ be a disk, the binary algorithm is simply $\psi(\mathcal{A}) = \mathcal{A} \circ \mathcal{B}$. The naturally lifted algorithm that operates on a fuzzy realization A of $\mathcal{A}$ is $\psi(A) = \mathcal{O}(A, B)$, where B is an appropriately chosen fuzzy subset of $\mathcal{B}$. We provide two examples.

Example 9 *Consider a noisy realization of a right-angled isoceless triangle shown in Fig. 17(a). To smoothen the corners, we can perform a fuzzy opening by* $C(r, h)$ *with* r *and* h *determining the amount smoothening. The roundness of corners in the opened image increases with increasing values of* r *and the flatness of membership values in the opened image increases with decreasing values of* h*. Figures 17(b) through 17(f) show the outcome of opening by structuring elements* $C(3, 0.7)$, $C(5, 0.7)$, $C(9, 0.7)$, $C(13, 0.7)$, *and* $C(17, 0.7)$ *respectively. We have used* $\lambda(x) = 1 - x$*. When* $r = 19$ *the opened image is* $C(r, 0.7)$ *(except for the negligible background), and for* $r \geq 20$*, the opened image contains only the (negligible) background as the circle no longer fits inside the triangle.*

Example 10 *Figs. 18(a) through 18(d) show the outcome of opening the image of Fig. 8(b) by the images* $C(5, 0.8)$, $C(5, 0.7)$, $C(9, 0.8)$, *and* $C(15, 0.8)$ *respectively. We have used* $\lambda(x) = 1 - x$*. Again note the rounding off of the sharp corners of the square and triangle in the opened image. Again, as the radius of the structuring element increases, the roundness of corners also increases. Since the structuring element* $C(15, h)$ *neither fits inside the square nor inside the triangle, the corresponding opened image only contains the circle.*

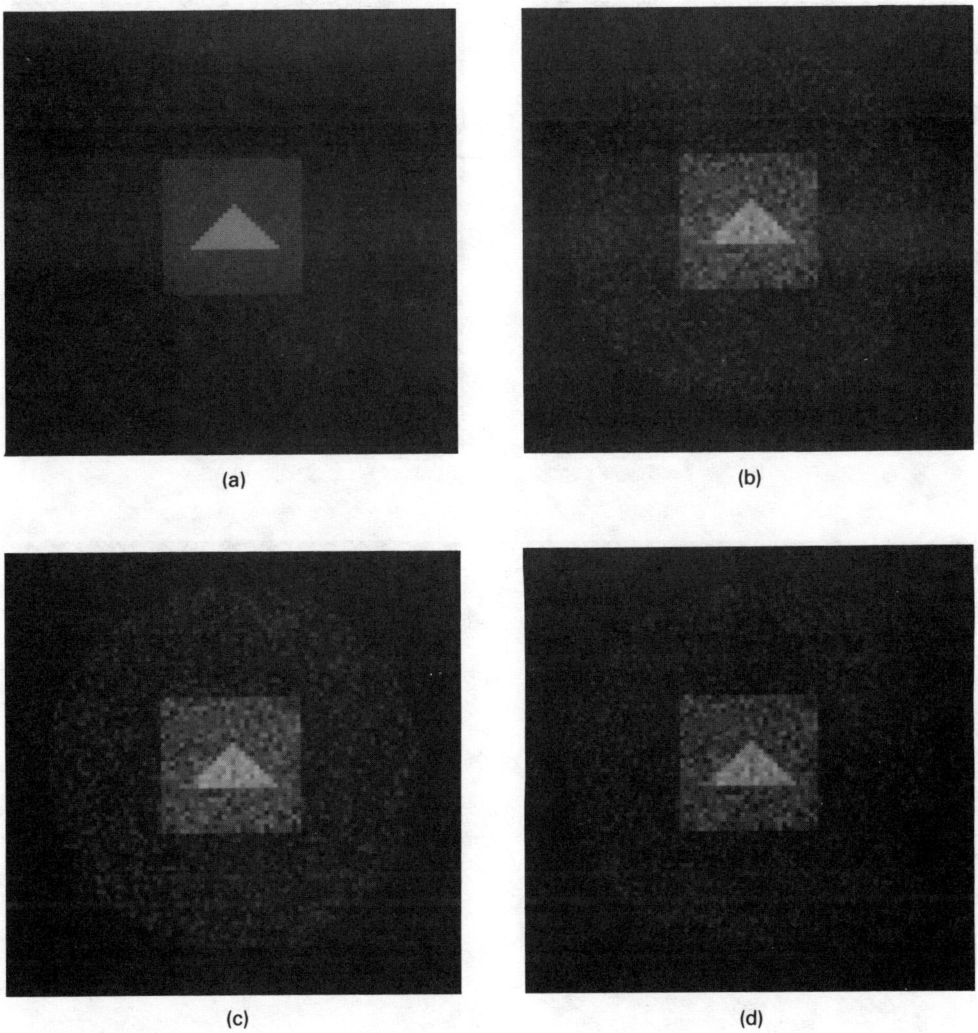

Figure 16: Illustration of peak detection via the fuzzy top-hat transform.

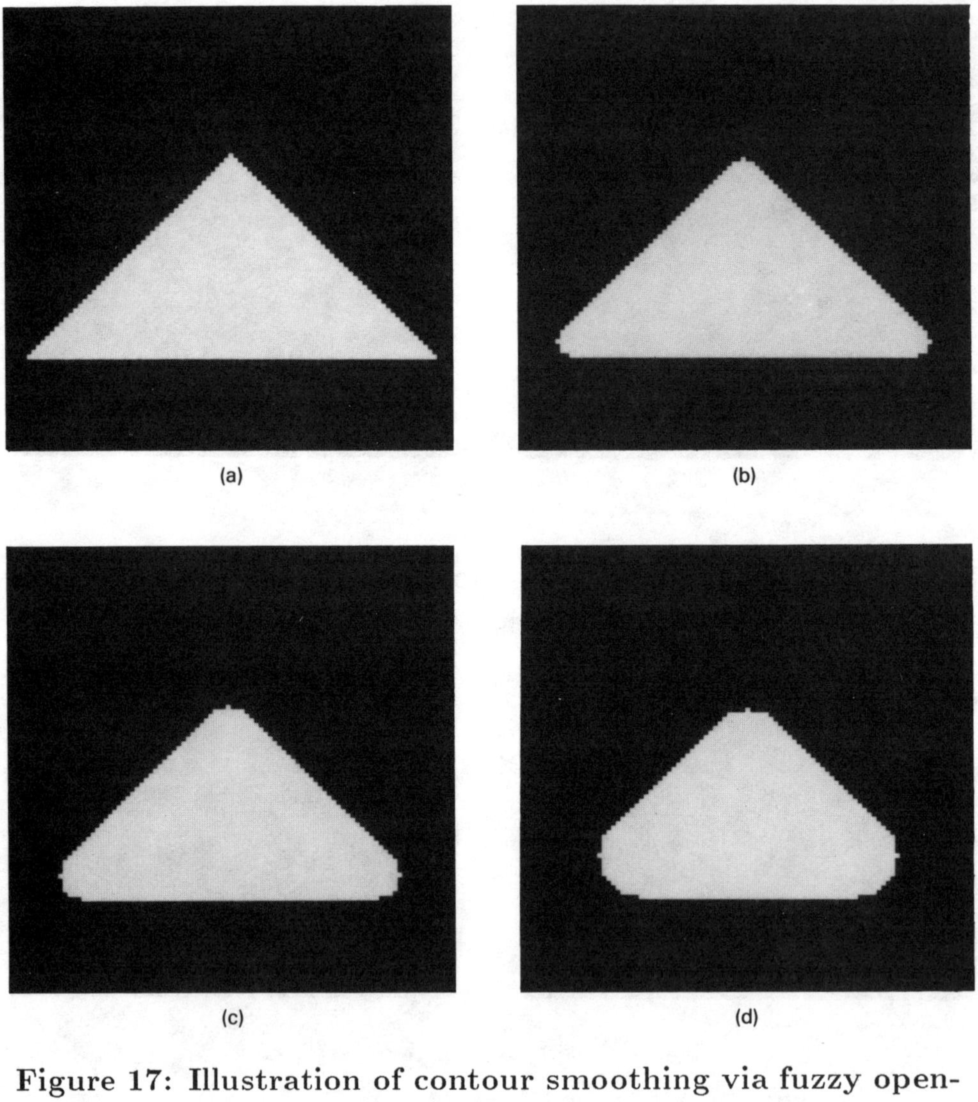

Figure 17: Illustration of contour smoothing via fuzzy openings. See Example 9 for details.

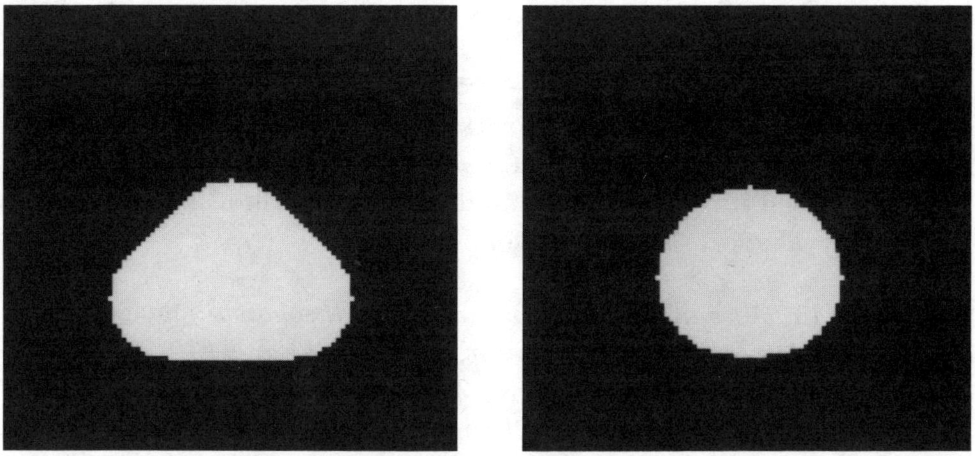

(e) (f)

Figure 17 (continued)

If we consider the triangle and square as the clutter, then opening by $C(15, h)$ *removes noise as well as the clutter. Removal of clutter and noise is discussed in detail in the next subsection.*

3.5 Filtering union noise by fuzzy opening

If S and N are normalized fuzzy realizations of the ideal binary images $\mathcal{S}$ and $\mathcal{N}$, respectively, and if the ideal image to be filtered is $\mathcal{S} \cup \mathcal{N}$, then the fuzzy image to be processesd is given by the fuzzy union $S \cup N$. As in the case of contour smoothing, the binary union is lifted to a fuzzy union, so that the fuzzy filter is defined by $\psi(S \cup N) = \mathcal{O}(S \cup N, B)$, where B is an appropriately chosen structuring element.

Rules of thumb

Unless there is a specific reason to preserve roundness in the image contour, the square-shaped family of images $S(l, h)$ provides us with the most useful structuring elements for noise removal. As a rule of thumb, we let $h \approx 0.8$ and $3 \leq l \leq 7$. When the noise is more widespread and when the objects are large, then and only then should one use larger values of l. If we use a large value of l while smoothing a smaller image, then we run the risk of blurring the image. In this sense l acts very much like the window size for median and moving-average type operators.

As mentioned earlier, when $n \to \infty$ in Eqs. (18) and (20) (or, when $n \to -1$ in Eq. (19)), the indicator treats "signal" and "noise" as the same. The net effect of this, so far as the opening is concerned, is that the opened image becomes a null image in the limiting situation. Hence, the value of n should be very small.

Example 11 *As our first example, consider the noisy image of Fig. 15(b) which does not have any clutter. Figs. 19(a) and 19(b) show the outcome of the median filter with* 3×3 *and* 5×5 *windows respectively. Fig. 19(c) shows the outcome of the moving-average filter with a* 3×3 *window. Figs. 19(d) and 19(e) show the outcome of fuzzy opening by structuring elements* $S(5, 0.8)$ *and* $S(7, 0.8)$*, respectively, when* $\lambda(x) = 1 - x$.

Figs. 19(f) through 19(j) show the outcome of opening by $S(5, 0.8)$ *when* $n = 1.25$, $n = 1.5$, $n = 1.75$, $n = 2$, *and* $n = 5$, *respectively, in Eq. (18). To get an intuitive feel for what is happening in these cases, simply note that we are trying to check if* $\gamma[S(5,0.8)]^c$ *fits inside (cf. Eq. (22)) the given noisy image (call it A). A simple calculation yields that for* $\lambda(x) = 1 - x^n$, $\gamma[S(5,0.8)]^c = S(5, 0.8^n)$. *Thus, for* $n = 2$, *any membership value of* 0.64 *or above in* $\gamma(A^c)$ *will be treated as* 0.64 *by the structuring element. Within the region of interest, the smallest membership value of* $\gamma(A^c)$ *is* 0.7127, *that is, within the region of interest, the structuring element* $\gamma[S(5,0.8)]^c$ *fits everywhere. That is precisely the reason why we only see a cylinder of "height"* 0.644 *in Fig. 19(i).*

As $n \to \infty$ *in* $\lambda(x) = 1 - x^n$, *we have* $\gamma[S(5,0.8)]^c \to \emptyset$. *Now,* $\mathcal{O}(A, \emptyset) = \emptyset$ *for any image A. Thus, as we increase the value of n beyond 2, we will find that the height of the cylinder keeps going down, and eventually, we will obtain a null image. The height of the cylinder in Fig. 19(j) is only* 0.3277 *(as it should since* $0.8^5 = 0.3277$.*)*

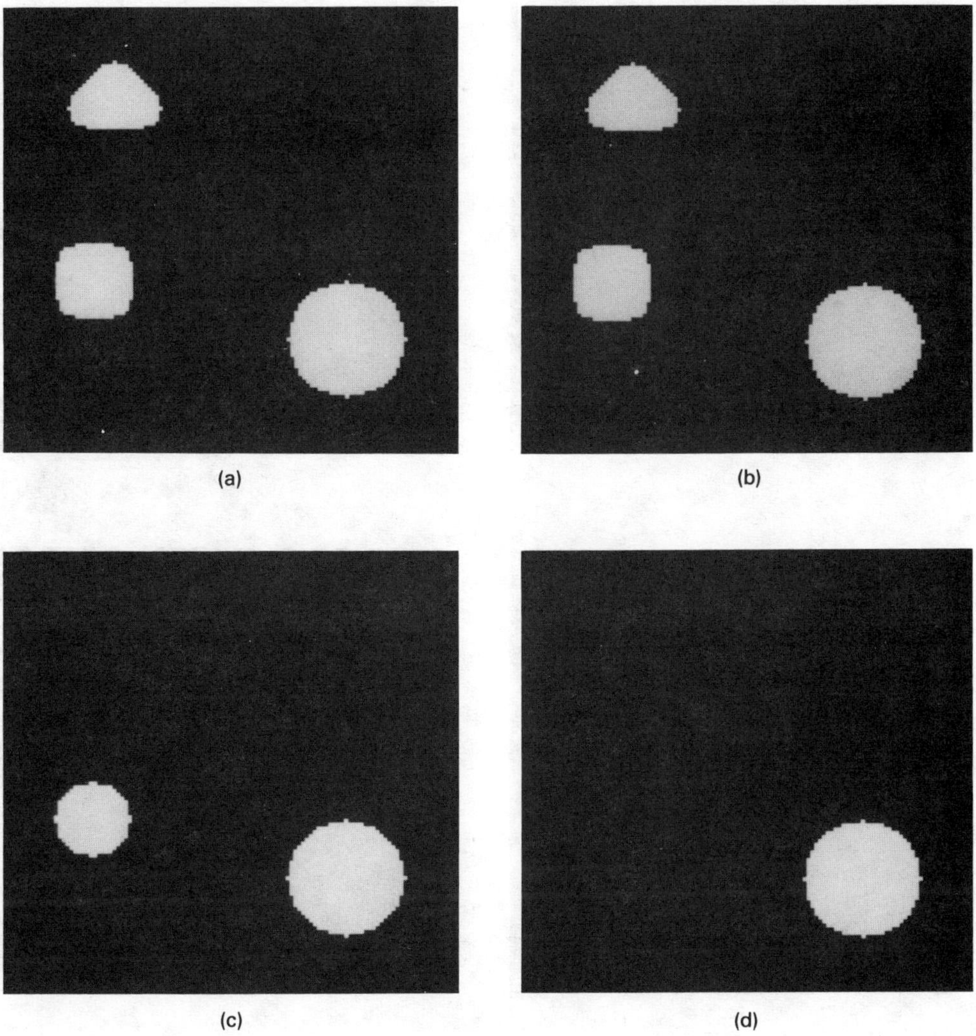

Figure 18: Illustration of contour smoothing via fuzzy openings. See Example 10 for details.

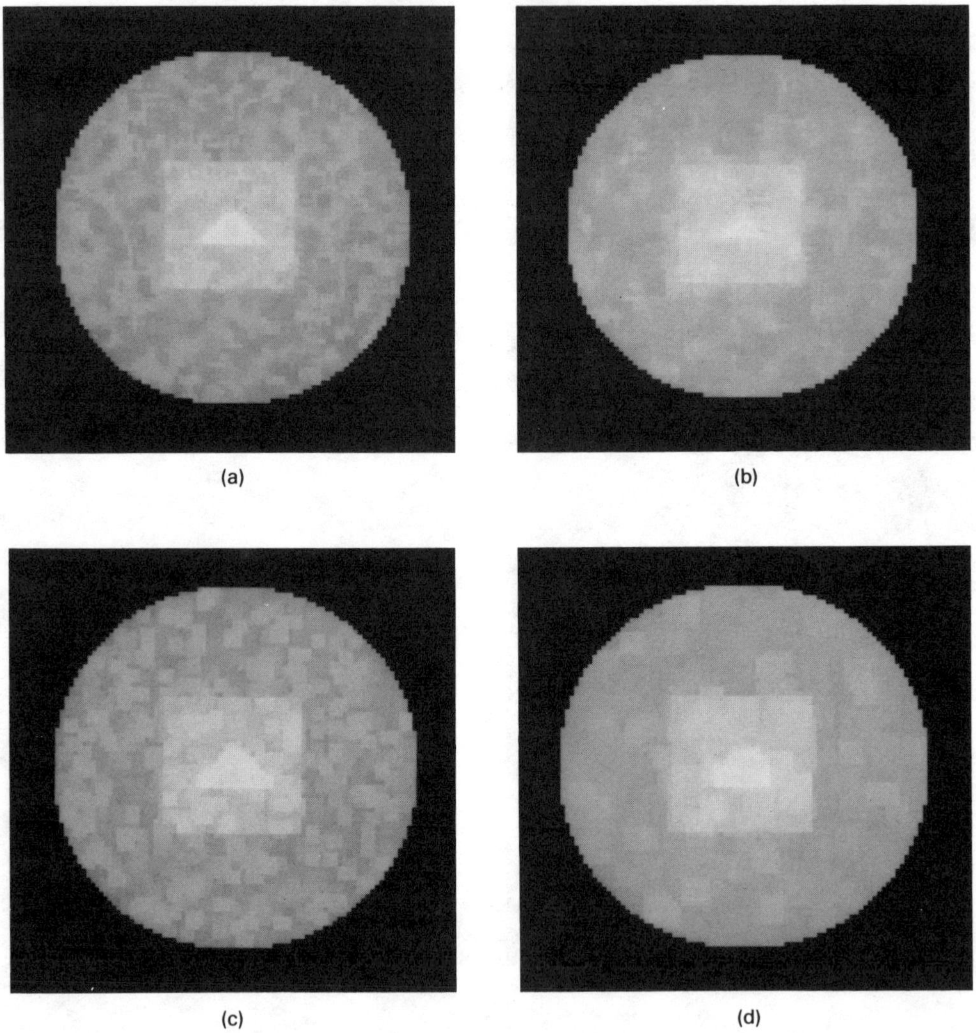

Figure 19: Illustration of random noise removal via fuzzy openings. See Example 11 for details.

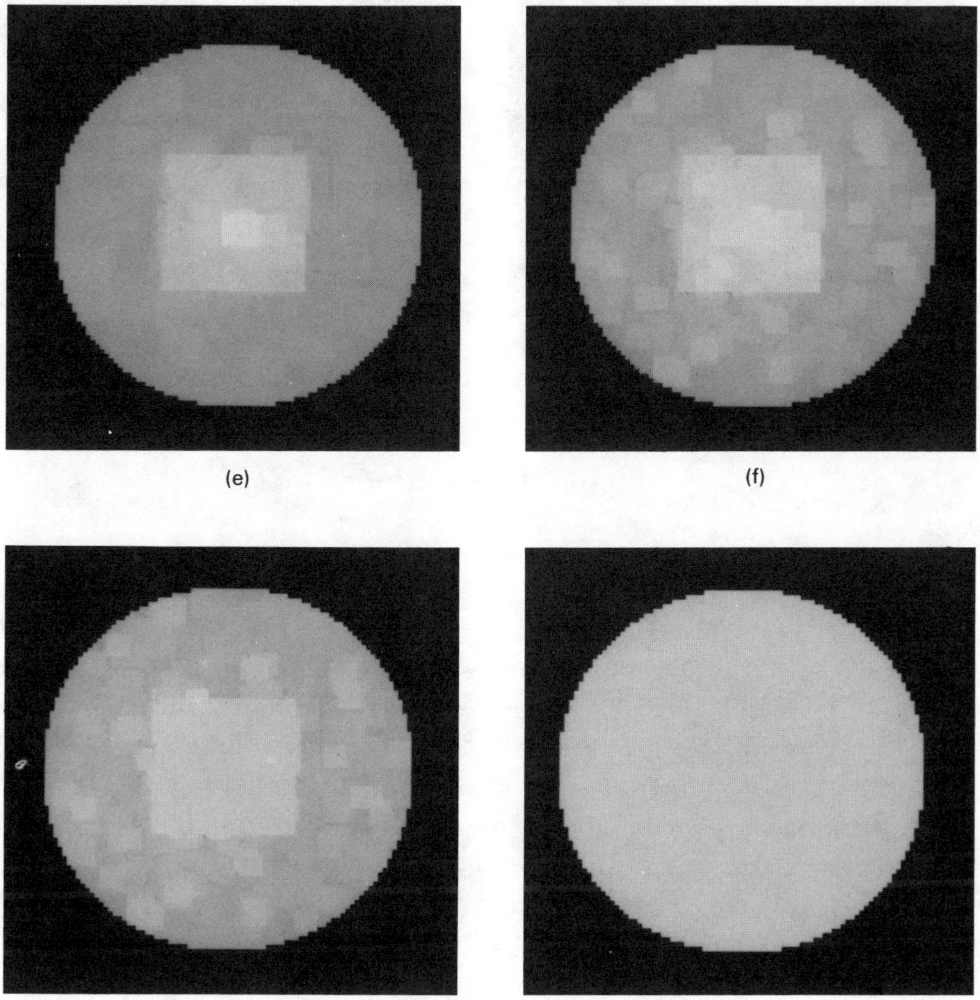

(e) (f) (g) (h)

Figure 19 (continued)

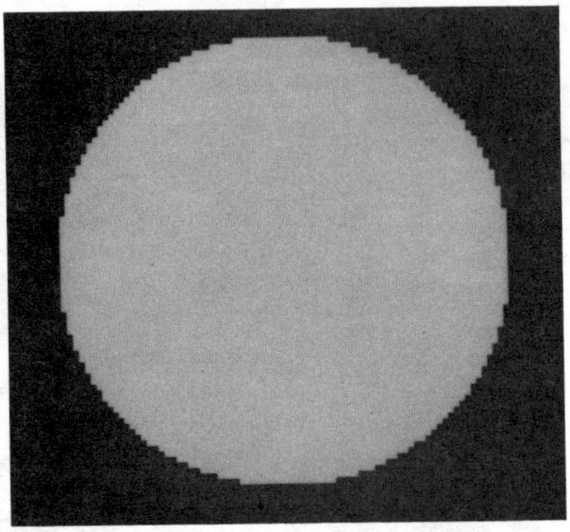

(i)

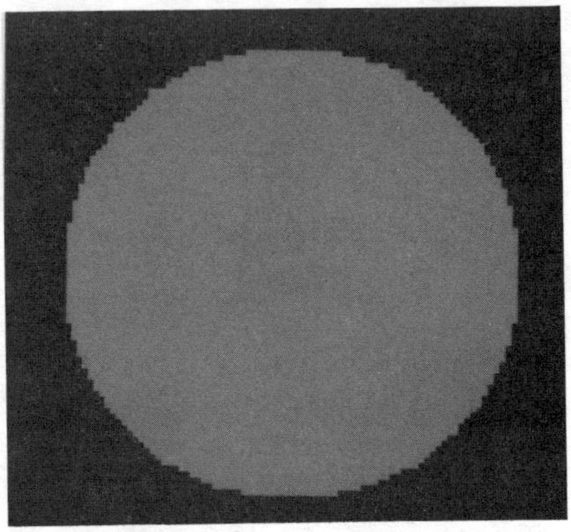

(j)

Figure 19 (continued)

Example 12 *Reconsider the camera image of a joystick that was shown in Fig. 12(a). For λ defined via Eq. (20), we obtain the best noise removal results when $n = 1.8$ and the structuring element is $S(3, 0.7)$. The fuzzy opening is shown in Fig. 20(a). Likewise, in Figs. 20(b) and 20(c) we show the fuzzy opening by $S(5, 0.8)$ when $n = -0.2$ in Eq. (19), and by $S(3, 0.8)$ when $n = 1.1$ in Eq. (18), respectively. For comparison purposes, Figs. 20(d) through 20(f) show the fuzzy opening by $S(5, 0.8)$ when $n = 1.8$ in Eq. (20), by $S(3, 0.8)$ when $n = -0.2$ in Eq. (19), and by $S(5, 0.8)$ when $n = 1.1$ in Eq. (18), respectively. Finally, Figs. 20(g) and 20(h) show the outcome of median and moving-average filters respectively. The window was assumed to be 3×3.*

Our last example considers the task of removing clutter as well as random noise.

Example 13 *Suppose that the ideal image only has a 31×31 square with its center at the origin. In Fig. 21(a) we show a (normalized)* *noisy gray-scale realization of this image. We have also added clutter consisting of ten randomly sized rectangles at random locations. The height and width of these rectangles is assumed to be uniformly distributed between* 5 *and* 14.

To remove the clutter we must choose the size of the structuring element in such a way that the structuring element does not fit inside the clutter but fits inside the object of interest. We can therefore choose $S(l,h)$ *with* $l \geq 15$*, as an appropriate candidate for the structuring element. In Figs. 21(b) through 21(d) we show the outcome of fuzzy opening by* $S(15, 0.8)$ *when* $n = 1$ *in Eq. (18),* $n = -0.1$ *in Eq. (19), and* $n = 1.6$ *in Eq. (20) respectively. As can be seen there, the random noise as well as the clutter has been removed from the image.*

4 Conclusion

A paradigm for the natural lifting of binary morphological algorithms to fuzzy algorithms has been presented and implemented. The key aspect of the paradigm is that fuzzy realizations of binary images are processed in a manner consistent with their binary natures, thereby allowing subsequent processing to proceed in a manner conducive to whatever decision procedure is to be employed. Contour smoothing and filtering clutter, edge detection, peak detection, and object detection have been illustrated via fuzzy opening, fuzzy morphological gradient, fuzzy top-hat, and fuzzy hit-or-miss transform, respectively. These applications indicate robustness in the presence of noise as well as the insensitivity of the filters to small changes in the structuring elements.

5 Acknowledgement

We are grateful to Yidong Chen for drawing the diagrams and for taking the photographs of all images.

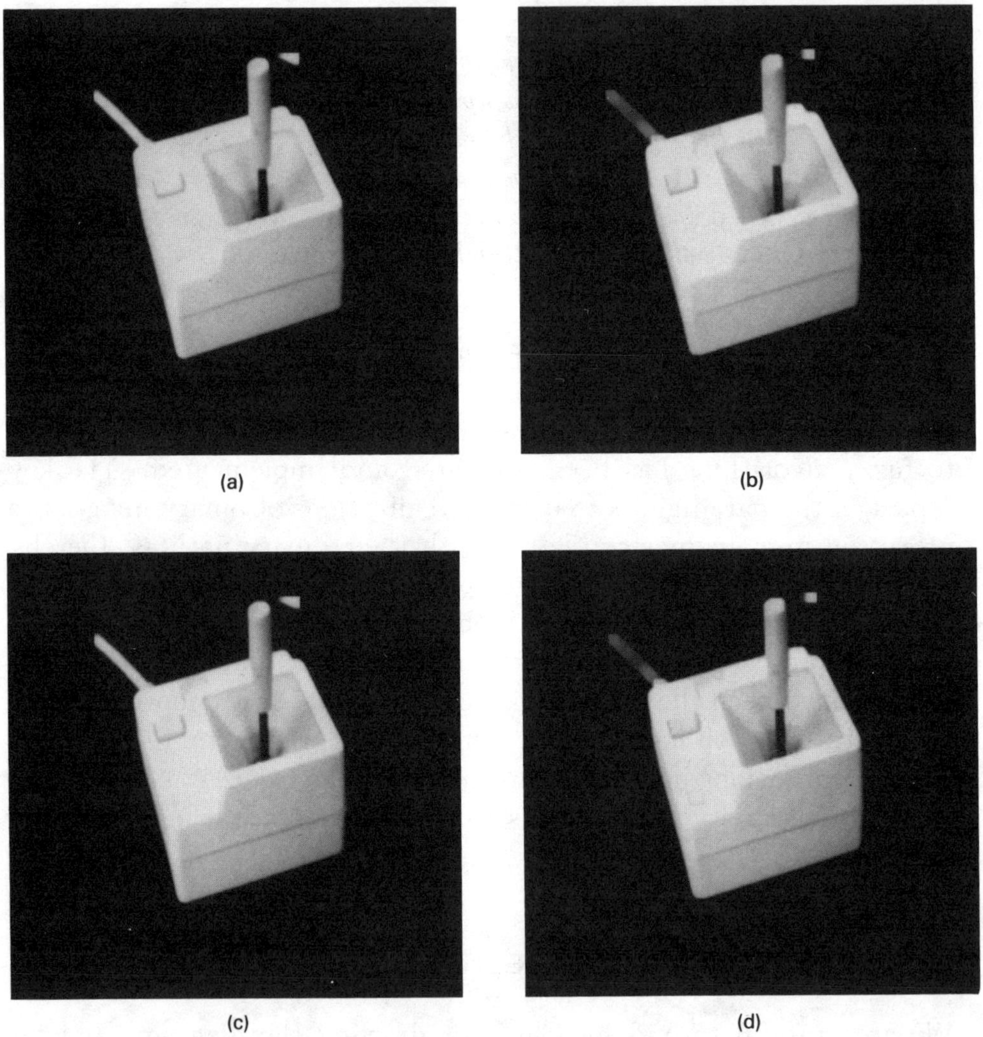

Figure 20: Illustration of random noise removal via fuzzy openings. See Example 12 for details.

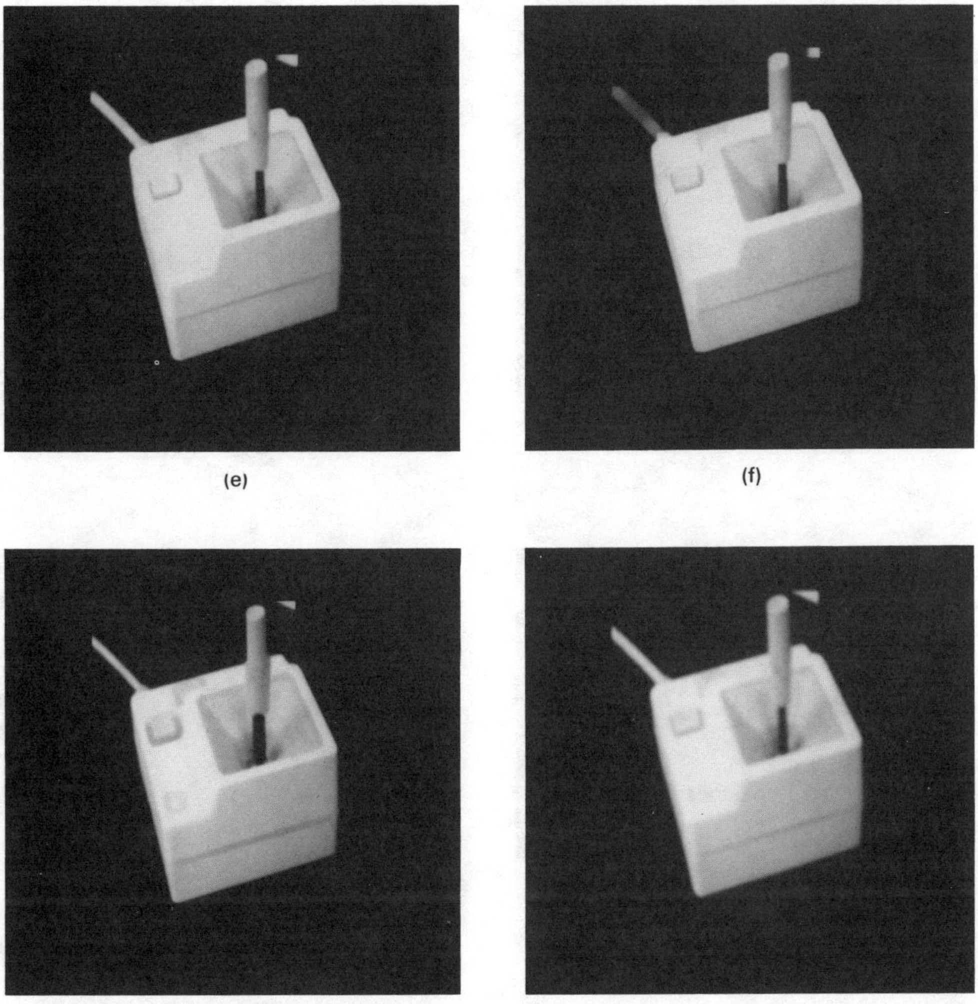

(e) (f) (g) (h)

Figure 20 (continued)

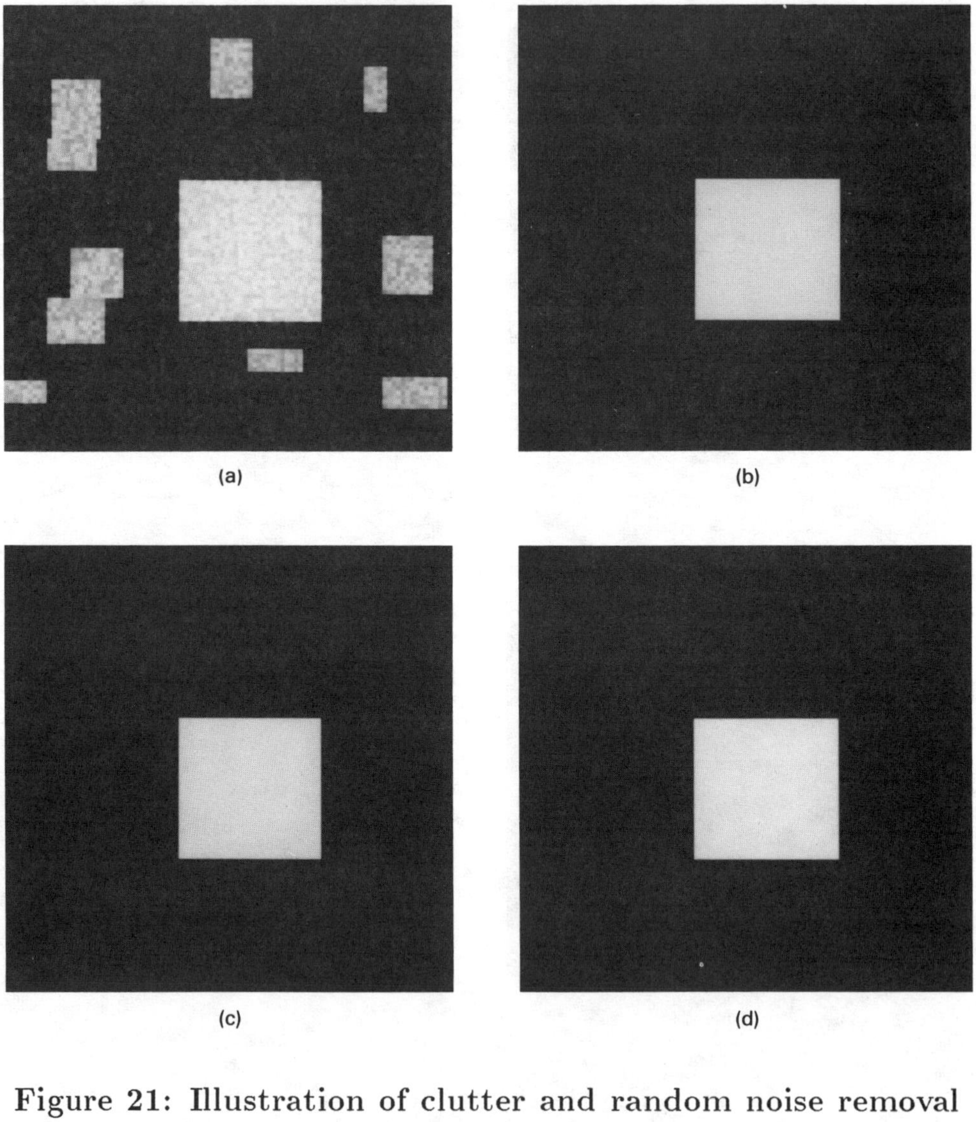

Figure 21: Illustration of clutter and random noise removal via fuzzy openings. See Example 13 for details.

6 Bibliography

1. J. Astola, L. Koskinen, and Y. Neuvo, "Statistical Properties of Discrete Morphological Filters," in *Mathematical Morphology in Image Processing*, E.R. Dougherty (editor), Marcel Dekker, 1992.
2. G.J.F. Banon and J. Barrera, "Decomposition of Mappings Between Complete Lattices by Mathematical Morphology — Part I: General Lattices," Report INPE-5346-NTC/300, Instituto Nacional de Pesquisas Espacials, Sao Jose dos Campos, Brazil, 1991.
3. T. Crimmons and W. Brown, "Image Algebra and Automatic Shape Recognition," IEEE Trans. on Aerospace and Electronic Systems, vol. 21, no. 1, January 1985.
4. E.R. Dougherty, *An Introduction to Morphological Image Processing*, SPIE Press, 1992.
5. E.R. Dougherty, "Optimal Mean-Square N-Observation Digital Morphological Filters — Part I: Optimal Binary Filters," *Computer Vision, Graphics, and Image Processing — Image Understanding*, vol. 55, no. 1, 1992.
6. E.R. Dougherty, "Optimal Mean-Square N-Observation Digital Morphological Filters — Part II: Optimal Gray-scale Filters," *Computer Vision, Graphics, and Image Processing — Image Understanding*, vol. 55, no. 1, 1992.
7. E.R. Dougherty and R.P. Loce, "Efficient Design Strategies for the Optimal Binary Digital Morphological Filter: Probabilities, Constraints, and Structuring-element Libraries," in *Mathematical Morphology in Image Processing*, E.R. Dougherty (editor), Marcel Dekker, 1992.
8. E.R. Dougherty and D. Sinha, "Computational Gray-scale Mathematical Morphology," submitted. (Morphological Imaging Laboratory Report MIL-17-92, Rochester Institute of Technology)
9. E.R. Dougherty, D. Sinha, and P. Sinha, "Fuzzy Morphological Filters," Proc. of SPIE, vol. 1825, Boston, November 1992.
10. E.R. Dougherty and D. Zhao, "Model-Based Characterization of Statistically Optimal Design for Morphological Shape Recognition via the Hit-or-miss Transform," *Journal of Visual Communication and Image Representation*, pp. 147-160, vol. 3, no. 2, 1992.
11. C.R. Giardina, "Advanced Imaging Techniques for Hierarchies of Intelligent PGM's," Proc. of NATO AGARD Guidance and Control Conf., 1982.

12. C.R. Giardina, "The Universal Imaging Algebra," *Advances in Electronics and Electron Physics*, vol. 67, pp. 121-182, 1986.
13. C.R. Giardina and E.R. Dougherty, *Morphological Methods in Image and Signal Processing*, Prentice-Hall, 1988.
14. V. Goetcherian, "From Binary to Grey Level Tone Image Processing using Fuzzy Logic Concepts," *Pattern Recognition*, vol. 12, 1980.
15. H. Heijmans, "Theoretical Aspects of Gray-level Morphology," *IEEE Trans. on Pattern Analysis and Machine Intelligence*, vol. 13, 1991.
16. H. Heijmans and C. Ronse, "The Algebraic Basis of Mathematical Morphology, Part I: Dilations and Erosions," *Computer Vision, Graphics, and Image Processing — Image Understanding*, vol. 50, 1990.
17. H. Heijmans and C. Ronse, "The Algebraic Basis of Mathematical Morphology, Part II: Openings and Closings," *Computer Vision, Graphics, and Image Processing — Image Understanding*, vol. 54, 1991.
18. A. Kauffman and M.M. Gupta, *Fuzzy Mathematical Models in Engineering and Management Science*, North-Holland, 1988.
19. G.J. Klir and T.A. Folger, *Fuzzy Sets, Uncertainty, and Information*, Prentice-Hall, 1988.
20. R.P. Loce and E.R. Dougherty, "Optimal Morphological restoration: The Morphological Filter Mean-Absolute-Error Theorem," *Journal of Visual Communication and Image Representation*, vol. 3, no. 4, December 1992.
21. G. Matheron, *Random Sets and Integral Geometry*, John Wiley, New York, 1975.
22. D. Schonfeld and J. Goutsias, "Optimal Morphological Pattern Restoration from Noisy Binary Images," *IEEE Trans. on Pattern Analysis and Machine Intelligence*, vol. 13, no. 1, 1991.
23. J. Serra, *Image Analysis and Mathematical Morphology*, Academic Press, 1982.
24. J. Serra (ed.), *Image Analysis and Mathematical Morphology, vol 2: Theoretical Advances*, Academic Press, 1988.
25. D. Sinha and E.R. Dougherty, "Fuzzy Mathematical Morphology,œ" *Journal of Visual Communication and Image Representation*, vol. 3, no. 3, 1992.
26. D. Sinha and E.R. Dougherty, "Fuzzification of Set Inclusion:

Theory and Applications," *International Journal of Fuzzy Sets and Systems*, April 1993.

27. D. Sinha and E.R. Dougherty, "Characterization of Fuzzy Minkowski Algebras," Proc. of SPIE, vol. 1769, San Diego, July 1992.

28. D. Sinha and E.R. Dougherty, "A General Axiomatic Theory of Intrinsically Fuzzy Mathematical Morphologies — Part I: Axioms and Primitive Operations," submitted. (Morphological Imaging Laboratory Report MIL-8-92, Rochester Institute of Technology)

29. D. Sinha and E.R. Dougherty, "A General Axiomatic Theory of Intrinsically Fuzzy Mathematical Morphologies — Part II: Minkowski Algebra and Representations," submitted. (Morphological Imaging Laboratory Report MIL-9-92, Rochester Institute of Technology)

30. D. Sinha and E.R. Dougherty, "Fuzzification of Set Inclusion," Proc. SPIE, vol. 1708, Orlando, April, 1992.

31. D. Sinha and C.R. Giardina, "Discrete Black and White Object Recognition via Morphological Functions," *IEEE Trans. on Pattern Analysis and Machine Intelligence*, vol. 12, pp. 275-293, 1990.

32. D. Sinha and H. Lee, "Invariant Characterization and Pseudocharacterizations of Discrete Multi-dimensional Sets using Mathematical Morphology," in *Mathematical Morphology in Image Processing*, E.R. Dougherty (editor), Marcel Dekker, 1992.

33. P. Sinha, D. Sinha and E.R. Dougherty, "Shape Detection via Fuzzy Morphology," Proc. SPIE, vol. 1904, San Jose, February 1993.

34. S. Sternberg, "Cellular Computers and Biomedical Image Processing," Proc. of US-France Seminar on Biomedical Image Processing, 1980.

35. S. Sternberg, "Grayscale Morphology," *Computer Vision, Graphics, and Image Processing*, vol. 35, pp. 333-355, 1986.

36. D. Zhao and D. Daut, "Morphological Hit-or-Miss Transformation for Shape Recognition," *Journal of Visual Communication and Image Representation*, vol. 2, no. 3, September 1991.

Analysis and Segmentation of Higher Dimensional Data Sets with Fuzzy Operators for Representation and Visualization

Wolfgang F. Kraske

Northrop Corporation

Abstract

The segmentation and representation of complex features in higher dimensional data sets is of paramount importance for machine recognition and human perception of image information. A multiresolution and multi–aspect data representation paradigm, a morphological skeleton, is used in this paper to provide a hierarchical framework for efficient representation and visualization of data for machine recognition and human perception of data features. The utilization of fuzzy operators establishes a basis within this framework for organizing packets, or fuzzy sets, of approximate information by minimum coverings or maximal subtense. 3–D and 2–D image data are used to demonstrate applications of these techniques on higher dimensional data sets.

Grayscale mathematical morphology provides an established basis and an algebra for fuzzy operators due to its representation of fuzzy maps over a set support. Specifically, the application of morphological operators in scale and orientation paradigms with tractable support shapes provides an ordered basis for topological analysis and user perception of data. To eliminate precision loss grayscale morphology utilizes only set operations requiring only computer addition and comparison.

1.0 Introduction

The paradigm of the morphological skeleton is introduced for hierarchical organization of fuzzy operators based on resolution. Operator smoothing and differencing operations in L_∞ are exploited to provide multiscale feature representations similar to inner product convolutional basis functions in the L^2 square integrable spaces.

This technique focuses on the specific dimensionalities, and scales of data features. A primary objective of this paper is to present an algorithm which preserves topological morphology and segments delicate and high resolution features to provide a comprehensible representation for machine recognition and human visualization [5, 21]. The fuzzy set algebraic techniques of grayscale mathematical morphology are the basis of this algorithm.

Mathematical morphology offers a multiscale or multiresolution approach for analyzing and segmenting higher dimensional data features across a spectrum of neighborhood scales and resolutions [16, 17, 22]. Set theory is the underlying concept of mathematical morphology which assumes the representation of data with an open topology such that data analysis with an algebra and topological base of sets is used in contrast to direct analysis of data points [20]. The extension to grayscale mathematical morphology with fuzzy set theory [22, 24] facilitates the analysis of data represented as collections of convex functions over an open topological base.

Matheron introduced the algebraic set analysis of mathematical morphology upon binary images to define probability distributions of size and shape spectra relative to a basis of kernel sets or structuring elements [20]. This approach to data representation, as a composition of sets extends to representations of grayscale or fuzzy mappings over sets. This approach deviates from the classical model of higher dimensional data sets perceived as a random and / or discrete collection of numbers or feature vectors to perform cluster analysis upon [8].

For direct applications the algebra of mathematical morphology provides an expression of feature dimensionality and morphology for analysis and classification. In particular the tractability of shape analysis allows the selective study of data features of particular dimensionality and size. Grayscale morphology extends this selectivity to local characteristics with the quantification of texture and gradient as well. The capability of measuring maximum and minimum size extents is also possible by the diametric comparisons of shapes of structuring elements at various scales and orientations. This capability of measuring local intensity and a global morphology of imagery may be particularly beneficial for the quantitative analysis and monitoring of physiological changes in the pathologies of tumors in biomedical imagery and and topographical changes for aerial reconnaissance of environmental data.

The idea of the spectral distribution of data relative to a basis, such as the Fourier transform is as well supported in the algebra of mathematical morphology with size and orientation distributions to describe images [20, 22]. These morphological distributions relate shape information to size and orientation spectra of image data just as the Fourier transform relates spectra of sinusoidal functions to images [17]. Furthermore, the spectral distributions relate to the complexity of describing images with morphological shapes. Simple shapes such as man-made machine parts typically have shape distributions consisting of only a few spectra whereas natural biological structures typically have spectra over a significant range with frequency distributions of the spectra conforming to exponential functions. This function dependence relates to the Hausdorff dimension [10] of fractals [1, 9]. Fractal dimensions are conveniently measured with the Minkowski content[6] in this paper. The shape characteristic of biological structures is postulated as both a fractal nature [1, 9] and a geometric nature [22, 23], in fact the Buffon Symposium of 1977 [23] addressed both the topics of fractal synthesis (Mandelbrot) and the morphological analysis (Serra) of biological structures. In retrospect it is therefore plausible from these mutual interests in biomathematics that morphological analysis and fractal synthesis of biological structures are interrelated by a common set theory. The synopsis of this discussion elucidates the philosophy of a broad application and the unifying prospect of morphological analysis suggested in this paper.

2.0 Methods

The primary goal of mathematical morphology has conventionally been the representation of data features by matching sets of particular shape size and orientation, the approach of this paper is to represent data features as deviations from sets. The selection of the appropriate dimensionality and shape of a structuring element or kernel set for morphological analysis is fundamental to the techniques of this paper. A rigorous development in henceforth developed from topological constructs to operator analysis and finally dimensional filtering.

2.1 Topology and Structuring Elements

The concepts of topology are fundamental to the analysis of image data relative to varieties of sets distinguished by shape, orientation, size and position. The topology used is founded by the norm or families of semi–norms on vector spaces. The text of Folland has been an indispensable reference in the development of this formalism [7].

2.1.1 Definition : A **vector space** X, over a field K, is closed for the following continuous linear mappings given any $\lambda \in K,\ x, y \in X$:

i) vector addition $X \times X \to X,\ (x, y) \to x + y\ ,\ \forall\, x, y \in X$.

ii) scalar multiplication $K \times X \to X,\ (\lambda, y) \to \lambda y$.

To begin, in particular, sets will be used to analyze images based on neighborhood properties over a vector space X. i.e.

2.1.2 Definition : A **topology** on X is a family $\mathcal{T}$ of subsets of X with the following properties :

i) $X,\ \emptyset \in \mathcal{T}$

ii) if the cardinality of A, $|A| \leq \aleph = |N|$, and $\{U_\alpha\}_{\alpha \in A} \subset \mathcal{T}$ then $\bigcup_{\alpha \in A} U_\alpha \in \mathcal{T}$.

iii) if $|A| < \aleph$ and $\{U_\alpha\}_{\alpha \in A} \subset \mathcal{T}$ then $\bigcap_{\alpha \in A} U_\alpha \in \mathcal{T}$.

note bene : $|A|$ denotes the cardinality of a set A, N denotes the natural numbers, $|N| = \aleph$.

Hence $\mathcal{T}$ on X defines an topological space $(\mathcal{T}, X)$. This paper will consider topological spaces with countable bases, defining a separable space.

2.1.3 Definition : A **neighborhood base,** $\mathcal{N}(E)$ of $E \subset X$ has the following properties :

given a set $\mathcal{V}(E) = \{A : E \subset A,\ A \in \mathcal{T}\}$,

i) $\mathcal{N}(E) \subset \mathcal{V}(E) \subset \mathcal{T}$,

ii) $\forall\, U \in \mathcal{V}(E)\ \exists\, V \in \mathcal{N}(E)$ such that $V \subset U$.

2.1.4 Definition : A **base** of a topological space $(X, \mathcal{T})$ is a family of sets $\mathcal{B} \subset \mathcal{T}$ such that $\forall\, E \subset X,\ \mathcal{V}(E) \subset \mathcal{B}$ and for any neighborhood base $\mathcal{N}$, of E, $\mathcal{N}(E) \subset \mathcal{B}$.

The consequences of the closure of a topology, $\mathcal{T}$, is considered a Borel σ–algebra is formed. The vector space over this algebra will be the closure of X. This paper will only consider closed vector spaces, denoted as X.

2.1.5 Definition : Given a topological space, $(X, \mathcal{T})$, the closure of $\mathcal{T}$ is a σ − **algebra**, $\mathcal{A}$, such that :

i) $\mathcal{T} \subset \mathcal{A}$, $\overline{X} \in \mathcal{A}$ i.e. $E \in \mathcal{A} \Rightarrow E \subset \overline{X}$

ii) $|A| \leq \aleph$, and $\{U_\alpha\}_{\alpha \in A} \subset \mathcal{A}$ then $\bigcup_{\alpha \in A} U_\alpha \in \mathcal{A}$.

iii) $|A| \leq \aleph$, and $\{U_\alpha\}_{\alpha \in A} \subset \mathcal{A}$ then $\bigcap_{\alpha \in A} U_\alpha \in \mathcal{A}$.

(note : If $\overline{X} = X$ then X is complete relative to $\mathcal{T}$, a Banach space)

To simplify analysis of images, restrictions on countable bases $\mathcal{B}$ of $\mathcal{T}$ must be considered. For the image data used in this paper X is an Euclidean vector space, either R^3 or $\mathbf{T}^3 = R^3 / \alpha\mathbf{Z}^3$, for image size α, if topology is continuous, or $\mathbf{Z}^3$ or $\mathbf{Z}^3 / n\mathbf{Z}^3$ if topology is discrete, for image size $n \in \mathbf{Z}$. The discrete topologies on $\mathbf{Z}^3$ or $\mathbf{Z}^3 / n\mathbf{Z}^3$ have a countable base of points and sets with finite collections of points which defines a separable space. For a continuous space it is necessary to know if a base can be developed from a countable set of scales and translations of an open set, E, a structuring element, with unit radius, $\mathbf{R}(E) = 1$, see below.

2.1.6 Definition : A **seminorm** on a vector space is a mapping $X \to [0, \infty)$ with the following properties :

i) $p(x + y) \leq p(x) + p(y)$, $\forall\ x, y \in X$.

ii) $p(\lambda x) \leq |\lambda| p(x)$, $\forall\ \lambda \in K$, $\forall\ x \in X$.

if (iii) $p(x) = 0$ iff $x = 0$ then p is a **norm** .

denote $d : X \to [0, \infty)$ as the **Euclidean norm.**

2.1.7 Definition : A **diameter,** of a set $E \subset X$ is

$$\mathbf{d}(E) = \sup\{d(x - y) : x, y \in E\},$$

2.1.8 Definition : A **radius,** of a set $E \subset X$ is

$$\mathbf{R}(E) = \inf\{x \in E : \sup\{d(x - y) : y \in E\}\} .$$

Another set property is now convenient to introduce, the concepts of open sets. Topological spaces in particular are constructions of open sets.

2.1.9 Definition : An **open ball** is a set $B_r(x)$ such that

$B_r(x) = \{y : x, y \in X,\ d(x, y) < r,\ r \in R,\ r > 0\}$,

where d is the Euclidean metric. $B_r(x) \subset X$.

2.1.10 Definition : The interior of a set, $\overset{0}{A}$, is the union of open balls contained in A,

$$\overset{0}{A} = \bigcup \left\{ x \in X : B_{\delta(x)}(x) \subset A, \text{ for some } \delta(x) > 0 \right\}.$$

2.1.11 Definition : An **open set,** U, is a set such that $\overset{0}{U} = U$.

Given the definition of an open set a more general concept is the definition of a neighborhood.

2.1.12 Definition : A **neighborhood,** A, of $E \subset X$, is a set such that $E \subset \overset{0}{A}$.

With open sets it is now possible to define compact sets, which will complete the introduction of primary topological concepts.

2.1.13 Definition (Heine – Borel) : $K \subset X$ is **compact** implies that given $K \subset \bigcup_{\alpha \in A} U_\alpha$, where $\{U_\alpha\}_{\alpha \in A}$ is an open collection, $\exists\, D \subset A,\ |D| < \aleph$, such that $K \subset \bigcup_{\alpha \in D} U_\alpha$.

2.1.14 Proposition : If $K \subset X$ is **compact** then K is bounded, i.e. $K \subset B_r(x)$, for some $r \geq 0,\ x \in X$.

2.1.15 Proposition : If $E \subset R^n$ is bounded and closed then E is compact.

In topological analysis it is convenient to have open sets with an additional bounding property shared by compact sets in vector spaces.

2.1.16 Definition : If $\overline{E}$ is compact and $E \subset X$ is open then E is **pre – compact** hence, $B_{r'}(x) \subset E \subset B_r(x)$, for some $r, r' > 0,\ x \in X$, such that $r' \leq \mathbf{R}(E) \leq r$.

Since the analysis of this paper is restricted to countable bases of X, be it R^3 or $\mathbf{T}^3$, the topological properties generated by the translates and scales of a pre–compact set E are established.

2.1.17 Proposition : A **simple base** of a topological space, $(X, \mathcal{T})$, is a base $\mathcal{E} \subset \mathcal{T}$ such that, $\exists\, E \subset X$, E is a pre – compact **structuring element,** $\mathcal{E}_E = \{E_\alpha(\beta) : \alpha \in \Omega,\ \beta \in A\}$, $E_\alpha(\beta) = \alpha E + \beta$, Ω is a countable subset of R, A is a countable subset of X.

Proof. assume $0 \in E$ otherwise choose $x \in E$ and consider $E' = E - x$ which contains $\{0\}$, also assume $R(E) = 1$ otherwise set $E' = \frac{1}{R(E)} E$.
Given $U \subset \mathcal{T}$, $\forall\, x \in U\ \exists\, B_r(x) \subset U$ and for some $r > 0$, $E_{\alpha(x)}(\beta(x)) \in \mathcal{E}$ such that $E_{\alpha(x)}(\beta(x)) \subset B_r(x)$, i.e. $d(x - \beta(x)) \leq \frac{r}{2}$ and $\alpha(x) \leq \frac{r}{2}$,
then $U = \bigcup_{x \in U} E_{\alpha(x)}(\beta(x))$, hence $\mathcal{E}_E$ is a base of $\mathcal{T}$.

The condition for a generating set E of a simple base $\mathcal{E}_E$ to be pre–compact can further be relaxed to include any set assuming the second condition of the Baire category theorem is satisfied, hence this theorem is stated below. First the definition of a dense set is required.

2.1.18 Definition : If $A \subset X$, then A is **dense** relative to X if $\overline{A} = X$.

2.1.19 Baire Category Theorem : Let X be a complete metric space.

i) If $\{U_n\}_{n \in N}$ is a countable collection of open dense subsets in X then $\bigcap_{n \in N} U_n$ is dense in X.

ii) if $X = \bigcup_{n \in N} E_n$ then for some $k \in N$, $\overline{E}_k$ is a neighborhood of some point in X. $\overline{E}_k$ is the closure of E_k .

The generating sets of the simple base of a topology are commonly referred to as structuring elements in mathematical morphology. To reduce complexity of analysis these structuring elements are further restricted to be convex as well as pre–compact.

2.1.20 Definition : $E \subset X$ is **convex** then $\forall\, x,\ y \in E$,

$$\lambda x + (1 - \lambda)y \in E,\ \forall\, \lambda \in [0,1].$$

Given a pre–compact convex set E several properties useful for further developments in functional analysis will be characterized. The following mathematical formalism for convex function spaces has been extended from that established by Conway [3].

2.1.21 Proposition : given a pre – compact and convex set E $\exists$ a unique point $c \in \overline{E}$ from which the radius of $\overline{E}$ can be defined, $E \subset B_{\mathbf{R}(E)}(q)$.
Denote this point c as the **center of Radius** of E.

Proof. consider $\mathbf{R}_x(E) = \sup\{d(x - y) : y \in E\}$,
since $\mathbf{R}(E) = \inf\{\mathbf{R}_x(E) : x \in E\}\ \exists\ x \in \overline{E}$
such that $\mathbf{R}_x(E) = \mathbf{R}(E)$.
Suppose $\mathbf{R}_p(E) = \mathbf{R}_q(E) = \mathbf{R}(E)$ with $q \neq t$, $q, t \in \overline{E}$.
Hence $\forall\, x \in E$, $d_t(x)$, $d_t(x) < \mathbf{R}(E)$, where $d_t(x) = d(t - x)$.
This implies that $E \subset B_{\mathbf{R}}(q) \cap B_{\mathbf{R}}(t)$,
But $E \subset B_{\mathbf{R}}(q) \cap B_{\mathbf{R}}(t) \subset B_{\mathbf{R}'}(\overline{x})$, $\overline{x} = \frac{t + q}{2}$.
$(R')^2 = R^2 - \left(\frac{t - q}{2}\right)^2 < R^2$,
$\overline{x} \in \overline{E}$ since E is convex and $\mathbf{R}(E) \leq \mathbf{R}_{\overline{x}}(E) < \mathbf{R}(E)$
with $q \neq t$, a contradiction. Therefore $q = t = c$.

Furthermore if E is has a sufficient number of axes of symmetry the unique center of radius is a member of the pre–compact structuring element E.

2.1.22 Proposition : if a pre – compact and convex set E has hyperplanes of symmetry $\{P_i\}_{i=1}^{N}$ with a unique point of intersection, c, this point is both in E and the center of radius of E, $\mathbf{R}_c(E) = \mathbf{R}(E)$.

Proof. given $P \in \{P_i\}_{i=1}^{N}$, By symmetry property if $x \in E$ then $\exists\ x' \in E$ such that x' is the reflection of x across P, i.e. $\bar{x} = \frac{(x + x')}{2} \in P$,

$\bar{x} = \varrho_P(x) = \inf\{d(x - y) : y \in P\}$.

Hence if c is center of radius of E, $\mathbf{R}(E) = \mathbf{R}_c(E)$, then the reflection of c across P, c' is also a center of radius of E $\mathbf{R}_{c'}(E) = \mathbf{R}_c(E)$, by the previous proposition 2.1.21 c is unique and $c = c'$, by symmetry this implies $c \in P$. But $P \in \{P_i\}_{i=1}^{N}$ is arbitrary hence $c = \bigcap_{i=1}^{N} P_i$ a unique point.

Now choose $x \in E$ with $x \neq c$ this implies that $x \notin P \in \{P_i\}_{i=1}^{N}$, from above x', and $\varrho_P(x) \in E$.

By induction $c = \left(\prod_{i=1}^{N} \varrho_{P_i}\right)(x) \in E$.

The structuring element or generating set, E, is selected to resemble the compact neighborhood form of the data structures of interest. For fractal data this is particularly convenient since the same primitive form and structure may be repeated over a broad range of scales. Hence the selected structuring element forms will locally resemble the features of interest in compact neighborhoods. Therefore, a compact 3–D sphere and 1–D arbitrarily oriented line segment are selected for biomedical imagery as structuring elements to represent the neighborhood of brain surface structures and vascular curvilinear structures respectively. The data analysis of complex image features with compact and convex structuring elements is easily implemented on parallel processing digital computers.

To demonstrate the shape of the approximate spherical structuring element, note that the structuring element in Figure 1 is eight sided and symmetrical in two sets of four directions. Note that the lines of symmetry of this structuring element require the center of radius to be within the structuring element, proposition 2.1.22. This shape possesses a quality of directional isotropy which is ideal for size and convexity analysis of 3–D image features however it is likewise not disposed to provide an analysis of the orientation of image features. Further analysis of the curvilinear features of 1–D forms, such as the arteries and veins of 3–D MRA acquisitions, requires oriented structuring elements resembling 1–D line segments such as elongated ellipses. Symmetry about a central point is common property of all structuring elements used in this paper and thus the coordinate system of the structuring element originates from this central point, or center of radius.

Since images are functions mapped over a complete vector space, a Banach space, X, [3,7] the development the above formalisms of set analysis must be extended to the representation of functions mapped over X. In discrete topologies with a countable base however, a sphere and other continuous space sets cannot exist in a digital representation. Hence the structuring element applied in this paper digitally approximates a sphere

A convex function, **g**, will be introduced in the next section, mapped over the structuring element domain to extend the topological basis to a functional basis. If **g** is non–constant over the basis support then the structuring element boundaries are fuzzified, permitting more precise approximation of a sphere and greater flexibility with generalization in data analysis.

Several generalizations of the set basis definitions are now presented for positive function mappings over the basis set, primarily, the structuring element diameter and radius will be that of the support.

2.1.23 definition : the **diameter** of a positive mapping $\mathbf{g}$ is $\mathbf{d(g)} = \sup\ \{d(x - y) : \mathbf{g}(x),\ \mathbf{g}(y) > 0\}$. Assume $\mathbf{d(g)} = 2$ for $\mathbf{g}$ unless otherwise stated.

2.1.24 definition : the **radius** of a positive mapping $\mathbf{g}$ is, $\mathbf{R(g)} = \inf\{\sup\ \{d(x - y) : \mathbf{g}(y) > 0\} : \mathbf{g}(x) > 0\}$. Assume $\mathbf{R(g)} = 1$ for $\mathbf{g}$ unless otherwise stated.

2.2 Fundamental Algebraic Operations

There are two primary neighborhood operations of mathematical morphology, which are related by set complementation and are critical for the analysis and development of the morphological representation. These operations are dilation and erosion. Conveniently the homogeneous cascade of these neighborhood operations increases the scale of the structuring element without precision loss. To perform these critical operations of dilation and erosion the structuring element must be translated over the image domain. To formulate these operations a definition of the functions and the bases which support the analysis of these functions must be derived. The approach of this development is to extend the developments of the last section to function bases which provide analysis of function or image mappings in higher dimensional vector spaces X. Convex fuzzy mappings are defined on the operators. Dilation is defined with properties proved and then its dual operation of erosion is introduced.

2.2.1 Definition : $\mathbf{g}$ is a convex fuzzy mapping over $\overline{E}$, $E \in \mathcal{E}_E$, where E is a compact convex generating set of $\mathcal{E}_E$.

$$\mathfrak{B}_{\mathbf{g}} = \{\mathbf{g}_A + \beta : \beta \in \Omega,\ A \in \mathcal{E}_E\}$$

For ease of notation $\alpha\mathbf{g} = \mathbf{g}_{\alpha E}$,

i.e. $\alpha\mathbf{g}\left(\frac{1}{\alpha}x\right) = \mathbf{g}_{\alpha E}(x) \quad \forall\ x \in E.$

hence, $\alpha\mathbf{g}(\alpha 0) = \mathbf{g}_{\alpha E}(0) = \mathbf{g}_E(0)\ \forall\ \varepsilon > 0.$

With the basis, $\mathfrak{B}_{\mathbf{g}}$, the objective is to approximate functions to arbitrarily precision in the space of L^∞ functions with Lebesque measure over the topological vector space $(X,\mathcal{T})$.

2.2.2 If $f : X \to K$ where $K = \mathbb{R},\ \mathbb{Z}$,

Consider $C^a f = \{x : x \in X,\ \|f\| > a\}$, furthermore

since f is a function if $a > b$, $C^a f \subset C^b f$,

then $\|f\|_\infty = \inf\{a \geq 0 : m(C^a f) = 0\}$,

alternately,

$$\|f\|_\infty = \sup\{a \geq 0 : m(C^a f) > 0\},$$

m is the Lebesque measure on $(X, \mathcal{A})$, if $f \in L^\infty$ then $\|f\|_\infty < \infty$.

note : to simplify notation assume $\|f\| = \|f\|_\infty$,

A more general class of functions can be admitted if these functions are considered integrable over pre–compact neighborhoods.

2.2.3 definition : $f \in L^\infty_{loc}$ implies $\forall$ precompact $U \subset X$,

$$\|f\|_U = \inf\{a \geq 0 : m(C^a_{U'}f) = 0,\ U \subset U' \in \mathcal{T}\},$$

where $C^a_{U'}f = U' \cap C^a f$.

Without loss of generality f can be restricted to a positive function over X. It can now be shown that the set base, $\mathcal{E}_E$, can approximate a function, f, arbitrarily closely in the L^∞ norm. The results will then be shown to extend to using the function base, $\mathfrak{B}_g$.

2.2.4 given $F \subset X$, the **measure** of F is :

$$m(F) = \inf\{m(U) : F \subset U \in \mathcal{T}\} .$$

Using the L^∞ norm there exists a mapping of $L^\infty{}_{loc}$ into the set of upper semi–continuous functions, $\mathcal{U}$, with function bases, $\mathfrak{B}_g$, supported by the the pre–compact topological bases, $\mathcal{E}_E$, of section 2.1.

2.2.5 definition : If $f \in \mathcal{U}$, $f : X \to K$, then
$f^{-1}((-\infty, a))$ is open in X $\forall\, a \in K$.

To facilitate further development a new function will be defined.

2.2.6 definition : $f^x_{\varepsilon g} = \left(f + \hat{g}^x_\varepsilon\right) \chi_{\text{supp}\,\hat{g}^x_\varepsilon}$, where χ_B is the characteristic function :

$$\chi_B(x) = \begin{cases} 1 & , \text{ if } x \in B \\ 0 & , \text{ if } x \notin B \end{cases} , \qquad \text{for } B \subset X .$$

The most fundamental operation for the image analysis developed in this paper is now introduced, the dilation mapping relative to a structuring element E and fuzzy map g.

2.2.7 definition : A **dilation** mapping, Γ, performs the following operation :

$$\Gamma^g_\varepsilon f(x) = \|f^x_{\varepsilon g}\| - \varepsilon, \quad \sup_{y \in \hat{E}_\varepsilon(x)} \hat{g}^x_\varepsilon(y) = \hat{g}^x_\varepsilon(0) = \varepsilon. \qquad (1)$$

$\hat{E}_\varepsilon(x)$ is the reflection of $E_\varepsilon(x)$ about x.
Likewise $\hat{g}^x_\varepsilon$ is the reflected mapping of $\hat{g}^x_\varepsilon$ about x with $\hat{E}_\varepsilon(x)$ the support of $\hat{g}^x_\varepsilon$.

note : the analysis of this paper will use symmetric $\hat{g}^x_\varepsilon$ such that, $\hat{g}^x_\varepsilon = g^x_\varepsilon$.

Denote Γ^g_ε as Γ_ε if the structuring element **g** is understood.

The introduction of the function mapping of dilation permits the proof of several interesting results for image analysis : extendibility, continuity and order preservation of $\mathcal{U}$.

2.2.8 proposition :

i) $\Gamma_\varepsilon f \geq \Gamma_{\varepsilon'} f$ if $\varepsilon \geq \varepsilon'$.

ii) $\Gamma_\varepsilon f$ (x) is continuous with respect to ε,
$\forall\ \varepsilon > 0,\ x \in X.$

proof :

(i) $\hat{g}^x_\varepsilon(y) \geq \hat{g}^x_{\varepsilon'}(y)\ \forall\ y \in X,$ if $\varepsilon \geq \varepsilon'$,

since **g** is convex,

hence, $f^x_{\varepsilon g} \geq f^x_{\varepsilon' g}$, for $f^x_{\varepsilon g} = \left(f + \hat{g}^x_\varepsilon\right) \chi_{\text{supp}\ \hat{g}^x_\varepsilon}$,

$$\Gamma_\varepsilon f(x) = \|f^x_{\varepsilon g}\| - \varepsilon\ , \quad \text{where } \hat{g}^x_\varepsilon(0) = \varepsilon\ ,$$

$$\geq \sup\left\{\|f^x_{\varepsilon g}\|_{\hat{A}_\varepsilon(x)},\ \|f^x_{\varepsilon g}\|_{\hat{A}_{\varepsilon'}(x)}\right\} - \varepsilon\ ,$$

$$A_\varepsilon(x) = E_\varepsilon(x) \setminus E_{\varepsilon'}(x),\ \varepsilon \geq \varepsilon'.$$

$$\geq \|f^x_{\varepsilon g}\|_{\hat{E}_{\varepsilon'}(x)} - \varepsilon$$

$$\geq \|f^x_{\varepsilon' g}\|_{\hat{E}_{\varepsilon'}(x)} - \varepsilon = \|f^x_{\varepsilon' g}\| - \varepsilon = \Gamma_{\varepsilon'} f(x)$$

Therefore, $\Gamma_\varepsilon f \geq \Gamma_{\varepsilon'} f$ since $\Gamma_\varepsilon f(x) \geq \Gamma_{\varepsilon'} f(x)\ \forall\ x \in X$.

(ii) $\Gamma_\varepsilon f$ is continuous from bellow :

$$\sup\{\Gamma_b f(x) + b : b < \varepsilon\}$$

$$= \sup\left\{\sup\left\{a \geq 0 : m\left(C^a f^x_{bg}\right) > 0\right\}\ ,\ b < \varepsilon\right\}$$

$$= \sup\left\{a \geq 0 : m\left(C^a f^x_{\varepsilon g}\right) > 0\right\}\ , \quad \text{since } f^x_{\varepsilon g} \geq f^x_{bg} \quad \text{for } \varepsilon > b\ ,$$

$$= \Gamma_\varepsilon f(x) + \varepsilon\ .$$

or ,

$$\Gamma_\varepsilon f(x) = \sup\{\Gamma_b f(x) : b < \varepsilon\}\ .$$

$\Gamma_\varepsilon f$ is continuous from above :

$$\inf\{\Gamma_b f(x) + b : b > \varepsilon\}$$

$$= \inf\left\{\inf\left\{a \geq 0 : m\left(F_a\left(C^a f^x_{bg}\right)\right) = 0\right\}\ ,\ b > \varepsilon\right\}$$

$$= \inf\left\{a \geq 0 : m\left(C^a_{\overline{E_b(x)}} f^x_{\varepsilon g}\right) = 0\right\}\ ,$$

$$\text{i.e. } \overline{E_\varepsilon(x)} = \bigcap_{b<\varepsilon} E_b(x)\ ,\ E_b(x) = \text{supp}\left\{f^x_{bg}\right\}\ ,$$

$$= \inf\left\{a \geq 0 : m\left(C^a_{\overline{E_b(x)}} f^x_{\varepsilon g}\right) = 0\right\}$$

$$= \|f(x)\|_{\overline{E_\varepsilon(x)}} = \sup\left\{\|f(x)\|_{E_\varepsilon(x)}\ ,\ \|f(x)\|_{\partial E_\varepsilon(x)}\right\}.$$

$$= \|f(x)\|_{E_\varepsilon(x)} = \Gamma_\varepsilon\ f(x) + \varepsilon\ .$$

or ,

$$\Gamma_\varepsilon\ f(x) = \inf\{\Gamma_b f(x) : b > \varepsilon\}\ .$$

By definition the basis functions of $\mathfrak{B}_g$ are bounded convex mappings over a pre–compact convex support. With definitions of a norm it is now possible to develop a norm with respect to the convex neighborhood properties of the support of **g**, i.e. E = sup **g** is convex.

2.2.9 definition : A set E is balanced if $\beta x \in E$, whenever $x \in E$, and $\beta \in [-1, 1]$.

A balanced set implies symmetry of the set about any hyperplane through the origin. Hence a pre–compact and convex structuring element set which is balanced contains its center of radius, which is as well a center of symmetry, as specified by proposition 2.1.22.

2.2.10 proposition : Given a pre – compact, convex and balanced set $E \subset X$, $\{0\} \in E$, then

$$p_E(x) = \inf\{\lambda : x \in \lambda E\} \text{ , is a norm on X.}$$

proof. Given a pre – compact and convex set, $E \subset X$, the three properties of the norm are to be demonstrated.

i) $p_E(x) = \inf\{\lambda > 0 : x \in \lambda E\} \geq 0 \ \forall \ x \in X$ since $\exists \ \delta > 0,\ 0 \in B_\delta(0) \subset E$, and $\dfrac{\delta}{d(x) + \varepsilon} x \in B_\delta(0)$,
or $x \in \dfrac{d(x) + \varepsilon}{\delta} E$ for any $\varepsilon > 0$ and $\forall \ x \in X$.
therefore $0 \leq p_E(x) \leq \dfrac{d(x)}{\delta} \ \forall \ x \in X$ and $p_E(x) = 0$ iff $x = 0$.

ii) $p_E(\beta x) = \inf\{\lambda > 0 : \beta x \in \lambda E\}$

$$= |\beta| \inf\left\{\frac{\lambda}{|\beta|} > 0 : x \in \frac{\lambda}{|\beta|} E\right\} = |\beta| p_E(x) \qquad \forall \ \beta \in K - \{0\}.$$

If $\beta = 0$, $\beta x = 0$ and $p_E(\beta x) = 0$ by (i).
n.b. that if $\beta \geq 0$, the balanced set condition can be relaxed.

iii) Given $x, y \in X$, $\varepsilon > 0$,

$$x' = \frac{x}{p_E(x) + \varepsilon} \text{ , } y' = \frac{y}{p_E(y) + \varepsilon} \in E \text{ since } p_E(x'),\ p_E(y') < 1.$$

Using the convexity of E with $\eta \in [0, 1]$,
then $\eta x' + (1 - \eta) y' \in E$, and $p_E(\eta x' + (1 - \eta) y') < 1$.
Now suppose

$$\eta = \frac{p_E(x) + \varepsilon}{p_E(x) + p_E(y) + 2\varepsilon} \text{ then by (ii)}$$

$p_E(x + y) < p_E(x) + p_E(y) + 2\varepsilon$, but ε is arbitrary,
hence $p_E(x + y) < p_E(x) + p_E(y) \ \forall \ x, y \in X$.

This norm in X, proposition 2.2.10, can now be extended to convex fuzzy mappings over these pre–compact, convex and balanced sets.

2.2.11 Theorem : Given a decreasing convex mapping, h, over $p_E(x)$, $h : [0, 1] \rightarrow [0, 1]$, $h(0) = 1$, the composition $g = h \bigcirc p_E$ is a convex mapping over $\overline{E}$, with the property ,

$$\varepsilon g\left(\frac{x}{\varepsilon}\right) + \eta g\left(\frac{y}{\eta}\right) \leq (\varepsilon + \eta) g\left(\frac{x + y}{\varepsilon + \eta}\right), \quad \forall \ \varepsilon,\ \eta > 0, \quad x \in \varepsilon E,\ y \in \eta E.$$

proof. Given a pre − compact and convex set $E \subset X$,

$$\varepsilon g\left(\frac{x}{\varepsilon}\right) + \eta g\left(\frac{y}{\eta}\right) = (\varepsilon + \eta)\left(\varepsilon' h\left(\frac{p_E(x')}{\varepsilon'}\right) + \eta' h\left(\frac{p_E(y')}{\eta'}\right)\right),$$

where $\varepsilon' = \dfrac{\varepsilon}{\varepsilon + \eta}$, $\eta' = 1 - \varepsilon'$, $x' = \dfrac{x}{\varepsilon + \eta}$, $y' = \dfrac{y}{\varepsilon + \eta}$,

now $\varepsilon' h\left(\dfrac{p_E(x')}{\varepsilon'}\right) + \eta' h\left(\dfrac{p_E(y')}{\eta'}\right) \leq h(p_E(x') + p_E(y'))$,

since h is convex.
By prop. 2.2.11 $p_E(x') + p_E(y') \geq p_E(x' + y')$,
combined with the decreasing property of h ,

$h(p_E(x') + p_E(y')) \leq h(p_E(x' + y')) = g\left(\dfrac{x + y}{\varepsilon + \eta}\right)$.

hence, $\varepsilon g\left(\dfrac{x}{\varepsilon}\right) + \eta g\left(\dfrac{y}{\eta}\right) \leq (\varepsilon + \eta) g\left(\dfrac{x + y}{\varepsilon + \eta}\right)$.

This property of g implies g is convex.
Since $p_E : \overline{E} \to [0, 1]$ and $h : [0, 1] \to [0, 1]$,
then $g = h \circ p_E : \overline{E} \to [0, 1]$.

2.2.12 corollary : Given **g** a convex mapping over E
$\Gamma_{\varepsilon'}\, \mathbf{g}_\varepsilon + \varepsilon' = \Gamma_\varepsilon\, \mathbf{g}_{\varepsilon'} + \varepsilon = \mathbf{g}_{\varepsilon+\varepsilon'}$.

proof. Since **g** is convex **g** is continuous, hence

$$\Gamma_{\varepsilon'}\, \mathbf{g}_\varepsilon(x) + \varepsilon' = \sup\left\{\varepsilon' g\left(\frac{y}{\varepsilon'}\right) + \varepsilon g\left(\frac{x - y}{\varepsilon}\right) : y \in E_{\varepsilon'}\right\}$$

$$= (\varepsilon' + \varepsilon) g\left(\frac{x}{\varepsilon + \varepsilon'}\right) = \mathbf{g}_{\varepsilon+\varepsilon'}(x), \text{ by theorem.}$$

Therefore $\Gamma_{\varepsilon'}\, \mathbf{g}_\varepsilon(x) + \varepsilon' = \mathbf{g}_{\varepsilon+\varepsilon'}(x)$.

2.2.13 Lemma : Given $\varepsilon' - \varepsilon = \delta > 0$ and $f \in L^\infty_{loc}$,
then $\Gamma_\varepsilon f(y) \leq \Gamma_\varepsilon f(y) + \hat{\mathbf{g}}^x_\delta(y) \leq \Gamma_{\varepsilon'} f(x) + \delta$, $\forall\, y \in E_\delta(x)$.

proof. If $f \in L^\infty_{loc}$ and $\varepsilon' > \varepsilon$ then $\Gamma_\varepsilon f(x) \leq \Gamma_{\varepsilon'} f(x)$,
by (i) prop. 2.2.8 .
Set $\varepsilon' - \varepsilon = \delta > 0$ and choose any $y \in E_\delta(x)$
then $\hat{\mathbf{g}}^y_\varepsilon + \hat{\mathbf{g}}^x_\delta(y) \leq \hat{\mathbf{g}}^x_{\varepsilon'}$ by prop. 2.2.11, hence $f^y_{\varepsilon g} + \hat{\mathbf{g}}^x_\delta(y) \leq f^x_{\varepsilon' g}$,
which implies $\|f^y_{\varepsilon g}\| + \hat{\mathbf{g}}^x_\delta(y) \leq \|f^x_{\varepsilon' g}\|$ and
$\Gamma_\varepsilon f(y) + \hat{\mathbf{g}}^x_\delta(y) - \delta \leq \Gamma_{\varepsilon'} f(x)$.

2.2.14 Theorem : $\Gamma_\varepsilon : L^\infty_{loc} \to \mathcal{U} \subset L^\infty_{loc}\ \forall\, \varepsilon > 0$.

proof. If $\Gamma_\varepsilon f \in \mathcal{U}$ for $f \in L^\infty_{loc}$ then
$A^a\, \Gamma_\varepsilon f = \{x \in X : \Gamma_\varepsilon f(x) < a\}$ is open in X.
Choose $x \in A^a\, \Gamma_\varepsilon f$ then $\Gamma_\varepsilon f(x) < a$:
Choose $\varepsilon' > \varepsilon$ then $\Gamma_\varepsilon f(y) \leq \Gamma_{\varepsilon'} f(x) + \delta$,

for any $y \in E_\delta(x)$ and $\varepsilon' - \varepsilon = \delta > 0$, by (i) lemma 2.2.13 .
Choose δ small enough such that $\Gamma_{\varepsilon'} f(x) + \delta < a$,
and hence $\Gamma_\varepsilon f(y) < a$, $\forall\, y \in E_\delta(x) \subset A^a\, \Gamma_\varepsilon f$.
Since $x \in A^a\, \Gamma_\varepsilon f$ was chosen arbitrarily this
implies that $A^a\, \Gamma_\varepsilon f$ is open and $\Gamma_\varepsilon f \in \mathcal{U}$.
Furthermore since $f \in L^\infty_{loc}$ is arbitrary it is shown
$\Gamma_\varepsilon : L^\infty_{loc} \to \mathcal{U} \subset L^\infty_{loc}\ \forall\, \varepsilon > 0$.

The now formally introduced image transformation of dilation not only projects the function space $L\infty$ into the upper semi–continuous functions with restrictions on the neighborhood mapping properties of these functions.

2.2.15 Definition : If $f \in \tilde{\mathcal{U}}$ (restricted $\mathcal{U}$) then $\lim_{\varepsilon \to 0} \Gamma_\varepsilon f = f$.

2.2.16 Lemma : Given $f \in \tilde{\mathcal{U}}$, $\varepsilon > 0$, then $\forall\, x \in X$,

$$\Gamma_\varepsilon f(x) = \sup\{f^x_{\varepsilon g}(y) \;:\; y \in E_\varepsilon(x)\} - \varepsilon \;.$$

proof. Suppose $f \in \tilde{\mathcal{U}}$ then by prop. 2.2.13 $\forall\, x \in X$,

$$\Gamma_\varepsilon f(x) + \delta \geq \sup\{\, \Gamma_{\varepsilon-\delta} f(y) :\; y \in E_\delta(x)\} \;,$$

hence $\Gamma_\varepsilon f(x) + \varepsilon \geq \sup\{\, f(y) :\; y \in E_\varepsilon(x)\}$.
But since $f^x_{\varepsilon g}(y) - \varepsilon \leq f(y)$,

$$\Gamma_\varepsilon f(x) \geq \sup\{\, f^x_{\varepsilon g}(y) - \varepsilon \;:\; y \in E_\varepsilon(x)\}$$
$$\geq \| f^x_{\varepsilon g}\| - \varepsilon = \Gamma_\varepsilon f(x) \;,$$

hence $\Gamma_\varepsilon f(x) = \sup\{\, f^x_{\varepsilon g}(y) - \varepsilon :\; y \in E_\varepsilon(x)\}$.

2.2.17 Lemma : Given $f \in L^\infty_{loc}$, $\varepsilon > 0$, then

$\Gamma_\varepsilon f = \Gamma_\varepsilon \hat{f}$, where $\hat{f} = \lim_{\varepsilon \to 0} \Gamma_\varepsilon f \in \tilde{\mathcal{U}}$.

proof. (a) If $f \in L^\infty_{loc}$ then by definition 2.2.7,
$\Gamma_\varepsilon f \geq f + \hat{g}^x_\varepsilon - \varepsilon$ on $E_\varepsilon(x)$ a.e.
where a.e. means almost everywhere except on a subset of $E_\varepsilon(x)$ of measure (definition 2.2.4) zero . Hence

$$\hat{f}(x) = \lim_{\varepsilon \to 0} \Gamma_\varepsilon f \geq \lim_{\varepsilon \to 0} \{f(y) + \hat{g}^x_\varepsilon(y) - \varepsilon,\ y \in E_\varepsilon(x)\} \text{ a.e. },$$

therefore $\hat{f} \geq f$ a.e. and $\Gamma_\alpha \hat{f} \geq \Gamma_\alpha f$ for all $\alpha > 0$.

(b) Now suppose $x \in X$ such that $\Gamma_\alpha \hat{f}(x) > \Gamma_\alpha f(x)$,
Then by lemma 2.2.16 , $\exists\, y \in E_\varepsilon(x)$ such that

$$\Gamma_\alpha \hat{f}(x) \geq \hat{f}^x_{\varepsilon g}(y) - \varepsilon = \hat{f}(y) + g^x_\varepsilon(y) - \varepsilon > \Gamma_\alpha f(x) \;,$$

but by lemma 2.2.13 $\hat{f}(y) + g^x_\varepsilon(y) \leq \Gamma_\alpha f(x) + \varepsilon$,
a contradiction,
Therefore (a) combined with (b) implies $\Gamma_\alpha \hat{f}(x) = \Gamma_\alpha f(x)$.

2.2.18 proposition : given the dilation mappings Γ_ε, $\Gamma_{\varepsilon'}$, constructed with the convex structuring element **g** , the composition map has the property :

$$\Gamma_{\varepsilon'}\Gamma_\varepsilon f = \Gamma_{\varepsilon'+\varepsilon} f \quad \forall f \in L^\infty_{loc} . \tag{2}$$

proof. Suppose $f \in L^\infty_{loc}$ then by lemma 2,2,17, $\forall$ x $\in$ X ,
$\Gamma_\varepsilon f(y) - \varepsilon = \sup\{ f^y_{\varepsilon g}(t) : \ t \in E_\varepsilon(t)\}$, $y \in E_{\varepsilon'}(x)$.
$\Gamma_{\varepsilon'}\Gamma_\varepsilon f(x) + \varepsilon' + \varepsilon$

$$= \sup\Big\{\sup\{ f^y_{\varepsilon g}(t) : t \in E_\varepsilon(y)\} + \hat{g}^x_{\varepsilon'}(y) : y \in E_{\varepsilon'}(x)\Big\}$$
$$= \sup\Big\{ f(t) + \sup\{\hat{g}^y_\varepsilon(t) + \hat{g}^x_{\varepsilon'}(y) : y \in E_{\varepsilon'}(x)\} \ : t \in E_\varepsilon(y)\Big\}$$
$$= \sup\{f(t) + \hat{g}^x_{\varepsilon'+\varepsilon}(t) : t \in E_\varepsilon(y)\}$$
$$= \Gamma_{\varepsilon'+\varepsilon} f(x) + (\varepsilon' + \varepsilon) .$$

Therefore $\Gamma_{\varepsilon'}\Gamma_\varepsilon f(x) = \Gamma_{\varepsilon'+\varepsilon} f(x) \ \forall \ x \in X$.

With the now standing dilation operator properties on $L^\infty{}_{loc}$ and the restricted upper semi–continuous functions of $\mathcal{U}$ a new set of functions can be derived. This class functions is called the restricted lower semi–continuous functions of $\mathcal{L}$. The generation of this dual function class is now introduced by the function transform operation of erosion, defined from the operation of dilation.

2.2.19 Definition : If $f \geq 0$, $f \in L^\infty_{loc}$ then $-f \in L^\infty_{loc}$, and the operation of **erosion** is : $\dot{\Gamma}_\varepsilon f(x) = -\Gamma_\varepsilon(-f)(x)$. (3)

The operation of erosion is dual to that of dilation and introduces many additional possibilities for morphological data analysis. Convex fuzzy mappings also pertain to erosion.

2.2.20 Proposition : If $f \geq 0$, $f \in L^\infty_{loc}$ then $\dot{\Gamma}_\varepsilon f \in \mathcal{L} \ \forall \ \varepsilon > 0$.

proof. Suppose $f \in L^\infty_{loc}$, then $-f \in L^\infty_{loc}$ and
$\Gamma_\varepsilon(-f) \in \mathcal{U} \ \forall \ \varepsilon > 0$.
Therefore, $A^a\Gamma_\varepsilon(-f)$ is open and $C^a\Gamma_\varepsilon(-f)$ is closed . Reversing polarity again then shows
$\{x \in X : -\Gamma_\varepsilon(-f)(x) > -a\}$ is open and $\dot{\Gamma}_\varepsilon f \in \mathcal{L}$.

Since dilation and erosion are duals, each property of dilation also has a dual property with erosion, which can be proved with this duality principle, hence proofs will be omitted.

2.2.21 Definition : If $f \in \tilde{\mathcal{L}}$, $f \geq 0$, $\lim_{\varepsilon \to 0} \dot{\Gamma}_\varepsilon f = f \in \tilde{\mathcal{L}}$.

2.2.22 proposition :

i) $\dot{\Gamma}_\varepsilon f \leq \dot{\Gamma}_{\varepsilon'} f$ if $\varepsilon \geq \varepsilon'$.

ii) $\dot{\Gamma}_\varepsilon f\,(x)$ is continuous with respect to ε, $\forall \ \varepsilon > 0$, $x \in X$.

2.2.23 proposition : given the erosion mappings $\dot{\Gamma}_\varepsilon$, $\dot{\Gamma}_{\varepsilon'}$, constructed with the convex structuring element **g**, the composition map has the following property similar to proposition 2.2.18 :

$$\dot{\Gamma}_{\varepsilon'}\,\dot{\Gamma}_{\varepsilon}\,f = \dot{\Gamma}_{\varepsilon'+\varepsilon}\,f \quad \forall f \in L^{\infty}_{\mathrm{loc}}\,. \tag{4}$$

If the finite values of X and g are binary valued these definitions of erosion and dilation correspond to their binary counterparts with no fuzzification.

The structuring elements of this paper are limited to arbitrary transpose invariant convex shapes which are selected to measure the size or orientations of image features as erosion and dilations are successively cascaded. An interesting property of dilation and erosion is that a homogeneous succession of these respective operations is equivalent to one operation with the composite dilation of the entire group of structuring elements, eq. (2), (4). Convex fuzzy mapping is a further extension of the operators. Thus the minimum diameter of a convex object is derived by a succession of erosions with a spherical structuring element until the object representation is deleted. Likewise the maximum diameter of a convex object may be ascertained with a succession of dilations with the same spherical structuring element.

With the operations of dilation and erosion the less destructive operations of opening and closing may be performed. These operations of closing and opening conform, respectively, to the definitions of maximum enclosure and minimum cover of a data mapping for unions of the structuring element as it is translated. Opening is of particular importance in the development of the morphological skeleton, a minimal image representation, which is defined from deviations of the image from the shape of the structuring element across a spectrum of scales.

2.2.24 Definition : opening , (5)

$$f_{\varepsilon g} = \Gamma_\varepsilon \dot{\Gamma}_\varepsilon f\,,\ \ \mathbf{g} \text{ is precompact} - \text{convex},\ \varepsilon > 0.$$

properties : anti–extensive $f_{\varepsilon g} \subset f$,

idempotent $(f_{\varepsilon g})_{\varepsilon g} = f_{\varepsilon g}$,

increasing $f \subset h \Rightarrow f_{\varepsilon g} \subset h_{\varepsilon g}$.

2.2.25 Definition : closing , (6)

$$f^{\varepsilon g} = \dot{\Gamma}_\varepsilon \Gamma_\varepsilon f\,,\ \ \mathbf{g} \text{ is precompact} - \text{convex},\ \varepsilon > 0.$$

properties : extensive $f^{\varepsilon g} \supset f$,

idempotent $(f^{\varepsilon g})^{\varepsilon g} = f^{\varepsilon g}$,

increasing $f \subset h \Rightarrow f^{\varepsilon g} \subset h^{\varepsilon g}$.

Implementation of morphological opening operations for diametric analysis has clearly provided a segmentation tool for the unobscured visualization of intracranial anatomy and internal organs from X–ray CT and magnetic resonance imagery [13, 14, 15]. In particular neurological surface anatomy is segmented from 3–D acquisitions due to properties of size and convexity across a spectrum of scales relative to the concave 2–D surface features of the enveloping extra–cranial tissues (Figure 3). For grayscale morphology, convexity also extends to the gray levels of the image data and thus requires that the features of the segmented object be the brightest or darkest within some scale of interest. Since erosion and dilation or opening and closing are related by set complementation it is equally possible to segment dark features by use of dilation and closing operations. This property is exploited to segment the ventricular system of the brain (Figure 9). For MRA studies the surface stripping properties of 3–D erosion is also exploited to eliminate extra–cranial capillaries for unobscured visualization of the intracranial arterial and venous tree. Unfortunately, most applications of mathematical morphology, unlike the methods introduced in this paper, do not extend beyond this level of morphological processing and hence destroy most of the high resolution detail of the segmented objects.

2.3 Morphological Skeleton

The representation of non–convex or concave topological characteristics, such as curvilinear features, over a spectrum of neighborhood scales is supported in mathematical morphology by a skeletonization technique. The morphological skeleton representation of an image is developed by the collection of successive shape deviations of an image set from a structuring element across a spectrum of scales. For arteries and veins this maps a uniform diametric representation when the circular structuring element is used.

It is apparent from the examples that morphological opening and closing operations destroy features, within an image, which deviate from the shape, size or orientation of the structuring element. To preserve these complementary features the morphological skeleton is used. Thus image features which deviate from the properties of size, shape and orientation of the structuring element are preserved within the morphological skeleton. The construction of the morphological skeleton of an image proceeds by differencing successively larger scales of morphological erosions until the image erodes to a monotone intensity, or successively larger scales of morphological dilations of an image from a largest scale, which just dilates the image to a monotone intensity.

Thus maximal features within a neighborhood scale (white) as well as minimal features within a neighborhood scale (black) are represented within the morphological skeleton.

To represent the white skeleton the anti–extensive operations of erosion and opening are performed.

2.3.1 Definition : $$S^{\varepsilon}_{\alpha}(g, f) = \left(\dot{\Gamma}_{\alpha} f\right) \setminus \left(\dot{\Gamma}_{\alpha} f\right)_{\varepsilon g} , \qquad (7)$$

$$d = \liminf_{\alpha \in \Re^{+}} \left\{\alpha : \inf(f_{\alpha g}) = \sup(f_{\alpha g})\right\} ,$$

$$S^{\varepsilon}(g, f) = \sum_{0 \le i \le 1+d} S^{\varepsilon}_{i\varepsilon}(g, f) ,$$ is the white skeleton.

Alternately for a black skeleton the extensive operations of dilation and closing are performed.

2.3.2 Definition : $$S^{\varepsilon}_{-\alpha}(g, f) = (\Gamma_{\alpha} f)^{\varepsilon g} \setminus (\Gamma_{\alpha} f) . \qquad (8)$$

$$b = \liminf_{\alpha \in \Re^{+}} \{\alpha : \inf(f^{\alpha g}) = \sup(f^{\alpha g})\},$$

For digital raster formats of pixels or voxels ε is equal to 1.

$$S^{\varepsilon}(g, f) = \sum_{0 \leq i \leq 1+b/\varepsilon} S^{\varepsilon}_{-i\varepsilon}(g, f) \text{ , is the black skeleton.}$$

2.3.3 Definition : $$S^{\varepsilon}(g, f) = \sum_{-1-b/\varepsilon \leq i \leq 1-d/\varepsilon} S^{\varepsilon}_{i\varepsilon}(g, f) \text{ ,} \tag{9}$$

the total skeleton.

An important feature of the skeleton is that the original image may be reconstructed without precision loss with dilations and additions of the positive morphological skeleton components $S_{\alpha}(X)$ or with erosions and additions of of the negative skeleton components $S_{-\alpha}(X)$. For grayscale images these equations are expressed in a recursive form :

2.3.4 Definition : $$f_{pg} = \Gamma_{\varepsilon} f_{(p+\varepsilon)g} + S^{\varepsilon}_{p}(g, f) \text{ , } f = f_{0 \cdot g} \text{ .} \tag{10}$$

white skeletonal reconstruction.

Likewise to represent the closings of the image the negative components of the skeleton are used.

2.3.5 Definition : $$f^{pg} = \dot{\Gamma}_{\varepsilon} f^{(p+\varepsilon)g} - S^{\varepsilon}_{-p}(g, f) \text{ , } f = f^{0 \cdot g} \text{ .} \tag{11}$$

black skeletonal reconstruction.

2.4 Fuzzy Connectivity

Connectivity is a characteristic of the morphological skeleton preserved relative to the corresponding image structures if a convex structuring element with finite surface curvatures is selected. The capacity of the morphological skeleton to preserve the connectivity of image structures is exploited by the algorithms of this paper. For digital imagery and fuzzy sets the actual definition of continuity, however, must be defined. The continuity of a skeleton representation is defined relative to the structuring elements used to make the morphological skeleton. To facilitate this definition the introduction of a modified Hausdorff metric provides the means of measuring the connectivity of image structures. The Hausdorff metric was devised to measure the distance and dissimilarity of sets, a nonlinear set analog to inner product correlation. This metric is precisely defined by mathematical morphology if a circular structuring element is utilized. A modified metric may likewise be devised with variously shaped structuring elements of relative convex and symmetric characteristics, Proposition 2.2.10. This metric likewise generalizes to a metric distance between functions with proposition 2.2.11.

Hence the modified Hausdorff metric between functional maps f and h in restricted $\mathcal{U}$,

2.4.1 Definition : Modified Hausdorff function metric (12)

$$d_g(f, h) = \inf\{\alpha : \sup(f \wedge h) < \inf(\Gamma_{\alpha} f \wedge \Gamma_{\alpha} h) \} \text{ .}$$

if f is connected by distance $\varepsilon > 0$, and element g then $d_g(f(x), f(y)) \leq \varepsilon$ for all points $x, y \in X$.

For the 3–D digital image applications of this paper connectivity is limited to a unity radial scale of the 3x3x3 basis structuring elements relative to the Hausdorff metric.

2.4.2 Proposition :
if an image f is distance εg connected then :

$$S_0^{\varepsilon}(\mathbf{g}, f) = f - f_{\varepsilon \mathbf{g},} \text{ and } f \text{ are } \varepsilon\mathbf{g} \text{ connected.} \tag{13}$$

proof. $S_0(\mathbf{g}, f)$, $f_{\varepsilon \mathbf{g},}$, are distance ε and element **g** connected , since $f = S_0(\mathbf{g}, f) + f_{\varepsilon \mathbf{g},}$, are distance ε and element **g** connected .

$$\text{by definition 2.3.1. Likewise } S_\alpha(\mathbf{g}, f), \left(\dot{\Gamma}_\alpha f\right)_{\varepsilon \mathbf{g},} \tag{14}$$

are distance ε and element **g** connected.

This proposition is exploited in morphological reconstruction techniques based on fuzzy connectivity.

The morphological skeleton is also a reduced representation of the image, since due to properties of opening no set interiors exist, which is ideal for image compression [18], and analysis of multiscale curvilinear features of images, independently from diametric features. For 3–D data sets the morphological skeleton consists of forms which are immeasurable to a 3–D structuring element such as fractals, 2–D surfaces, 1–D curves and 0–D point noise. For curvilinear feature segmentation the morphological skeleton is an ideal representation since arteries and veins are represented as 1+D curvilinear forms with uniform diameters of thickness immeasurable to the scales of the structuring element used. This is due to the dimensional filtering characteristic of opening since features of lesser dimensional measure are consistently covered by collections of structuring elements of larger euclidian dimensionality and thus filtered (Figure 5) and placed into the morphological skeleton. The ability of the morphological skeleton to represent features over a range of neighborhood scales and within a range of integer dimensional bounds permits the representation of fractal features or fractional dimension, which is demonstrated in the results section.

2.5 Segmentation of 3–D Features Based on Fuzzy Connectivity

Intracranial features of the brain are segmented from MRI acquisition based upon a morphological size filter of their large extent followed by a morphological reconstruction based upon fuzzy connectivity of smaller scale features from the grayscale skeleton components. This provides the restoration of the high resolution features, particularly brain surface and arterial features. The size filter is based upon cascaded erosions, eq. (4), with a spherical structuring element, Figure 2, until the extra–cranial tissues are deleted due to the lack of convexity and diametric properties of the scaled structuring element. Coincidentally, the spectral components of the skeleton can be formed from the difference of the openings of the successive erosions of the image, eq. (7), for algorithm optimization. The morphological reconstruction of the high resolution 3–D data features proceeds by selectively adding connected skeletal component features in an interative process from large to small scale based on connectivity with the modified Hausdorff metric, eq. (12). Fundamentally the skeletal features are added to the reconstruction based upon principals of fuzzy set connectivity proposition 2.4.2 [14, 15]. Additional requirements of the morphological skeleton relative to 1+D curvilinear features are addressed in the following methods section of this paper. The fundamental principal of segmenting image structures based on size and connectivity is demonstrated in Figure 3 and Figure 12.

2.5.1 Definition : Morphological reconstruction based upon fuzzy connectivity ,

$$R_d^{\varepsilon}(f) = \left(\Gamma_{\varepsilon} R_d^{\varepsilon}(f)\right) + \left[S_d^{\varepsilon}(g, f) \wedge \left(\Gamma_{\varepsilon} R_{d+\varepsilon}^{\varepsilon}(f)\right)\right] \tag{15}$$

$$R_{d \leqq d}^{\varepsilon}(f) = f_{dg} = S_d^{\varepsilon}(g , f) = \emptyset, \ .$$

$d = n\varepsilon < \mathbf{d} + \varepsilon$, $n \in N$. for digital images ε is 1.

$R_0^{\varepsilon}(f)$ is the completed reconstruction.

2.6 Morphological Scale Spectra of Skeleton Components

A similarity of the morphological set analysis to Fourier analysis is the ability to represent image data with spectral distributions. The frequency content of the morphological spectrum are however relative to tractable shapes rather than the more abstruse representation of frequency for sinusoidal function spectra in the Fourier distribution. The idea of the shape spectra of images was developed by Matheron [20] to define probabilistic size distributions of shapes for the study of 2–D binary sets on continuous domains. Serra furthermore applied this technique to study the porosity of petrographic and biological images [22, 23]. The development of a morphological frequency spectrum is based upon the measures of the components of the morphological skeleton. The frequency magnitude (PS_{α}) of each spectral point (α) of scale or orientation is therefore related to the volume measure of the morphological skeleton components.

2.6.1 Definition :

White spectral distribution,

$$PS_{\alpha}(\mathbf{g}, f) = \frac{d}{d\alpha} \left\|\dot{\Gamma}_{\alpha} f\right\|_1 = \lim_{\varepsilon \downarrow 0} \frac{1}{\varepsilon}\left[\left\|\dot{\Gamma}_{\alpha} f\right\|_1 - \left\|\dot{\Gamma}_{\alpha+\varepsilon} f\right\|_1\right] . \tag{16}$$

Black spectral distribution,

$$PS_{-\alpha}(\mathbf{g}, f) = \frac{d}{d\alpha} \|\Gamma_{\alpha} f\|_1 = \lim_{\varepsilon \downarrow 0} \frac{1}{\varepsilon}\left[\|\Gamma_{\alpha+\varepsilon} f\|_1 - \|\Gamma_{\alpha} f\|_1\right] . \tag{17}$$

$\|\cdot\|_1$ denotes the Lebesque measure on X, $\|f\|_1 = \int_X f \, dm$.

For digital raster formats of pixels or voxels, α is limited to the integer domain Z and ε is equal to 1. The spectral distribution of shape for digital images is furthermore approximated with the volume of the morphological skeleton, represented as the sum of grayscale integer voxels intensities :

2.6.2 Definition : Spectral distributions for digital data,

$$PS_n(\mathbf{g}, f) = \left\|\dot{\Gamma}_n f\right\|_1 - \left\|\dot{\Gamma}_{n+1} f\right\|_1 , \ \text{white} , \tag{18}$$

$$PS_{-n}(\mathbf{g}, f) = \|\Gamma_{n+1} f\|_1 - \|\Gamma_1 f\|_1 , \ \text{black} . \tag{19}$$

For Figures 3 and 4 the pattern spectra analysis of this continuous binary model of a MRI transverse section, through the brain, is exemplified qualitatively without high resolution features in Figure 6 for infinitesimal dilations and erosions. Presented in this spectral distribution is a positive spectral component where the encircling extra–cranial ring disappears due to erosion at a spectral scale and a negative spectral component where the dark ring representation of the skull disappears due to dilation at a negative spectral scale. For biomedical image acquisitions or synthetic aperture radar (SAR) clutter data with rich morphological content the behavior these scale spectral distributions relate a description of the more complex and fractal nature of biological structures, particularly for the neurological and the vascular systems. The preservation of this nature is an extremely important attribute in the visual representation and human evaluation of biomedical imagery.

A more general interpretation of the spectral distribution is related to the Minkowski content [6], μ, which can be generalized to functions over an image domain with the structuring element base $\mathfrak{B}_{\mathbf{g}}$. The result provides a means of measuring fractal dimension, s, however in a more restricted sense than the Hausdorff dimension [6, 10].

2.6.3 Definition : Minkowski Content μ is a scalar function mapping dependent on $s \in [0, n]$ of $f \in L^{\infty}_{loc.}$ such that $|\log \mu(s)| < \infty$.

white lower bound,

$$\lim_{\varepsilon \downarrow 0} \frac{\left\|\check{f}\right\|_1 - \left\|\dot{\Gamma}_\varepsilon f\right\|_1}{\varepsilon^{n-s}} = \mu_-(s), \quad \lim_{\varepsilon \downarrow 0} \dot{\Gamma}_\varepsilon f = \check{f} \in \tilde{\mathfrak{L}}. \tag{20}$$

black upper bound,

$$\lim_{\varepsilon \downarrow 0} \frac{\left\|\Gamma_\varepsilon f\right\|_1 - \left\|\hat{f}\right\|_1}{\varepsilon^{n-s}} = \mu_+(s), \quad \lim_{\varepsilon \downarrow 0} \Gamma_\varepsilon f = \hat{f} \in \tilde{\mathfrak{U}}. \tag{21}$$

2.7 Size Filtering with Morphological Scale Spectra and Skeleton Components

A further similarity of the morphological size spectral distributions to the Fourier transform is the ability to perform frequency filtering of the spectral components. Hence it is possible to perform band–pass, band–reject, high–pass or low–pass filtering of the shape size composition of an image. In this manner a spectral filter $F(\alpha)$ may be defined over the size spectral domain. Specifically size filtering introduces a new post–processing methodology for segmenting static tissues from MRA acquisitions.

For the 3–D MRI images the discriminating feature of static tissues of the brain is a convexity property across a broad range of neighborhood scales relative to the limited radius of intracranial arteries and veins. Filtering this convexity feature over a broad range of scales permits the selective elimination of static tissues by limiting the scale of morphological reconstruction to within the radius of arteries and veins. In effect this technique is a selective feature scale filtering and enhancement technique. To perform this preprocessing operation a size spectral filter is defined.

2.7.1 Definition : $F(\alpha) = 1$ for $0 \leq \alpha \leqq v$, and $F(\alpha) = 0 \; v \leq \alpha$.

i.e. Morphological reconstruction based upon a limited range of diametric scale v, v is the approximate maximum diameter of arteries and veins.

With this scale filter the convexity properties of images over a pattern spectrum are selectively enhancement or suppression to assist visualization and image analysis.

2.7.2 Definition :

$$f_{\alpha g}^{*v} = \Gamma_{\varepsilon} f_{(\alpha+\varepsilon)g}^{*} + F_v(\alpha) S_\alpha(g, f)$$

$$\text{where } S_\alpha^*(g, f) = F(\alpha) S_\alpha(g, f) = \sum_{\alpha \leqq v} F(\alpha) S_\alpha(g, f) . \qquad (22)$$

$f^{*v} = f_{0g}^{*v}$, is the limited scale reconstruction.

2.8 Curvilinear Feature Segmentation

The curvilinear features of images are discriminated and enhanced from other concave features, such as surfaces and noise, due to their behavior relative to 1–D structuring elements. This analysis exploits maximum comparison, by opening, and minimal comparison, by closing, of 1–D line segments oriented, at various angles and scales of length [17, 18]. The morphological skeleton representation of the image is used due to the reduced dimensional representation of curvilinear forms of the arteries and veins as 1+D curves of uniform diameter. In particular a hierarchical representation by cascading maximum orientations extends the scale of interest for selection of arterio–venous curvatures. This technique may likewise be extended to analyze 2–D surface features with 2–D planar shaped structuring elements at various angles of orientation. This capability of dimensional form filtering extends to a broader application for the operation of erosion, and complimentarily the operation of dilation, a capability of analyzing dimensional forms and representing data features in the skeleton of a hence lesser dimensional form to that of the structuring element. Such as for a spherical structuring element the skeleton contains points, curves, surfaces. Figure 5 lists a chart of the discriminated forms based of the structuring element dimensionality.

Curvilinear analysis differs from the size and convexity analysis of the preceding segmentation schemes primarily due to the reduced dimensionality of the structuring element from 3–D sphere to a 1–D segment and hence the greater sensitivity to the orientations of curvilinear structures (Figure 8). Erosion with this structuring element eliminates impulse noise as well as curve and surfaces points which do not have parallel tangents to the segment. Utilizing the supremum over all orientations of opening with the 1–D structuring element filters noise and define the appropriate tangent direction of curvilinear features eq. (23). The infimum over all orientations of closings with the 1–D structuring element defines the tangent direction, Figure 7, and enhance the curvilinear features of the MRA acquisitions, eq. (23). The length of the 1–D structuring element is also scaled for improved sensitivity to the curvatures of curvilinear forms, Figure 8.

2.8.1 Definition : The median operations are :

$$\text{erosion, } \Gamma_r^{\ell} f = \bigvee_{\omega \in \Omega} \Gamma_r^{\ell(\omega)} f , \quad \text{dilation, } \overset{\bullet\,\ell}{\Gamma}_r f = \bigwedge_{\omega \in \Omega} \overset{\bullet\,\ell(\omega)}{\Gamma}_r f ,$$

$$\text{opening, } f_{r\ell} = \bigvee_{\omega \in \Omega} f_{r\ell(\omega)} , \quad \text{closing, } f^{r\ell} = \bigwedge_{\omega \in \Omega} f^{r\ell(\omega)} . \qquad (23)$$

$\ell(\omega)$ – denotes the 1 – D oriented structuring element at solid angular ω within the domain Ω.

$\wedge$ denotes an infimum over X of a collection of functions .

$\vee$ denotes an supremum over X of a collection of functions .

Equations of (23) describe modified median filter operations.

for digital imagery:

ℓ_i– denotes the 1–D structuring element at orientation i,
$i \in \{0, \ldots, 12\}$, denotes the 13 possible digital orientations of a 3–voxel segment within a 3x3x3 voxel matrix.

Curvilinear or reduced dimensional analysis further offers a means of morphological skeletonization over a range of scales, relative to the lengths of the variously oriented structuring elements. The analysis of this skeleton demonstrates curvature information about the curvilinear forms of veins and arteries. Intuitively, this analysis is reasonable since smaller curvatures, as apposed to larger, should be less sensitive to changes in the scale length of the 1–D structuring element. For digital imagery curvilinear forms have the finite dimensions of voxels which permits curvilinear analysis over a broad range of scales.

2.8.2 Definition : Oriented skeleton ,

$$S_r^{\varepsilon}(\ell, f) = \left(\dot{\Gamma}_r^{\ell} f\right) \setminus \left(\dot{\Gamma}_r^{\ell} f\right)_{\varepsilon\ell} , \tag{24}$$

$$S_{-r}^{\varepsilon}(\ell, f) = -\left(\Gamma_r^{\ell} f\right)^{\varepsilon\ell} \setminus \left(\Gamma_r^{\ell} f\right) . \tag{25}$$

Curvilinear analysis is only performed upon the morphological skeleton, formed from eq. (7) which exclusively contains < 3–D dimensional forms. This restriction upon curvilinear analysis is imposed since the varying diameters of arteries and veins as well as static tissues of the brain, in the original MRA data set, are 3–D forms which are invariant to the oriented 1–D structuring elements used for curvilinear analysis (Figure 5).

Hence the curvilinear reconstruction is based upon a limited scale of the morphological skeleton to eliminate static tissues from the image. The connectivity criteria is also used to eliminate extra–cranial tissue features of the morphological skeleton for an analysis focused upon 1–D curvilinear features. An enhancement of the composite skeleton is thus possible with a technique, eq. (26) similar to the intracranial reconstruction process, eq. (15).

2.8.3 Definition :

$$\begin{aligned} CL1_r^{\varepsilon}(\ell, S^*) = {} & \left[\, \Gamma_{\varepsilon}^{g}\, CL1_{r+\varepsilon}^{\varepsilon}(\ell, S^*)\right] \\ & + S_r^{\varepsilon}(\ell, S^*) \wedge \left[\, \Gamma_{\varepsilon}^{g}\, CL1_{r+\varepsilon}^{\varepsilon}(\ell, S^*)\right] , \end{aligned} \tag{26}$$

$$CL1_r^{\varepsilon}(\ell, S^*) = \emptyset, \quad r \geq \mathbf{r} .$$

Further enhancement can be provides with median closing,

$$\mathbf{CL1}_{\mathbf{r}}^{\varepsilon}(\ell, S^*) = \bigvee_{0 \leqq r \leqq \mathbf{r}} \left(\left[CL1_r^{\varepsilon}(\ell, S^*)\right]\right)^{r\ell} \tag{27}$$

For 3–D digital MRA studies equation (26) requires r = 2 to approximate arterio–venous curvatures and ε = 1 voxel spacing. A simpler approach is to perform only one opening and one closing operation on the skeleton for filtering and enhancement respectively. This approach however does not preserve image features well when r becomes larger due to the lack of a connectivity criteria.

$S \equiv S^*(g, f)$, is th composite skeleton of the image
f derived with a spherical structuring element, **g**,
of diametric scale $\leqq$ the maximum diameter of arteries and veins.

2.8.4 Definition :

$$CL2(\ell, S^*) = S^* \wedge (S^*_{r\ell})^{r\ell}, \text{ where } S^* \equiv S^*(g, f) \, . \tag{28}$$

Equation (27) yields noise suppression and enhancement of the limited scale morphological skeleton, typically $r = 1$ or 2 for 3–D digital MRA studies. In general to retrieve the true diametric representation of the arteries and veins the results of equations (26, 27) are used to reconstruct a filtered representation of the image.

2.8.5 Definition :

$$R^*_r(f) = \left[\, \Gamma^g_\varepsilon \, R^*_{r+\varepsilon}(f)\right] + S^*_r \wedge CLn(\ell, S^*) \, , \; S^*_r \equiv S^*_r(g, f) \, . \tag{29}$$

$R^*_0(f)$ is the limited scale reconstruction for $r = 0$.

3.0 Results

Grayscale morphological analysis of a spin proton image section through the brain was used to illustrate the elementary aspects of morphological description. In Figure 10 an example of segmenting the intracranial brain tissue from the extra–cranial tissues of the scalp is presented, Figure 10c in comparison with Figure 10b. The method utilized is similar to the idealized version shown in Figure 3, however, the fuzzy connectivity reconstruction technique, eq. (15), was used to restore small scale morphological features of the brain after size filtering. This same technique was extended to 3 dimensions to segment intracranial features for unobscured surface visualization of the brain, Figure 12.

To demonstrate a morphological skeleton and its behavior the skeletal components S_α of the image were derived with eq. (7) up to the maximum diameter d which eroded the image to a monotone value. Two 3x3 pixel structuring elements were used in a cascaded succession of erosions to generate the spectrum of necessary scales, eq. (4). The shapes of the structuring elements were selected to approximate a circle in a discrete topology, $\mathbf{Z}^2 / n\mathbf{Z}^2$, one of a diamond shape and the other of an octagonal shape. It should be emphasized that the reconstruction of the image from its morphological skeleton components, eq. (10), is lossless relative to the original image. The sum of these skeletal components, eq. (8), are visible in Figure 10a. This Figure illustrates a 1–D curvilinear form of the composite skeleton which is a reduced representation of the 2–D morphology of the imaged brain section. Furthermore the integral volume of the skeletal components eq. (18), over a spectrum of diameters produces a size spectral distribution, Figure 11, which demonstrates a Minkowski content or fractional dimension of the brain section relative to a logarithmic scale. In particular the first slope over a range of the smallest structuring element diameters indicates a 1.65 Hausdorff dimension which is basic to the morphology the brain gyri represented in this 2–D section. By selecting a particular set of "bones", Figure 10d, from the skeleton it is possible to independently reconstruct the image features of the bones, Figure 10e, or to see the image with the features of the bones missing, Figure 10f. This example particularly represents the natural and fuzzy representation of image features with grayscale morphology. The technique introduced for the analysis of the morphological skeleton of MRA acquisitions, eq. (26, 27, 28), provides a means of selecting these skeleton bones based features of one dimensional curvilinear forms. These broad capabilities of the morphological skeleton to represent imagery without distortion, preserve connectivity, isolate diametric features, and to furthermore represent the natural or fractal nature of imagery posed the morphological skeleton as an ideal and extremely attractive representation for arteries and veins, from MRA acquisitions.

Figure 13 demonstrates a magnetic resonance angiography study of the carotid arteries of the neck. The arteries are selected relative to flow velocity directions toward the extremity of the head from the body. An undesirable artifact from this imaging technique is that arterial flow oblique to the selected flow direction is distorted resulting in a staircase effect. The objective with this study is to enhance the arteries within the 3–D volume of the data. The unadulterated data is initially observed with a transparency projection algorithm, performed through a succession of projection angles. The first method to exploit the curvilinear features of arteries and veins is to represent arterio–venous features in a uniform diametric representation. This representation is rendered by the morphological skeletonization of arteries and veins which defines a central axis of the arteries i.e.. Figure 13b. The arteries of the skeleton representation are approximately of a single voxel diametric width to simplify curvilinear analysis of the arterio–venous features. This single voxel representation allows features of curvature and length to be analyzed independent of the diametric scale of the artery or vein. The limited scale morphological reconstruction, eq. (22), was used to enhance this study by eliminating static tissue over a broad range of scales followed by application of eq. (26, 28) to eliminate capillary features, Figure 14c versus Figure 14f. Of particular importance it the fact that the diameters of the arteries and veins are unchanged by this scale selective technique. Additionally, the "stair case" artifact has been eliminated from the oblique flow arterial acquisitions.

The visualizations of Figure 14 were acquired at a courser resolution than Figure 13, however these scans encompass the broader range of the carotid arterial tree which feeds the base of the brain. The basal arterial features of the brain were enhanced, in Figure 14a, by the elimination of posterior and anterior coronal slices for a region of interest (ROI) study. To eliminate the expert intensive and computer interactive techniques of ROI selection, the automated methods of this paper can be used to eliminate extra–cranial and sub–dura capillaries for unobscured visualization of arterial tree. Once again eq. (22) is eliminates image features of diametric dimensions larger than the arteries and veins, thus only a limited scale morphological skeleton reconstruction was used. Curvilinear segmentation of the limited scale morphological skeleton with eq. (26, 27) then eliminates 0–D point noise and limits arterial curvatures with line segments of various lengths and orientations. Thus segments of 3 and 5 voxels in length, or radius, $r = 1$ and 2, were used with maximum values of orientation in 13 directions, eq. (23). This segmented skeleton then provides a basis to morphologically reconstruct the enhanced data volume for reprojection and visualization. Figures 14b and 14e show examples of the filtered morphological skeleton axinometrically projected along 2 different angles of direction. Figures 14c and 14f are the projections of the filtered MRA volume. Note the preservation of small diametric features between projections of the original volume 11a and the filtered volume 11c, and note the radical elimination of the obscuring capillaries from Figure 14d demonstrated in Figure 14f.

Figure 15 demonstrates a study for visualizing the arteries and veins of the head. Again a directional flow field was selected to enhance blood flows through the transverse plane towards the superior of the head. Since arteries and veins have no particular direction in the head both types of flow are realized by this imaging method. The elimination of extra–cranial veins as well as capillaries is the objective for this study, particularly since intracranial arteries and veins are of paramount importance for planning neurological surgeries. To facilitate this objective, eq. (15) has shown the capability of segmenting intracranial features for unobscured 3–D visualization of the brain, Figure 12, and now may be used to include the segmentation arteries and veins within the proximity of the static brain tissues. Figures 15a and 15b demonstrate two projected directions of the MRA study, extra–cranial capillaries obscure the visualization of the intracranial arteries and veins and must be eliminated. Equation (15) was used to segment the static tissues of the brain based upon a size and connectivity criteria, which permitted the unobscured visualization of brain anatomy, Figure 14, Figure 15e,f. The static tissue representation defined with this method was then used to mask the intracranial arteries and veins for further curvilinear enhancement eq. (15) and limited scale filtering eq. (22), and eq. (26, 27) which resulted in Figure 12c,d. Comparing Figure 15a,b with Figure 15c,d. we observe that the intracranial arteries and veins are segmented successfully without the overlying extra–cranial features.

In Figure 16 a complex scene is analyzed consisting of foliated trees, tree stump, road and telephone poles. The original image is demonstrated in the upper left. In the upper left a 2–D structuring element is used, operator **g** in eq. 21, to develop a morphological skeleton, and then reconstruct the image based on fuzzy connectivity of the skeleton components, eq. 25. note the preservation of foliage details and elimination of manmade features such as the road sign and lane strip on the road. In the lower left a 3x3 median opening operation , consisting of 4 1–D oriented structuring elements, was employed in a succession of 20 cascaded operations [12]. This operation preserved the 1–D features of the image, including man–made features of the sign and lane strip on the road, and eliminated the features of leaves from the trees. The leaves have a characteristic dimension less than one in the image and are filtered by the 1–D oriented operations of the median erosion, eq. (23). In the lower right a median dilation, eq. (23), consisting of four 1–D oriented structuring elements, was used on the eroded image to fill the textual features segmented by median erosion. This median operation of erosions followed by dilations results in a more convex representation of image features such as leaves.

In Figure 17, a synthetic aperture radar image is analyzed consisting of several specular targets in a speckle noised clutter background. The original image is demonstrated in the upper left. Successively morphological median dilation operations are used to fill the fractal features of the clutter and leave the impulsive specular details of the man–made targets unaltered [12].

4.0 Conclusion

Mathematical morphology has provided a powerful segmentation tool for the unobscured presentation of 3–D brain anatomy [2, 4] and now can eliminate noise and static tissues for a clear demonstration of the intracranial arterial tree from MRA acquisitions. In particular the fine structures of arteries and veins are preserved by the curvilinear feature segmentation with mathematical morphology.

A theoretical development of a battery of morphological analysis techniques have been developed and applied to biomedical and remote sensing image data. These techniques analytically capture the dimensionality, fractal nature and fuzzy set structure of features from the image acquisitions. A variety of structuring elements, digitally familiar to spheres, planes, and lines defined over a 3–D domain, have been utilized over a spectrum of scales and orientations to segment features of similar or complementary dimensions and shape. In particular, the complementary dimensional features of fractals are preserved as deviations in the morphological skeleton across a wide range of scales relative to the diameter of the convex structuring elements.

The image analysis of mathematical morphology has demonstrated an ability to segment 3–D biomedical image features of magnetic resonance imagery acquisitions and 2–D natural clutter features from remotely sensed SAR images for an unobscured visualization of features of interest from axinometric sum projections. In comparison with intensity histogram and thresholding techniques the distributions of shape spectra, eq. (16, 17), of data are thresholded and mapped as a new analysis and visualization tool, e.g. eq. (22), just as intensity histograms are thresholded and mapped.

Three progressively improving methods for segmentation of the curvilinear features of arteries and veins from MRA acquisitions were introduced. A successive application of these methods illustrated an evolutionary refinement of algorithms to segment intracranial arteries and veins from MRA studies. One of these methods alternately provided a means of segmenting brain tissues from the involvement of extra–cranial tissues eq. (15). Alternately, a technique established a means of suppressing static tissue features from MRA projections eq. (22), and enhancing curvilinear features eq. (26, 27, 28). This technique successfully addressed the segmentation requirements of the pecùliar curvilinear nature of MRA acquisitions, a resplendent example is the comparison of Figures 14d and 14f, and a testament of its accuracy is exemplified in the preservation and enhancement of delicate curvilinear features, comparison of Figure 14a and 14c.

The premise of imagery as a representation over a Borel field topology in mathematical morphology facilitates the analysis of imagery with fuzzy set algebra and offers a catalyst for further advancement into the field of data interpolation and reprojection. In particular, MRI acquisitions and SAR image data are a representation of natural organisms with a property of self similarity at various scales [9] which is consistent to a fractal description [1]. The utilization of the algebra of mathematical morphology makes possible the direct analysis and segmentation of fractal image features relative to spectral distributions of size and shape from the morphological skeleton, as presented in this paper, eq. (16, 17), Figures 10,11. This capability of fractal representation establishes a basis for the most literal and informative presentation and segmentation of biomedical imagery.

Bibliography

1. M. Barnsley, *Fractals Everywhere,* Academic Press, San Diego (1988).
2. H.E. Cline, "Connectivity algorithms for MR angiography," *Society of Magnetic Resonance In Medicine: 9th Annual Conf.* **1**, 61 (1990).
3. J. B. Conway, *A Course in Functional Analysis,* Springer–Verlag, New York (1990).
4. K. S. Denison, "The maximum intensity projection as a segmentation tool," *Society of Magnetic Resonance Imaging* **8**, 73 (1990).
5. R.A. Drebin, "Volume rendering," *Computer Graphics* **22**, 65 (1988).
6. K. Falconer, *Fractal Geometry : Mathematical Foundations And Applications*, Wiley, New York (1984).
7. G. B. Folland, *Real analysis : modern techniques* and their applications, Wiley, New York (1990)
8. G. Gerig, "Medical imaging and computer vision," *M.DAGM–Symp. Mustererkennung–Informatik Fachberichte* **279**, 425–430 (1989).
9. A.L. Goldberger, "Fractal description of nature," *Scientific American* **262**, 44–50 (1990).
10. F. Hausdorff, "Dimension und ausseres Mass," *Mathematische Annalen* **79**, 157 (1919)
11. X. Hu, et. al., "Volumetric rendering of multimodality, multivariable medical imaging data," *Chapel Hill Workshop on Volume Visualization* **1**, 45–50 (1989).
12. W.F. Kraske, "Resolvable Analysis and Synthesis of Synthetic Aperture Radar Imagery," *SPIE Intelligent Robots and Computer Vision 11 : Biol. Neural Nets & 3–D Methods* **1826**, 317–323 (1992).
13. W.F. Kraske, "VOXAR 3–D Curvilinear Feature Description of MR Angiographic Acquisitions for Graphic Presentation and Analysis," *Soc. of Mag. Res. In Med. : 9th Ann. Conf.* **1**, 59 (1990).
14. W.F. Kraske, "3–D Morphological description and visualization of brain anatomy from magnetic resonance imagery," *Radiological Society of North America* **173**, pp. 337 (1990).
15. W.F. Kraske, "Morphological description in 3–D volumetric visualization," *Proc. Chapel Hill Workshop on Volume Visualization* **1**, 59 (1989).
16. P. Maragos, "A representation theory for morphological image and signal processing," *IEEE Trans. Pattern Analysis and Machine Intelligence* **11**, 586–596 (1989).
17. P. Maragos, "Pattern spectrum and multiscale shape representation," *IEEE Trans. Pattern Analysis and Machine Intelligence* **11**, 701–716 (1989).
18. P. Maragos, "Morphological skeleton representation and coding of binary images," *IEEE Trans. Acoustics, Speech, Signal Processing* **34**, 1228–1240 (1986).
19. P. M. Margosian, "Statistical segmentation and region growing methods," *Society of Magnetic Resonance Imaging* **8**, 73 (1990).
20. G. Matheron, *Random Sets and Integral Geometry,* Wiley & Sons, New York (1975).
21. D.R. Ney, "3–D Volume Visualization in Medicine," *IEEE Computer Graphics and Applications* **10**, 24–32 (1990).
22. J. Serra, *Image Analysis and Mathematical Morphology, Acad. Press, New York* (1982).
23. J. Serra, R.E. Miles, *Buffon Symposium on Geometrical Probability in biomathematics,* Springer–Verlag, **23**, Berlin (1978).
24. L.A. Zadeh, *Fuzzy Sets and their Applications to Cognitive and Decision Processes,* Academic Press, New York (1975).

figure 1

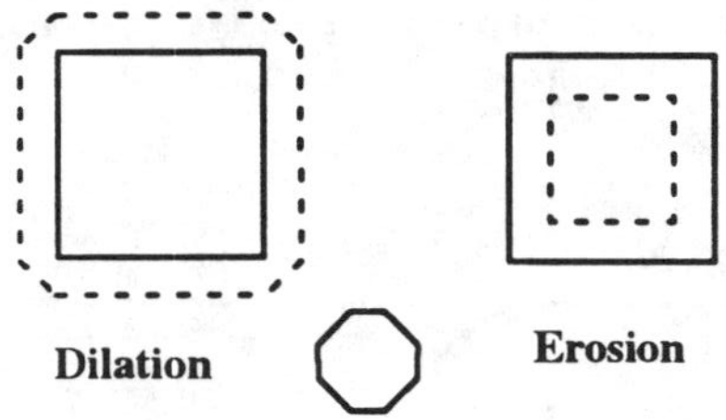

Examples demonstrating dilation and erosion of a square oject relative to the Minkowski addition and subtraction with the given structuring element.

figure 2

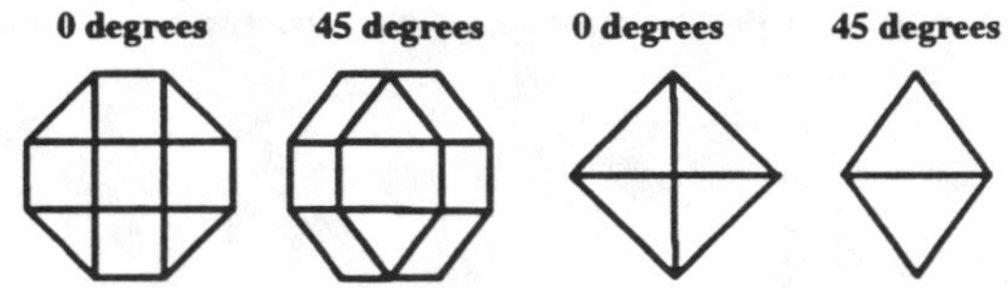

Axinometric projections of the sets used as structuring elements to develop the 3–D grayscale skeletons, a lossless representation of digital imagery. These operators were used alternately to erode and open the 3–D imagery for discrimination of spectral components of the skeleton. Likewise these operators are utilized for morphological reconstruction algorithms.

figure 3

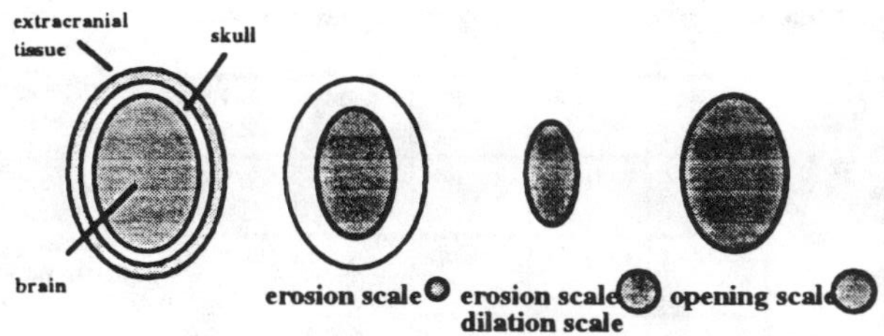

Extra–cranial tissues are deleted due to concavity beyond a particular scale, demonstrated by erosion and opening

figure 4

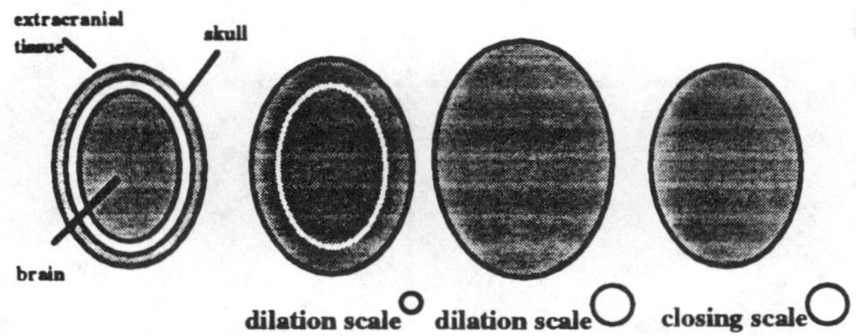

Skull tissues are deleted due to concavity beyond a particular scale, demonstrated by dilation and closing

figure 5

image form

structuring element form	point	curve	surface	volume
point	point	curve	surface	volume
segment	null	curve	surface	volume
disk	null	null	surface	volume
sphere	null	null	null	volume

Filtering of dimensional forms by morphological erosions relative to the structuring element form. The forms filtered to null are complementarily represented in the morphological skeleton

figure 6

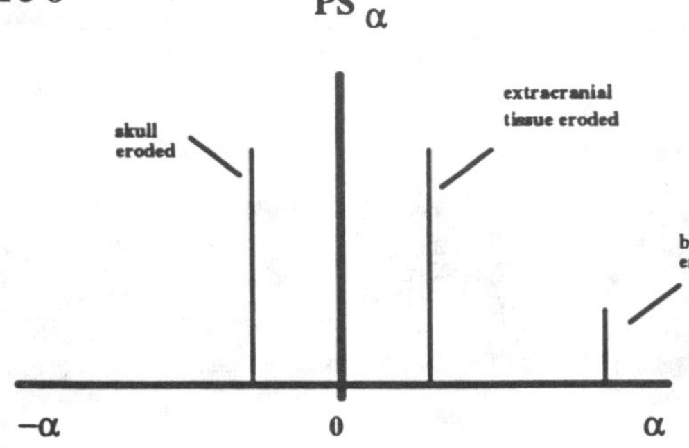

Pattern spectrum of figures 3 and 4
negative spectrum corresponds to dilations of figure 4

figure 7

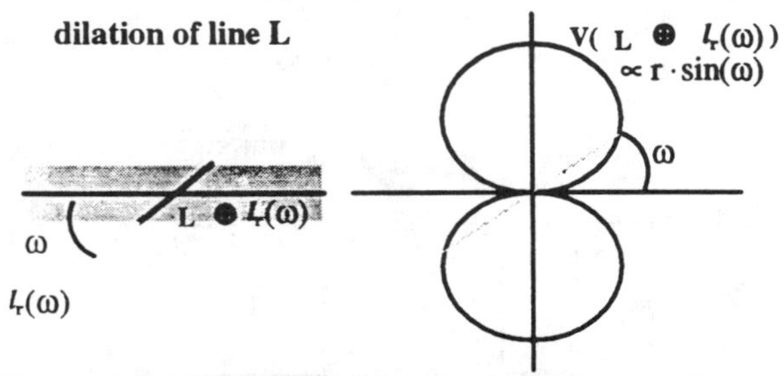

The oriented spectral distribution demonstrates the change in diameter of the line (L) relative to dilations with an oriented structuring element $\ell_r(\omega)$
The line diameter is unchanged with $\ell_r(\omega)$ and L parallel

figure 8

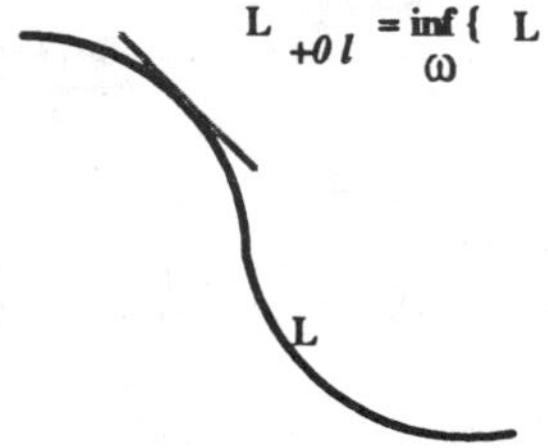

$$L_{+0l} = \inf_{\omega} \{ L \oplus \ell_r(\omega) \}$$

Tangent to curve defined by minimum of oriented spectral distribution at a point of the curve

Sensitivity to structuring element length indicates curvature

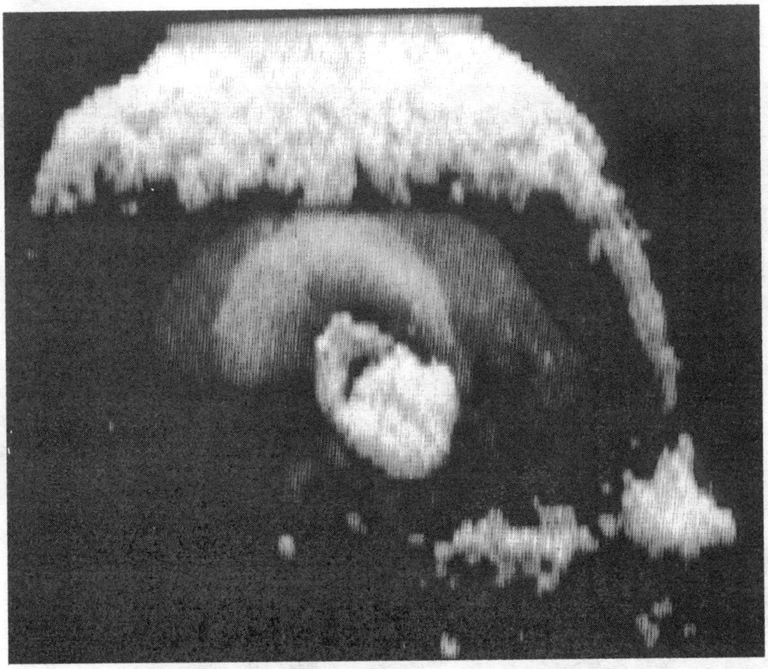

Figure 9 : Ventricles and blood sinuses of the brain. Low contrast ventricular system is demonstrated from overlying sub–dura cerebral–spinal fluid (CSF) by enhancement of dilation. Blood sinuses, inner brain hematoma, and hemorrhaging superior to the brain are demonstrated by enhancement of erosion.

(a) (b) (c)

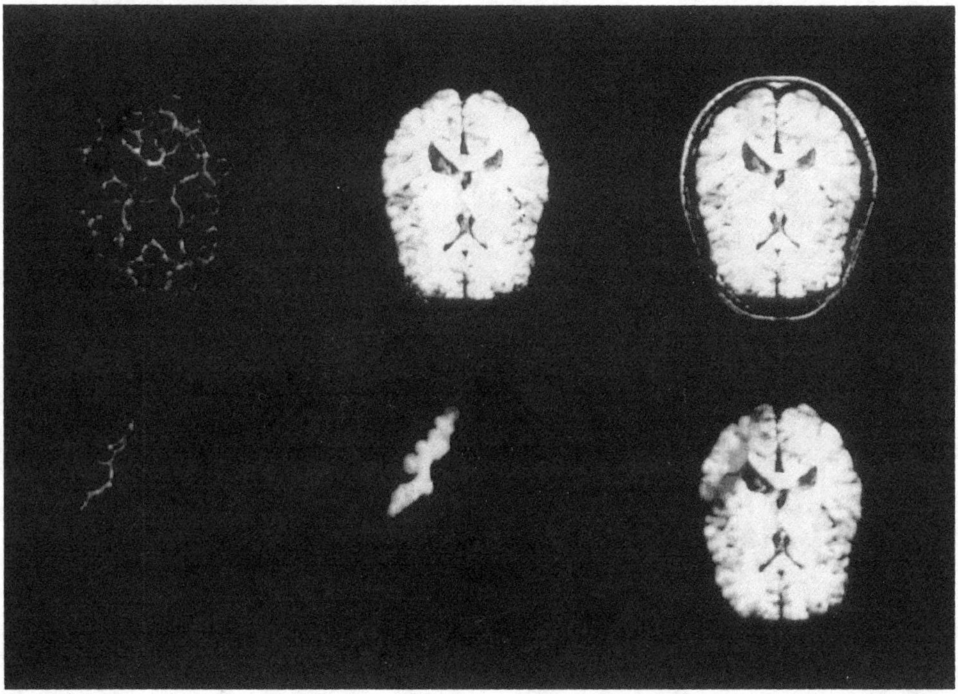

(d) (e) (f)

Figure 10 : An array of image processing operations on a proton spin acquisition of a brain section (a) Morphological skeleton with a selected enhancement of a "bone" (b) Segmented neurological tissue (c) Original MRI scan section (d) Selected feature bone of the skeleton (e) Morphologically reconstructed gyri from selected "bone" of skeleton (f) Morphologically reconstructed brain tissue without gyri bone feature from skeleton. Note natural continuity of texture on the boundaries where features are removed.

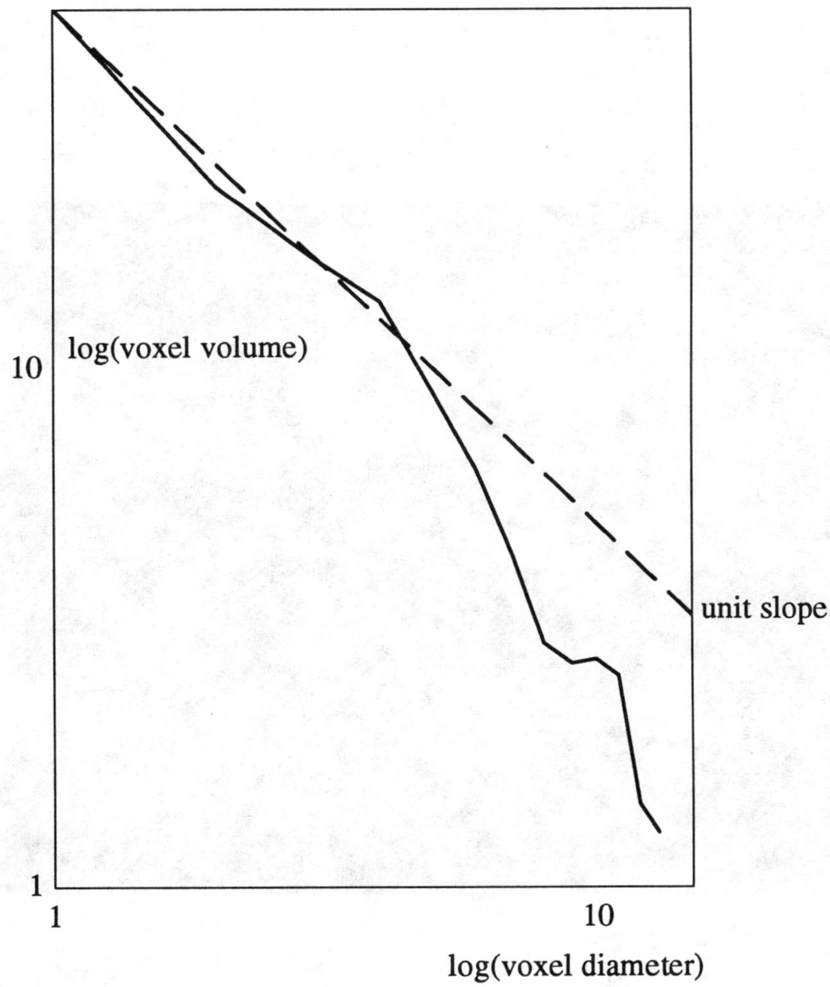

Figure 11 : Log–log distribution of size spectra relative to the volume measure of each spectral component of the morphological skeleton. In the smallest scale region of the distribution the slope corresponds to the fractal dimension of the gyri of the brain scan section of Figure 10. Utilizing the first eight voxel diameters for a least means square quadratic fit estimates the gyri to have an 1.65 dimensional representation within the 2–D scan section.

(a) (b)

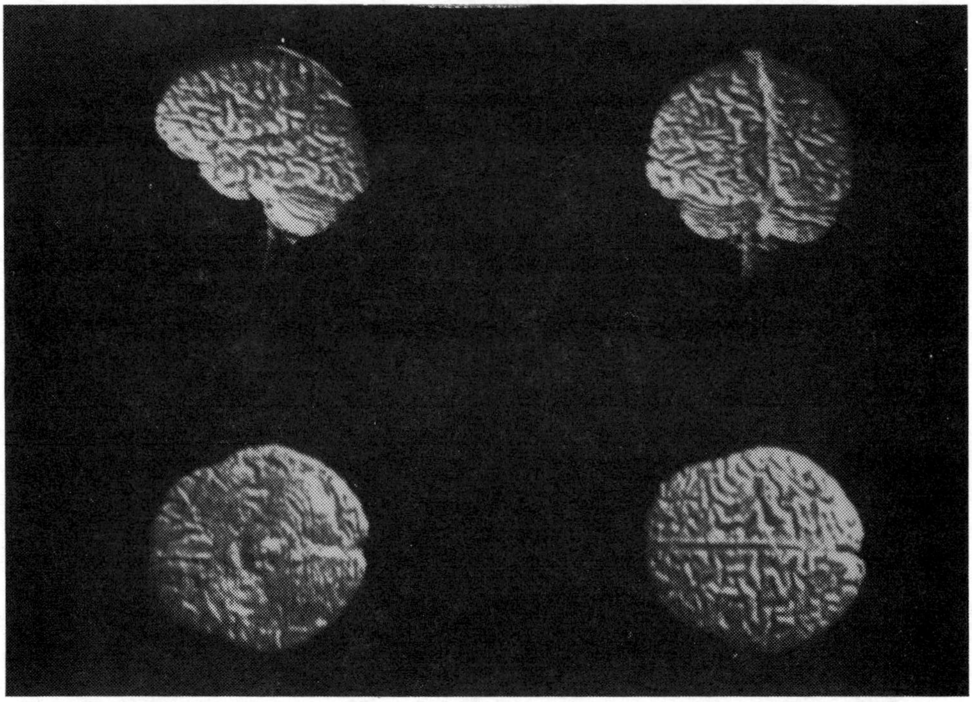

(c) (d)

Figure 12 : Sum projections (SP) of a 3–D morphologically segmented and reconstructed MRI brain scan. Brain surface is grayscale gradient enhanced and shaded with a single light source at a direction from infinity. (a) Lateral sagittal SP (b) Near posterior coronal SP (c) Inferior transverse SP (d) Superior transverse SP. The method of visualization is based on a voxel based surface enhancement [5, 21]. The classification of internal brain anatomy has also been approached with multi–modal fusion techniques [8, 11].

(a) (b)

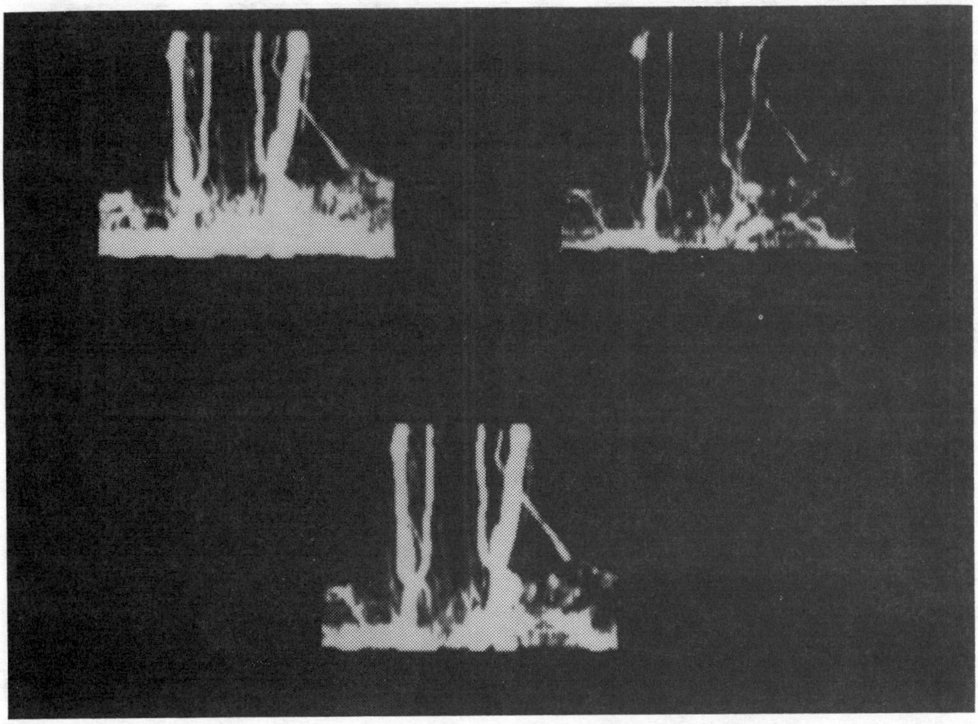

(c)

Figure 13 : Coronal sum projections (SP) of variously 3–D processed MRA acquisition of the carotid arteries, phase saturated. (a) Surface enhanced SP (b) SP of the morphological skeleton demonstrating curvilinear features, note the uniform diametric representation (c) SP of the 3–D acquisition.

(a) (b) (c)

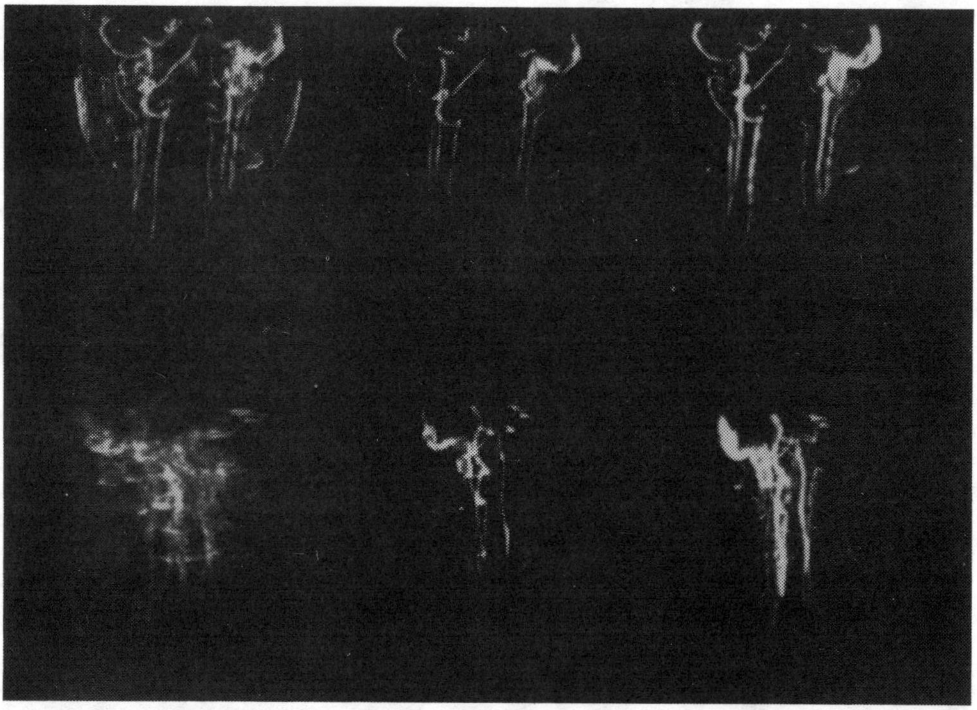

(d) (e) (f)

Figure 14 : Demonstration of variously 3–D processed MRA acquisitions with coronal sum projection (SP), top row is coronal sum projection (SP), bottom row is near lateral SP (a, d) SP of original data, posterior and anterior sections removed to reveal basal arteries, note the obscuring facial capillaries in (d). (b, e) SP of curvilinear (CV) enhanced limited morphological skeleton (LMS). (c, f) SP of morphologically reconstructed data from CVE limited morphological skeleton, note the preservation of delicate arterial features and the filtering of facial capillaries.

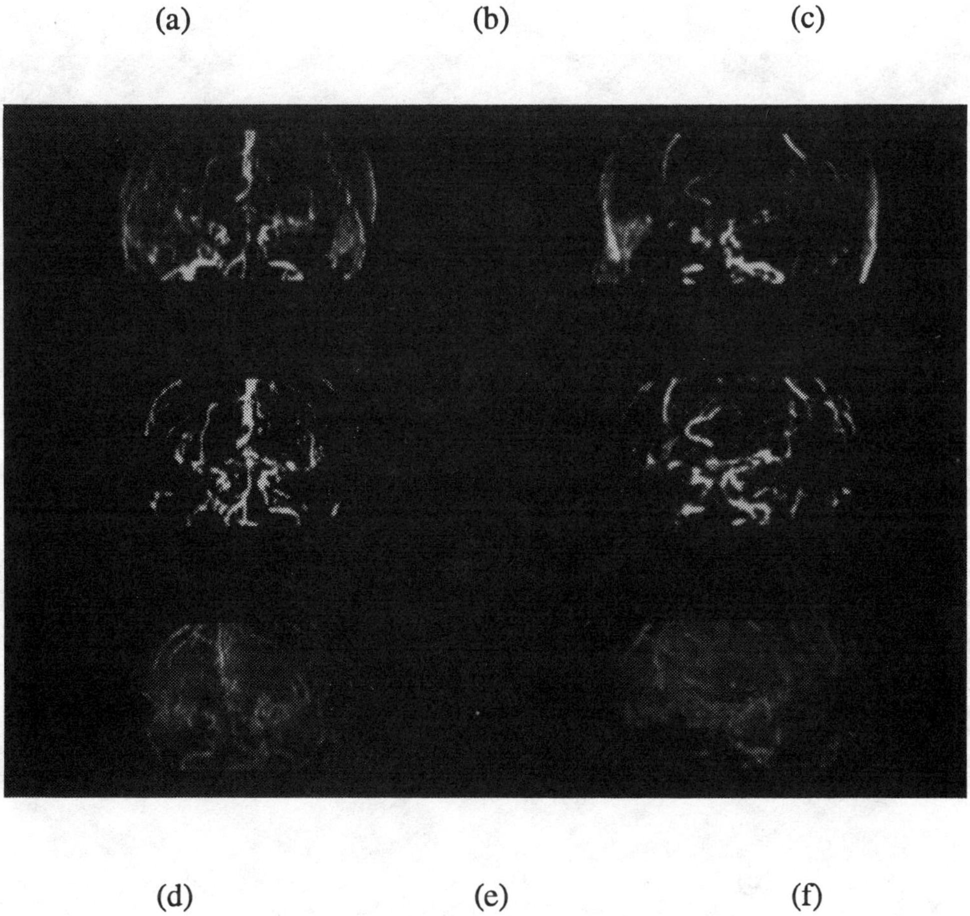

Figure 15 : Sum projections (SP) of blood vascular and static brain tissue features, left column is an anterior coronal SP, right row is a near left lateral SP (a, b) SP of MRA acquisition with limited scale reconstruction. (c, d) SP of CV enhancement of intracranial blood vasculature. Segmentation of intracranial vasculature was realized with segmented static brain tissues. (e, f) Fused surface shaded visualization of vasculature and static brain tissues.

(a) (b)

(c) (d)

Figure 16 : Median filtering on 2–D imagery provides a means of removing fractal dimensional features and "filling" textures for an alternate texture mapping. The original image (a) is decomposed into median erosion skeleton components and reconstructed with eq. (26) (b). In (c) the original image is processed with 20 cascaded operations of median erosion. In (d) the image of (c) is processed with 20 median closing operations.

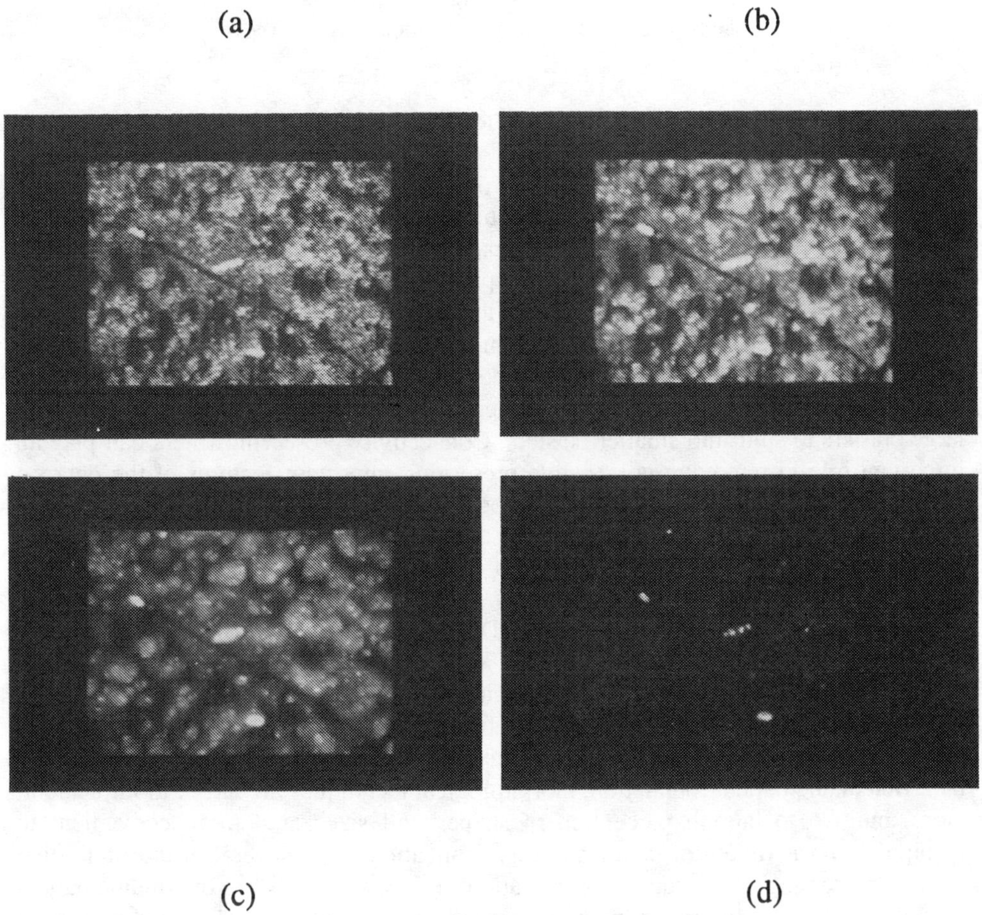

Figure 17 : Median filtering on 2–D imagery provides a means of removing fractal dimensional features and "filling" textures for an alternate texture mapping. The original synthetic aperture radar (SAR) image (a) is median closed with 8 operations in (b). 32 operations in (c) 128 operations in (d).

Clustering analysis of multidimensional fuzzy sets using mathematical morphology.

Alok Kher and Sunanda Mitra

Computer Vision and Image Analysis Laboratory,
Department of Electrical Engineering,
Texas Tech University, Lubbock, TX-79409.

ABSTRACT

The tools of mathematical morphology originally developed for image analysis can be generalized for structural analysis of the general class of multidimensional fuzzy data sets. In this chapter we discuss morphological tools useful for clustering analysis of fuzzy sets. Morphological techniques can perform cluster segmentations that are stable with respect to relative scale changes of the axes of the multidimensional space. We briefly explain a few simple approaches to convert data sets from continuous spaces to discrete spaces where the morphological algorithms can be applied. Morphological filters can be designed to eliminate noise which can cause large clusters to split into smaller clusters. Connectivity preserving filters can play an important role in removing noise while preserving important features of the data set which may be vulnerable to filtering processes. Morphological tools can segment complex data sets such as those consisting of shell type of clusters. We have reviewed several morphological algorithms useful for clustering of binary and fuzzy data sets.

1. INTRODUCTION

The tools of mathematical morphology were first developed in the late sixties by Matheron[1] for structural analysis of binary images. However, the subject has become popular in the image analysis community only in the last few years. Despite their rich mathematical foundation, morphological techniques are easier to implement, more intuitive to human perception of shapes and sizes, and more convenient to manipulate for a variety of image analysis applications than several of the traditional signal processing techniques. Application oriented books[2,3] on mathematical morphology have been written recently for readers with general background in science or engineering. These treatments are easier to understand than the mathematically rigorous classical treatments[4,5]. Several review papers[6,7,8] may also be very useful for gaining an insight in this area.

In binary morphology, the binary images under investigation are represented by crisp sets. The morphological operations are built on the basic set operations such as union, intersection, set difference, set translation, etc. In gray level morphology, gray level images are represented by fuzzy sets, where the gray levels are considered as membership values. With this approach many of the algorithms developed originally for binary images are generalized for gray level images. These algorithms may be used in several other areas of analysis of fuzzy data sets besides gray level images. Most of the numerical morphological algorithms are applicable to fuzzy sets in digitized spaces. Thus, fuzzy sets in continuous spaces need to be converted to digitized spaces in order to apply the morphological operations. The present chapter focuses on the applications of gray scale morphological algorithms to the problem of clustering of fuzzy data sets. Morphological filtering may be efficiently used for preprocessing tasks such as leveling the potential noise clusters with their backgrounds and filling small holes due to insufficient information. Morphological algorithms have already been applied for efficient segmentation of disjoint or overlapping binary[9] and fuzzy clusters[10,11]. These approaches implicitly use the notion of shapes, the concept more intuitive to human perception. Unlike the algorithms using distance norms for clustering, the morphology based algorithms are tolerant to axes scaling of multidimensional data sets.

Preprocessing becomes a difficult task when the clusters under consideration comprise complex components such as shell structures instead of solid regions. Simple filtering operations could destroy these thin regions. More complex filters, namely connectivity preserving filters are needed for the preprocessing of such complex data sets. Median closing and opening introduced by Matheron and Serra[5], and further studied by Maragos[12], are the most commonly used[13,14] connectivity preserving filters. These filters, however, result in partial loss of connected regions. Another class of filters called comparative filters, also introduced by Serra[5] and used by Safa and Flouzat[15] for the filtering of radar imagery, may be used for complete preservation of connected regions. However, these filters do not remove noise efficiently. We have developed connectivity preserving filters which can preserve connectivity completely, and at the same time remove the noise efficiently. We have used these filters to remove speckle noise from synthetic aperture radar (SAR) images[16].

The next section discusses the basics of mathematical morphology. Section three discusses the issues involved in the conversion of a data set from Euclidean space to digital space. Section four discusses the morphology-based algorithms used for clustering tasks. Section five deals with the concept of the connectivity of fuzzy sets. Section six discusses various connectivity preserving filters. Section seven described how the scattered regions ignored in the preprocessing and clustering steps are assigned to various clusters. Section eight summarizes the role of morphological algorithms in the preprocessing and clustering tasks of fuzzy data sets.

2. PRELIMINARIES OF MATHEMATICAL MORPHOLOGY

We first discuss the four basic morphological operations defined for binary sets: erosion, dilation, opening, and closing and the extension of these operators to fuzzy sets. Next, a few important properties of these operations are outlined. The information on each of the above topics is available in all texts on mathematical morphology[2,3,4]. We mention a few tricks useful for faster implementations of the basic operations of erosion and dilation. We also discuss the "Alternating Sequential Filters[4,7]" useful for preprocessing of the data sets prior to the application of clustering algorithms.

2.1 Morphological operations for binary sets:

A binary set used in the present context is the set of points in a digitized space. An example is the set of white pixels of a black and white picture. A typical basic morphological operation is of the form "$A * B$", where "A" is the given data set and "B" is a small set called structuring element which operates on A. A point in each of these sets is assigned as the origin of that set.

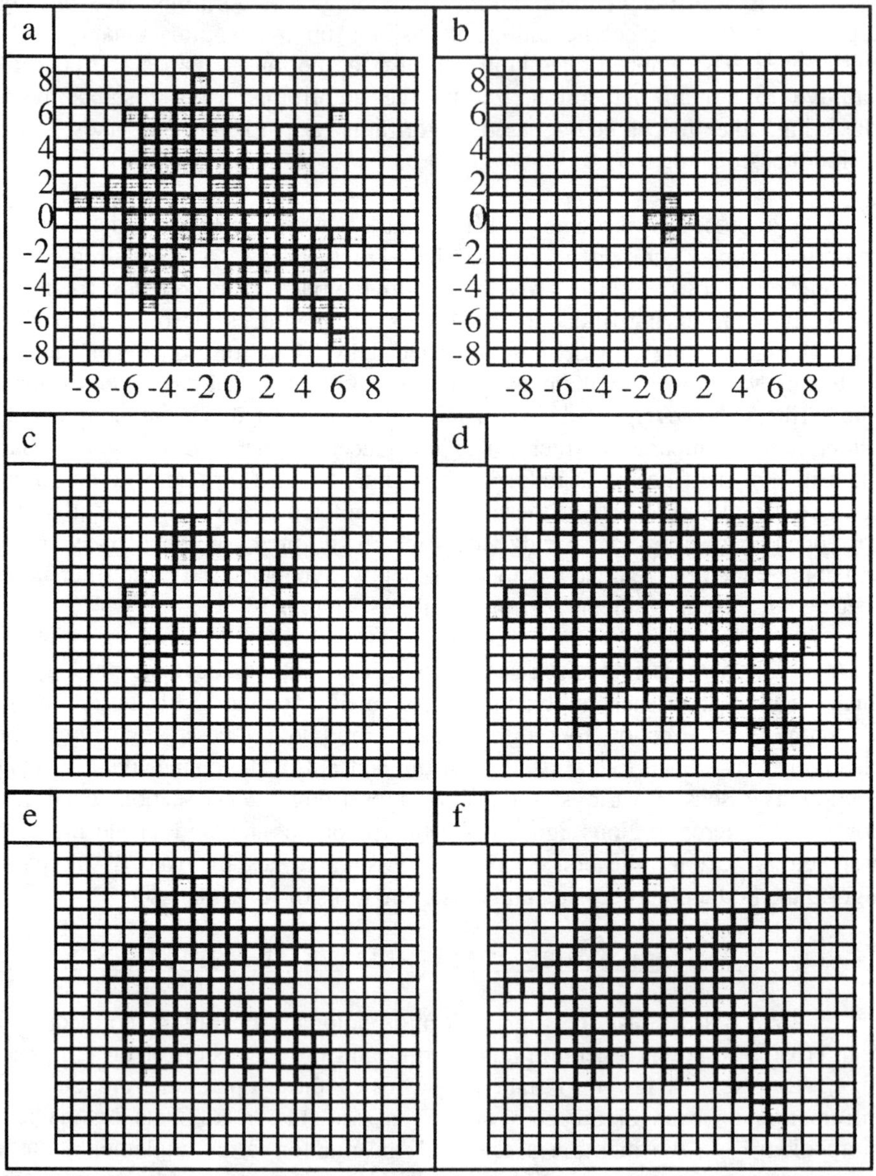

Figure 1: Basic morphological operations on a binary set in a 2-D discrete space. (a) Binary set "A" (defined by the shaded region), (b) Structuring element B, (c) $A \ominus B$, (d) $A \oplus B$, (e) $A \circ B$, (f) $A \bullet B$

Erosion of a binary set "A" with a structuring element "B", is defined as

$$A \ominus B = \cup \, (x: x+B \subset A) \qquad (1)$$

The term "x" represents a point in the given space and "x+B" represents the translation of set "B" which causes the new origin of "B" to coincide with point "x". In words, the erosion of a binary set A by a structuring element B is the set of points such that when the origin of "B" is positioned at these points, the entire structuring element "B" lies inside the binary set "A". Thus, for a structuring element which contains its origin, the eroded set lies inside the original set with the structuring element acting as an "inner marker". Figure 1 shows basic morphological operations. Figure 1(a) shows binary set "A" in a two dimensional (2-D) discrete space. Figure 1(b) shows structuring element "B". This structuring element is called the nearest neighbor mask in 2-D[1]. The nearest neighbor mask in a multidimensional space consists of origin the (0,0,0,...) and the cells ($\pm$1,0,0,...), (0,$\pm$1,0,...), (0,0,$\pm$1,...) and so on. Figure 1(c) shows erosion of set "A" with structuring element "B".

Dilation of a binary set "A" with a structuring element "B" is defined as:

$$A \oplus B = \cup\, (x + B : x \subset A) \qquad (2)$$

In words, dilation of a binary set "A" with a structuring element "B" is the union of translations of the structuring element "B" when the origin of "B" is positioned at all points of the set "A". Thus for a structuring element containing its origin, the dilated image contains the original image with the structuring element acting as an "outer marker". Figure 1(d) shows dilation of set "A" with structuring element "B".

Dilation can be expressed in terms of erosion as follows

$$A \oplus B = ((A^c \ominus (-B))^c \qquad (3)$$

Where

$$-B = \{ -x : x \in B\} \qquad (4)$$

Thus, dilation of A by B leads to erosion of A^c by -B. Here A^c is the set complement of A in the given space.

Two very useful operations derived from erosion and dilation are opening and closing. Opening is given by

$$A \circ B = (A \ominus B) \oplus B = \cup\, (x + B : x + B \subset A) \qquad (5)$$

Physically, opening is the union of all the structuring element translates that can fit in the given binary image. Thus, opening removes the parts of the set where the structuring element can not fit. Closing is equivalent to opening of the complement of the image. Mathematically, closing can be expressed as

$$A \bullet B = (A \oplus B) \ominus B \qquad (6)$$

As illustrated in figure 1(e), opening may be used to remove thin elements of the set under investigation. Similarly, as shown in figure 1(f), closing may be used to fill the thin regions missing from the set. If the thin elements are the effects of noise or insufficient data, the operations of opening and closing act as filters which improve the data set.

A useful operation in several morphology algorithms is the operation of set difference given by

$$P/Q = \{x : x \in P,\ x \notin Q\} \qquad (7)$$

Thus, the set difference of two sets P and Q is the set of points that are included in P but not included in Q.

Some books[1,2] define the erosion as $A \ominus B = \cup\, (x: x - B \subset A)$, and the opening and the closing as $A \circ B = (A \ominus B) \oplus (-B)$ and $A \bullet B = (A \oplus B) \ominus (-B)$ respectively. Any one notation may be followed consistently.

2.2 Morphological operations for fuzzy sets:

Let $\mu_X(x)$ represents membership function value of point x for the fuzzy set X. Then, basic concepts of union, intersection, subset, and set difference can be generalized to fuzzy sets in terms of membership functions as follows:

Union: $R=P \cup Q \Rightarrow \mu_R(x) = \max(\mu_P(x), \mu_Q(x))$ (8)

Intersection: $R=P \cap Q \Rightarrow \mu_R(x) = \min(\mu_P(x), \mu_Q(x))$ (9)

Sub-set: $P \subset Q \Rightarrow \mu_P(x) < \mu_Q(x)$. (10)

The basic morphological operations of erosion and dilation may be generalized for fuzzy sets as follows:

Erosion: $C=A \ominus B \Rightarrow \mu_C(x) = \min((\mu_A(y) : y \in (B+x))$ (11)

Dilation: $C=A \oplus B \Rightarrow \mu_C(x) = \max((\mu_A(y) : y \in (B+x))$ (12)

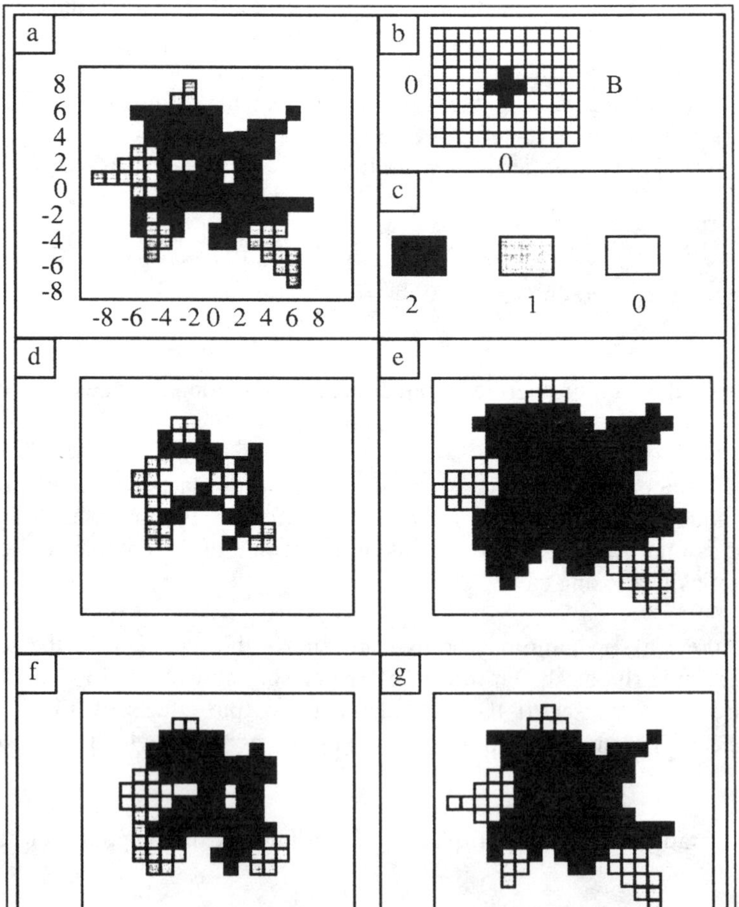

Figure 2: Basic morphological operations on a fuzzy set in a 2-D discrete space.
(a) Fuzzy set A, (b) Structuring element B, (c) Membership value scheme,
(d) $A \ominus B$, (e) $A \oplus B$, (f) $A \circ B$, (g) $A \bullet B$

Here, "A" is a fuzzy set with membership values "$\mu_A(x)$". The structuring element "B" is still a binary or crisp set. It is possible to use a fuzzy set for "B". However, most of the fuzzy morphological algorithms use crisp structuring elements, and so, in the present chapter we will stick to the crisp (also called "flat") structuring elements. The operations involving fuzzy structuring elements may give undesirable effects when the scale of membership function is changed. With the help of definitions 8 through 12, most of the operations defined for binary morphology can be extended to gray scale or fuzzy morphology. For example, the operations of opening and closing for fuzzy morphology are defined by equations 5 and 6 respectively, using the definitions 11 and 12 for the erosion and dilation operations. The morphological filtering operations mentioned in the rest of the chapter will be applicable to either binary, or fuzzy data sets, unless mentioned otherwise.

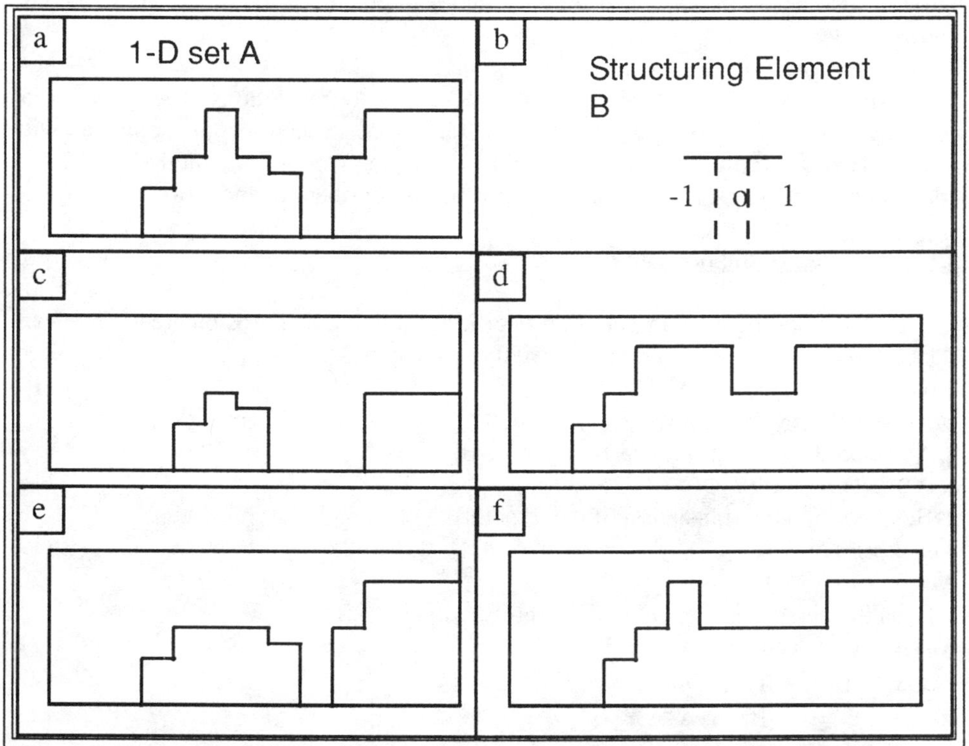

Figure 3: Basic morphological operations on a fuzzy set in a 1-D discrete space. (a) Fuzzy set A, (Plot of membership function versus spatial co-ordinates.) (b) Structuring element B, (c) $A \ominus B$, (d) $A \oplus B$, (e) $A \circ B$, (f) $A \bullet B$

The fuzzy operations of erosion, dilation, opening, and closing commute with the operation of thresholding. Let X_T represent a crisp set derived from the fuzzy set X, such that $X_T = \{x : \mu_X(x) > T\}$. With this notation it can be shown that

$$(A*B)_T = A_T*B \tag{13}$$

where * represents any of the four basic operations: erosion, dilation, opening, and closing. The operation A*B is defined for the fuzzy set A, whereas the operation A_T*B is defined for the binary set A_T. Thus, the above property establishes a relationship between the binary and the fuzzy morphological operations. A fuzzy opening may be used to reduce high membership values of thin regions which can not accommodate the structuring element, such that these regions merge with their background. On the same lines, fuzzy closing may be used to increase low membership values of thin regions to level them with their background. Figure 2 illustrates morphological operations on a 2-D fuzzy set. For simplicity, the points in the 2-D space under consideration are assumed to take membership values of zero, one, or two. Figure 3 illustrates morphological operations on a 1-D fuzzy set.

2.3 Important properties of morphological operations

In this section we mention some of the most important properties of the binary or fuzzy morphological operations. All morphological operations are increasing and are invariant with respect to translations of the data sets on which they operate. Thus we have:

$P \subset Q \Rightarrow (P * B) \subset (Q * B)$ (* is increasing) and (14)

$(A+x) * B = (A * B) + x$ (* is invariant with respect to translations of A) (15)

where X+x represents translation of set X, such that its new origin coincides with point x. Here '*' represents either dilation or erosion. We first mention properties of dilation and erosion. Next, we discuss the properties of opening and closing.

2.3.1 Properties of dilation and erosion:

Some of the most important properties of dilation and erosion are as follows:

a. Extensivity of dilation and antiextensivity of erosion:

If $o \in B$, $(A \ominus B) \subset A \subset (A \oplus B)$ (16)

where o is the origin of the co-ordinate system.

b. Associativity of dilation: $A \oplus (B \oplus C) = (A \oplus B) \oplus C$ (17)

c. $A \ominus (B \oplus C) = (A \ominus B) \ominus C$ (18)

d. Behavior under a translation of the structuring element:

$A \oplus (B+x) = (A \oplus B) + x$ (19)

$A \ominus (B+x) = (A \ominus B) - x$ (20)

e. Equalities involving set unions and intersections:

$A \oplus (B \cup C) = (A \oplus B) \cup (A \oplus C)$ (21)

$(P \cup Q) \oplus B = (P \oplus B) \cup (Q \oplus B)$ (22)

$A \ominus (B \cup C) = (A \ominus B) \cap (A \ominus C)$ (23)

$(P \cap Q) \ominus B = (P \ominus B) \cap (Q \ominus B)$ (24)

f. Ordering involving set unions and intersections:

$A \oplus (B \cap C) \subset (A \oplus B) \cap (A \oplus C)$ (25)

$(P \cap Q) \oplus B \subset (P \oplus B) \cap (Q \oplus B)$ (26)

2.3.2 Properties of opening and closing:

Some of the important properties of opening and closing are as follows:

a. Extensivity of closing and antiextensivity of opening:

$A \circ B \subset A \subset A \bullet B$ (27)

b. $D \circ B = D \Rightarrow (A \circ D) \circ B = (A \circ D)$ and $(A \bullet D) \bullet B = A \bullet D$ (28)

c. Idempotence of opening and closing:

$(A \bullet B) \bullet B = A \bullet B$ and $(A \circ B) \circ B = A \circ B$ (29)

This property is a spatial case of property c, since $B \circ B = B$.

d. Behavior under a translation of the structuring element:

$A \circ (B+x) = A \circ B$ and $A \bullet (B+x) = A \bullet B$ (30)

Thus opening and closing are independent of the origin of the structuring element.

2.4 Fast implementation of dilation and erosion:

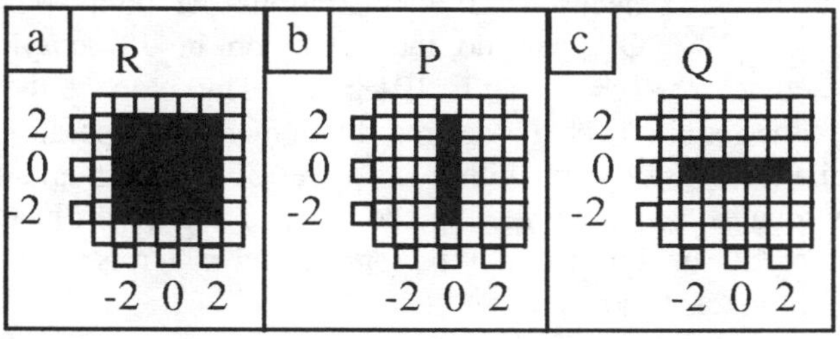

Figure 4: $A \oplus R = (A \oplus P) \oplus Q$ and $A \ominus R = (A \ominus P) \ominus Q$ since $R = P \oplus Q$. Here A is any given data set in 2-D.

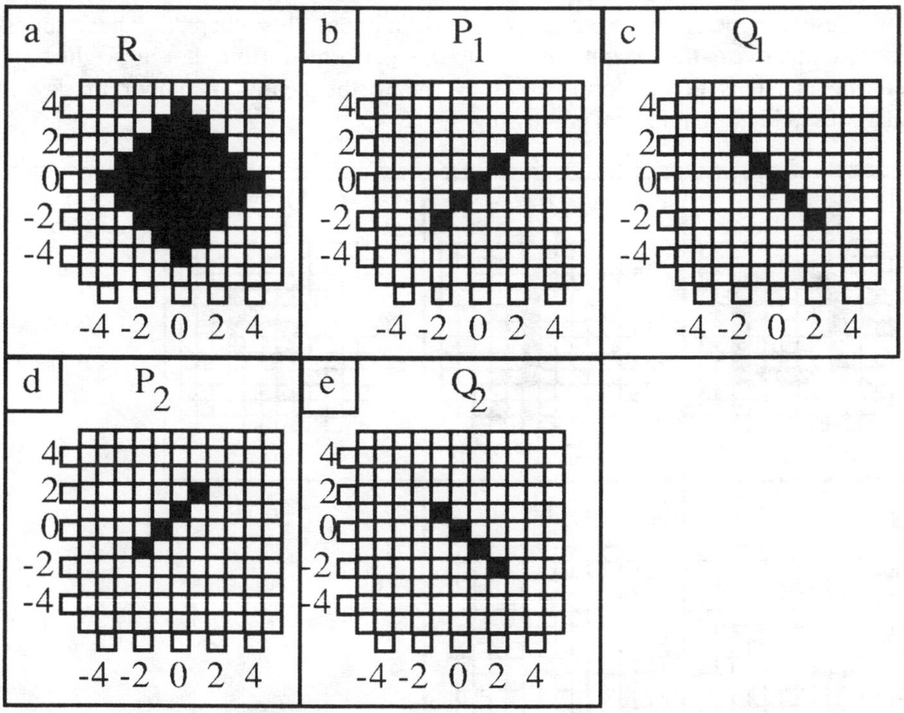

Figure 5. $R = (P_1 \oplus Q_1) \cup (P_2 \oplus Q_2) \Rightarrow A \oplus R = ((A \oplus P_1) \oplus Q_1) \cup ((A \oplus P_2) \oplus Q_2)$, and $A \ominus R = ((A \ominus P_1) \ominus Q_1) \cap (A \ominus P_1) \ominus Q_1$, where A is any data set in 2-D.

Equations (17) and (18) imply that a dilation / erosion, with a bigger structuring element can be replaced by multiple dilations / erosions with smaller structuring elements. This property is useful to reduce computational times required for the dilation and erosion operations. In figure 4, a dilation/erosion with structuring element "R", is equivalent to a dilation/erosion with "P" followed by a dilation/erosion with "Q". Dilation/erosion with "R" requires 25 max/min operations per point, whereas dilation/erosion with "P" followed by dilation/erosion with "Q" requires only 10 max/min operations. Consider a generalized cube in "k" dimensional space with sides of length "n", parallel to the axes of the co-ordinate system. A dilation/erosion with this cube requires "n^k" max/min operations per point. Successive dilations/erosions with the adjoin edges of the cube can do the same job with only "kn" max/min operations per point. This trick may be used to significantly cut down computational times of clustering algorithms operating on data in multidimensional spaces[9].

Figure 5 illustrates another example of representing a dilation/erosion with a 2-D structuring element in terms of dilations/erosions with 1-D structuring elements. A dilation/erosion with "R" in figure 5 needs 41 ax/in operations per point. The structuring elements "P_1", "Q_1", "P_2", "Q_2" can do the same job in 18 ax/min operations per point. Note that 25+16=41 and 10+8=18. Thus, in either representation, the processing time required for the desired dilation/erosion operation in figure 5 is equivalent to a dilation/erosion each with two squares of lengths 5 and 4 respectively, the sides of the squares being parallel to the axes of the co-ordinate systems. The structuring element "R" in figure 5 may be represented in terms of the nearest neighbor mask "B" in figure 1 as follows:

$$R=B^4=B\oplus B\oplus B\oplus B \tag{31}$$

where

$$B\oplus B\oplus B\oplus B=(((B\oplus B)\oplus B)\oplus B) \tag{32}$$

We will use the notation in equation 32 throughout this chapter. Thus, unless mentioned otherwise the operations are to be performed from left to right. The structuring elements of the form "B^k", "B" being the nearest neighbor mask, are popularly used in morphological processing.

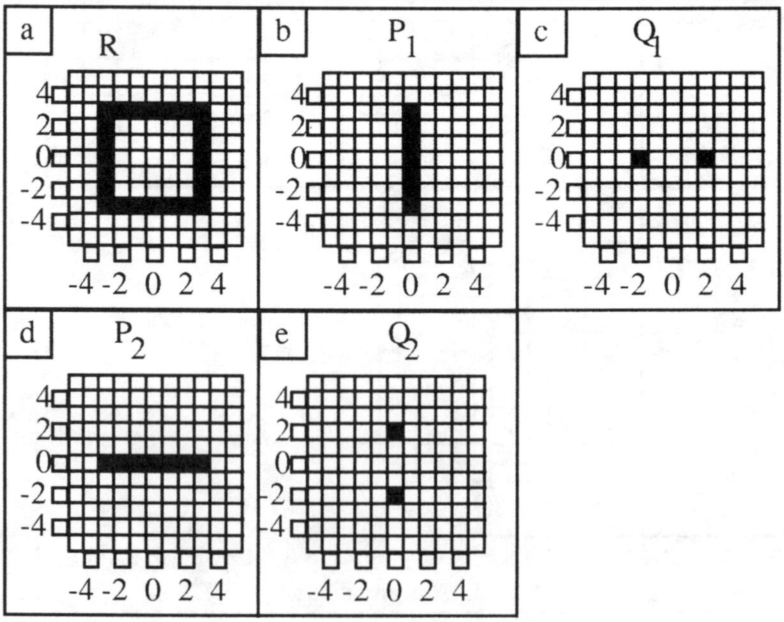

Figure 6: $R=(P_1\oplus Q_1)\cup(P_2\oplus Q_2) \Leftrightarrow A\oplus R=((A\oplus P_1)\oplus Q_1)\cup((A\oplus P_2)\oplus Q_2)$, and $A\ominus R=((A\ominus P_1)\ominus Q_1)\cap(A\ominus P_1)\ominus Q_1$, where A is any data set in 2-D.

Properties of dilations and erosions may also be used for faster implementation of morphological operations with certain types of non-convex structuring elements[16]. Figure 6 shows a rectangular non-convex structuring element R. This kind of structuring elements are used to implement certain types of connectivity preserving filters which are described later in this chapter. As illustrated in figure 6, the dilation/erosion with "R" may be expressed in terms of dilations/erosions with structuring elements "P_1", "Q_1", "P_2", and "Q_2" to cut down the computational time from 24 max/min operations per point to 18 max/min operations per point.

2.5 Alternating Sequential filtering (ASF) :

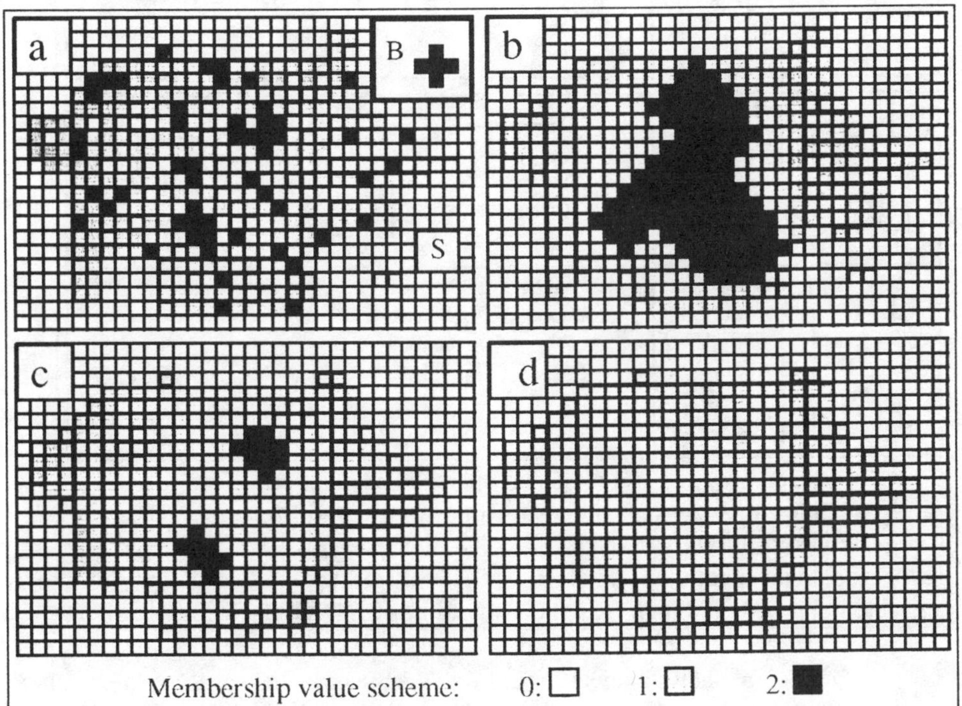

Figure 7: Alternating Sequential Filters
(a) Noisy data set S and structuring element B, (b) $S \bullet B^2 \circ B^2$, (c) $S \bullet B \circ B$,
(d) $ASF(S \bullet B \circ B)^2 = A \bullet B \circ B \bullet B^2 \circ B^2$

Although simple opening and closing operations act as filters which remove small noisy regions, practical filters usually consist of a sequence of several opening and closing operations. These filters are called "alternating sequential filters" or ASFs[5,7]. The general form of kth order ASF of set A with structuring element B may be given by

$$ASF(A \wedge B \vee B)^k = A \wedge B \vee B \wedge B^2 \vee B^2 ..... \wedge B^k \vee B^k \quad (33)$$

where the operations are performed from left to right. Here "$\wedge$" and "$\vee$" are called "algebraic closing" and "algebraic opening" operations. An operation is an algebraic opening if it is increasing (equation 14) and satisfies the conditions:

$A \vee B \subset A$ (antiextensivity) and (34)

$(A \vee B) \vee B = A \vee B$ (idempotence) (35)

Similarly, an operation is an algebraic closing ($\wedge$) if it is increasing and satisfies the conditions:

$A \subset A \vee B$ (extensivity) and (36)

$(A \wedge B) \wedge B = A$ (idempotence) (37)

Morphological closing ($\bullet$) and opening ($\circ$) operations are examples of algebraic closing ($\wedge$) and algebraic opening ($\vee$) respectively.

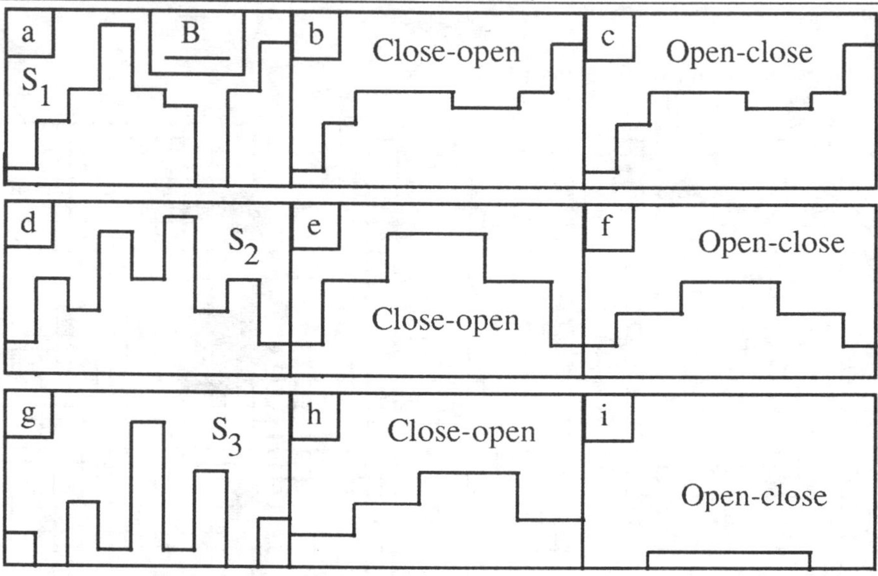

Figure 8: Comparison of close-open and open-close operations
(a)Signal "S_1" with a low spatial density of noise; structuring element "B". (b) $S_1 \bullet B \circ B$, (c) $S_1 \circ B \bullet B$, (d)Signal "S_2" with high spatial density of noise, but high signal to noise ratio, (e) $S_2 \bullet B \circ B$, (f) $S_2 \circ B \bullet B$,(g)Signal "S_3" with low signal to noise ratio and high spatial density of noise, (e) $S_3 \bullet B \circ B$, (f) $S_3 \circ B \bullet B$,

Figure 7(a) shows a noisy set S. Figure 7(b) shows the result of the combined operation of morphological closing followed by morphological opening (morphological close-open operation) with structuring element B^2. Figure 7(c) shows the result of morphological close-open operation with structuring element B. Figure 7(d) shows the result of second order ASF of "S" with "B". The ASF here is based on morphological closing and opening operation. Notice the superior filtering in 7(d) compared to filtering in either 7(b) or 7(c). Note that in equation 33, an algebraic closing operation with B^m is performed prior to the corresponding algebraic opening operation for each value of m=1,k. An N dimensional fuzzy set can be equivalently represented by a lower bounded N dimensional function, the lower bound being zero. For such functions, the morphological operation of close-open retains the signal information better than the operation of open-close[16]. Figure 8 compares performances of close-open and open-close operations for various 1-D signals. For signal S_1 with low spatial density of noise, the morphological close-open and open-close filters give identical results. For signal S_2 with high spatial density of noise, but high signal to noise ratio, the signals resulting from close-open and open-close operations have approximately the same shape. However, for lower bounded signal S_3 with high spatial density of noise and low signal to noise ratio, only the close-open operation preserves the signal shape, whereas the open-close operation destroys the signal information. Hence in all our filters, we will perform the closing operation prior to the opening operation. Although we described the general notions of algebraic closing and opening in this section, we will use the terms closing and opening for morphological closing and opening operations throughout this chapter. In section 6 we will describe more examples of algebraic closing and opening.

The brief introduction in this section was intended to give the reader a flavor of the tools of mathematical morphology. In our discussion on the clustering algorithms, and the connectivity preserving filters, we have used the basics developed in this section.

3. DATA CONVERSION FROM EUCLIDEAN SPACE TO DIGITIZED SPACE

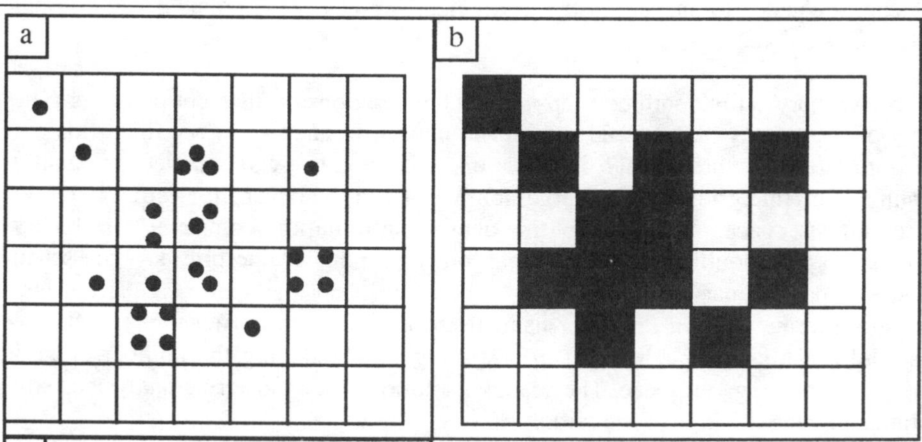

c

1	0	0	0	0	0	0
0	1	0	3	0	1	0
0	0	2	2	0	0	0
0	1	1	2	0	4	0
0	0	4	0	1	0	0
0	0	0	0	0	0	0

Figure 9: Binary and fuzzy sampling
(a) A binary data set in a 2-D continuous space, (b) Result of binary sampling, (c) Result of fuzzy sampling.

In the last section, morphological operations were defined for binary and fuzzy sets in digitized (or discrete) spaces. Several fuzzy sets of interest are defined in continuous spaces. As a first step towards converting the data from a continuous space to a discrete space, a grid must be establish. Figure 9(a) shows a binary data set in a 2-

D continuous space. A grid used to convert this data into discrete space is also shown. Two approaches may be used to convert the data into discrete space. One approach is to assign a point to each cell which contains at least one point of the original data set in the continuous space. This approach is used by Postaire, Zhang, and Lecocq-Botte[9] in their work on "cluster analysis by binary morphology". We will call this sampling as binary sampling. Figure 9(b) shows result of the binary sampling of the data set in figure 9(a) for the given grid. Thus, the binary sampling converts a binary data set in a continuous space to a binary data set in a discrete space. Another approach is to convert a binary set in a continuous space to a fuzzy set in the discrete space where each cell is assigned a membership value equal to the number of points of the original set lying inside that cell. We will call this sampling the fuzzy sampling. Figure 9(c) shows the result of fuzzy sampling on the binary data set in figure 9(a). Fuzzy sampling may also be used to convert a fuzzy data set in a continuous space to a fuzzy data set in a discrete space. In this case, the membership value of a cell in the discrete space can be equated to the sum of the membership values of the points of the original set lying inside that cell.

A binary set in continuous space may have regions of different densities. The density of points may play crucial role in the clustering analysis. Very fine grids are needed to preserve the density information in the data sets obtained by binary sampling. The fine grids demand large number of cells to cover the desired range in the continuous space. To preserve the density information completely in binary sampling, no cell should contain more than one data point of the original data set in continuous space. Binary sampling with "not so fine" grid may cause significant loss of density information. On the other hand fuzzy sampling can efficiently encode the density information for much larger grid spacing, thus reducing the number of cells needed to cover the given space. The spatial resolution of the clustering analysis is still determined by the fineness of the grid.

The fuzzy sampling scheme described above converts a binary or a fuzzy data set in a continuous space to a fuzzy data set in a discrete space. Fuzzy morphology algorithms may be applied on the data set in the discrete space. In a typical data analysis task the final result of the analysis may need to be represented in continuous space. For example, in a clustering problem, the final result needs to tell the cluster associated with each point in the continuous space. In short, the above approach of using discrete spaces for intermediate processing, needs to establish a correspondence between points in continuous space and cells in discrete space. Let the ranges of x and y co-ordinates of a 2-D continuous space range from "x_{min}" to "x_{max}" and "y_{min}" to "y_{max}" respectively. A grid of size "m x n" will split this space into cells of size "d_x x d_y" where

$$d_x=(x_{max} - x_{min})/m \text{ and } d_y=(y_{max} - y_{min})/n. \tag{38}$$

Let the cells be denoted by ordered pairs (i,j) where "i" and "j" are integers ranging from 0 to "m-1" and 0 to "n-1" respectively. Then, the cell corresponding to a point "P(x,y)" is given by

$$i=\text{int}((x-x_{min})/d_x) \text{ and } j=\text{int}((y-y_{min})/d_y) \tag{39}$$

where the operator "int" finds the largest integer, smaller than its argument.

In the subsequent discussion we will deal with discrete spaces. We will use the term "point" to refer to a cell in the given discrete space.

4. DISTANCE MAJORS AND CLUSTERING

The topological notion of distance provides a major of nearness of points in the given space. Euclidean distance is the most common distance norm. The Euclidean distance between point A and B with co-ordinates (x_a,y_a) and (x_b,y_b) in a 2-D space is given by

$$d(A,B)=((x_a-x_b)^2+(y_a-y_b)^2)^{1/2}. \tag{40}$$

In general "d(A,B)" represents a topological distance between points "A" and "B" if and only if it satisfies the triangular inequality

$$d(A,C)\leq d(A,B)+d(B,C). \tag{41}$$

Many of the algorithms[17,18] that use a distance norm for clustering, handle the data sequentially. The first couple of points are designated as centroids of different clusters. A new data point may start a new cluster if its distances from the centroids of the pre-existing clusters are larger than some predetermined limit such as the average inter cluster distance between the pre-existing clusters. Otherwise, the data point is assigned the pre-existing cluster which corresponds to the cluster centroid nearest to the point. When a new point is added to a cluster, its centroid is recalculated. The algorithms also allow merging of small clusters into big clusters. However, it is assumed that eventually, after processing sufficient data points, the number of clusters and the locations of their centroids will be stable. Many of the algorithms, directly or indirectly assume a priori knowledge of the number of clusters present in the given data set.

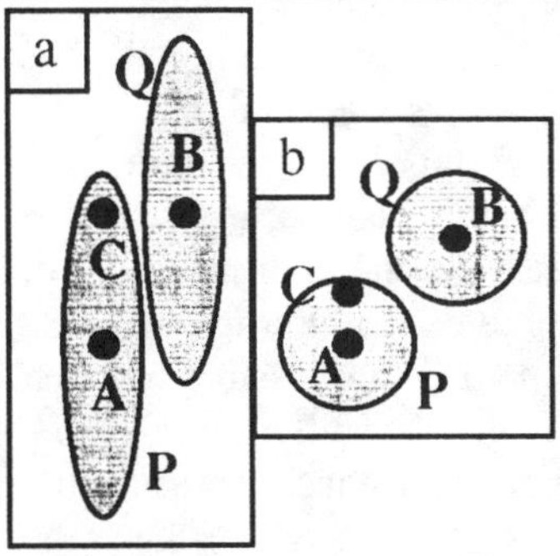

Figure 10: Scale dependence of clustering processes based on the distances from the cluster centroids.
(a) d(B, C) < d(A,C)
(b) d(A,C)<d(B,C)

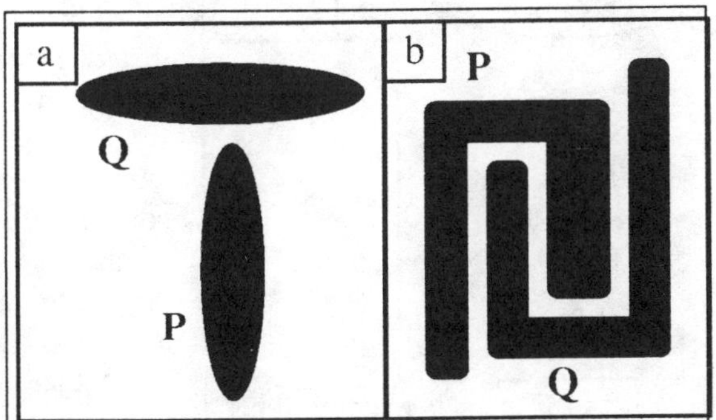

Figure 11: Examples of clusters which could not be identified based on the distances from their centroids for any given axes scaling.

Figure 10(a) shows a binary data set in a continuous 2-D space. The data set has two distinct clusters "P" and "Q". Now, let us assume that we know that the centroids of these clusters are located at "A" and "B" respectively. A clustering algorithm based on Euclidean distance norm will classify the point "C" as a member of cluster "Q" instead of cluster "P", as d(B,C)< d(A,C). Such an algorithm will not be able to find the cluster centroids to begin with. Figure 10(b) shows the same data set as figure 10(a) after the axes are scaled by certain factors. The clustering algorithms based on Euclidean distance norm will now give the expected results. For a general data set, it is difficult to guess the appropriate scaling factors for axes without a priori information. In certain cases it may be possible to estimate the axes scaling factors from consistent information about relative density of data points along various axes in different regions. In large variety of problems such as those illustrated in figure 11(a) and 11(b), it is not possible to find any scaling which will convert the data sets in the forms that can be correctly segmented using clustering algorithms based on Euclidean distance norm. In fact no distance norm may correctly work on these data sets.

Neural networks have been widely applied to solve complex clustering problems. Some of these approaches need enough training-data. The clustering problems addressed in this chapter do not assume access to any training data. Many of the self learning neural networks on the other hand use distance norms as a part of the learning schemes. The schemes to update weights in neural networks are many times computationally expensive on sequential machines. For example in a perceptron network, the number of neuron layers is the same as the maximum number of decision surfaces that could be constructed with that network[19]. Thus for a complex clustering task, a network with several neural layers is needed. Hence, high computational time is required for tuning the weights involved in these multilayer networks. The hardware implementations of neural networks to achieve parallel computations are cumbersome in many cases because of the analog circuit elements involved in these efforts.

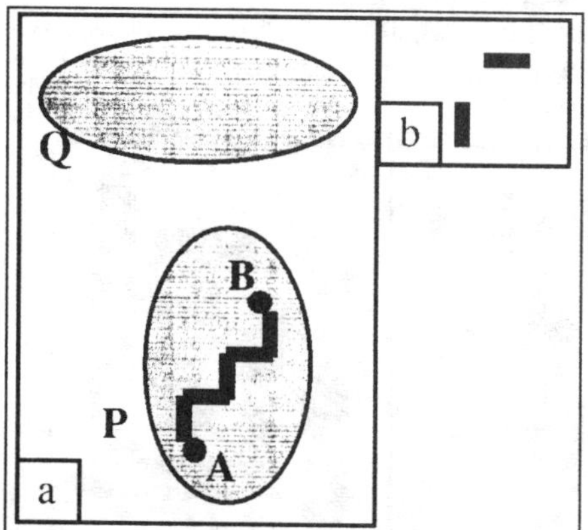

Figure 12: (a) Data sets in continuous space: Any two points of a cluster could be connected by at least one path. A path is a sequence of links such that the end points of each link belong to the cluster. (b) Links of the maximum allowable length.

As illustrated in the rest of the chapter, mathematical morphology offers elegant ways to overcome the problems faced by the clustering techniques based on distance norms. A note of caution is due here. Figure 12 shows the same data set as figure 11(a). As illustrated in figure 12, any two points in the same cluster can be connected by at least one path made out of small links, where each link joins two data points. Such a path is not possible between any two points in the different clusters, since the minimum distance between the two clusters is still larger than the maximum allowable length of a link. This approach is tolerant to axes scale changes as long as the minimum distance between the two clusters is greater than the maximum allowable link length. Our simplest morphological algorithm in the next section approximately implements the above idea. The notion of distance is indeed used here on successive neighborhoods. Thus, when we categorize morphological algorithms as being different from the distance based algorithms, we are referring to global distances such as those from cluster centroids. In general any clustering algorithm has to use the concept of distance in one way or the other. The size of structuring element and the fineness of the grid used for data conversion from continuous space to discrete space implicitly use the concept of distance norm over local neighborhoods. The same two factors also determine the spatial resolution of the morphological algorithms.

5. MORPHOLOGICAL ALGORITHMS FOR CLUSTERING ANALYSIS

In the present section we will discuss several mathematical morphology algorithms useful for clustering analysis. We will discuss different categories of clustering problems based on various attributes of the data sets. Practical problems may comprise one or more of these categories. The algorithms for binary morphology are discussed first for the sake of simplicity. The extensions to fuzzy morphology are discussed later. The section deals with the data sets in discrete space

5.1 Binary morphology algorithms for clustering analysis:

5.1.1 Segmentation of non-overlapping connected clusters:

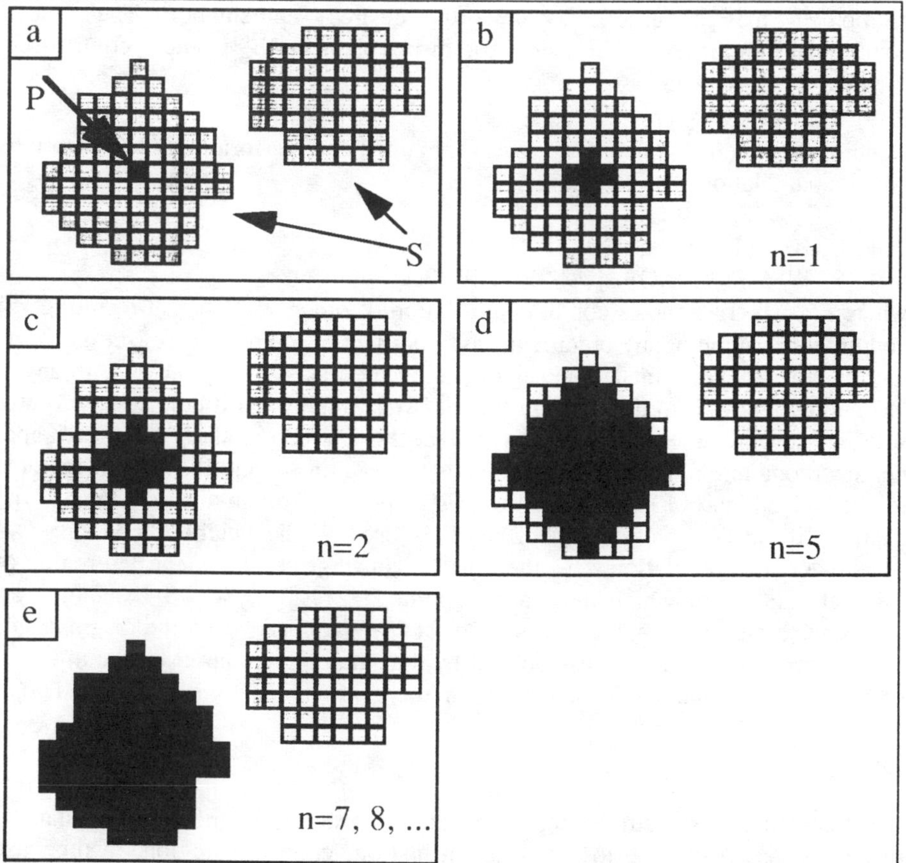

Figure 13: Segmentation of a binary data set into non-overlapping clusters. (a) Data set S and an arbitrary point P, (b, c, and d) $P(\oplus)(B,S)^n$ for n=1,2, and 5 respectively. (e) $P(\oplus)(B,S)^n$ for n>6. (B is the nearest neighbor mask in 2-D.)

The simplest case of clustering is when each cluster is a connected region and it is disconnected from any other cluster. In short, the clusters are non-overlapping. In a connected region, it is possible to find a path between any two points in the region such that all the points on the path also belong to the region. A path is a sequence of points, such that any two consecutive points in this sequence are neighbors of each other in the discrete space. A neighbor of a point may either be a nearest neighbor or a simple neighbor. Nearest neighbors of a point are the points under the nearest neighbor mask when the center of the mask is located at the given point. Simple neighbors of a point are the points inside a generalized cube of length three units when the center of the cube is located at the given point. The dimension of the generalized cube is same as the dimension of the co-ordinate system. Thus, depending on the definition of "neighbor," either the nearest neighbor mask, or the generalized cube of length three units may be used for the clustering process. In rest of the chapter we will use the general term "neighborhood mask" to refer to either the nearest neighbor mask or the generalized cube of length three units. An operation called conditional dilation[1,2,3] may be used to achieve the desired segmentation. The first order conditional dilation of set "F" with structuring element "B" inside a control set "G" such that $G \supset F$ is given by

$$F(\oplus)(B,G) = (F \oplus B) \cap G \tag{42}$$

Conditional dilation of order "k" is obtained by performing "k" iterations of conditional dilation. Thus we have:

$$F(\oplus)(B,G)^1 = F(\oplus)(B,G)$$
$$F(\oplus)(B,G)^2 = \{F(\oplus)(B,G)^1\}(\oplus)(B,G)$$
$$F(\oplus)(B,G)^j = \{F(\oplus)(B,G)^{j-1}\}(\oplus)(B,G) \quad (j=3,k) \tag{43}$$

where $F(\oplus)(B,G)^k$ denotes conditional dilation of order "k". The set resulting from a conditional dilation of any order is always confined inside the control set (G) in which the operation is performed. The segmentation process starts by picking up any point "P" and assigning it to the first cluster set "S_1". Conditional dilation is performed on set "S_1" several times until the set converges. The original data set "S" and the neighborhood mask "B" in 2-D are used as the control set and the structuring element respectively. Figure 13 illustrates various steps involved in the processing. In this example the 2-D nearest neighbor mask is used as the neighborhood mask. After several conditional dilation steps the set "S_1" converges to the connected region of the original data set in which the starting point "P" resides. Set "S_1" obtained above, represents a cluster. A point "Q" outside set "S_1" is picked from the original data set and the second cluster S_2 is established by the same process used to obtain "S_1". The process is continued until each point in the original data set is assigned to some cluster.

For the data set in continuous space, the fineness of the spatial grid used for the digitization has a strong impact on the connectivity of the resulting data set in discrete space. A crude grid can result in just one connected region. In this case the clustering technique discussed in the present section will indicate that this data set has only one cluster. On the other hand the technique may assign each point to a different cluster if the digitization grid is excessively fine. One way to handle this problem is to perform clustering for various sizes of grids and determine the range of grid sizes over which the number of clusters obtained is relatively stable[9]. The final clustering is performed by selecting a grid in this range. Morphological openings and closings may be performed as noise filtering stages prior to the above clustering processes.

5.1.2 Segmentation of overlapping clusters:

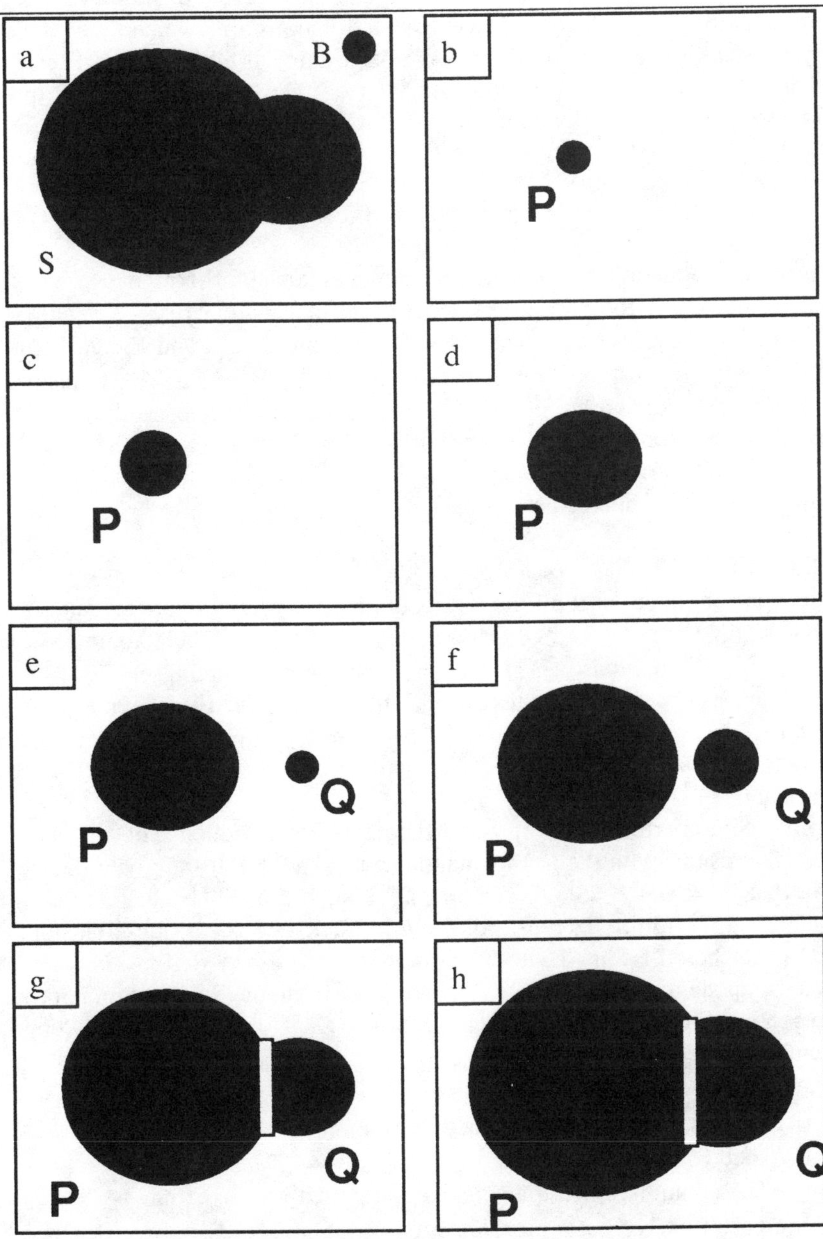

Figure 14: Binary watershed algorithm.
(a) Binary data set S and structuring element B, (b) S^6, (c) S^5, (d) S^4, (e) S^3, (f) S^2, (g) S^1, (h) S where $S^k = S \ominus B^k = S^{k-1} \ominus B$.

When distinct clusters in the continuous space are very close to each other, a very fine grid needs to be used to keep these clusters from touching each other in the digitized space. Thus a large number of cells are needed to cover the space under consideration. The large number of cells demands a long processing time. There are situation when it is impossible to find any grid that can prevent different clusters from touching each other in the discrete space. These situations arise when there is a region of nearly constant low density between two cluster cores of high density. Hence, these clusters are connected in the digitized space. The present section discusses segmentation of binary data sets into clusters that may be connected to each other. It is assumed that the region connecting the clusters is narrower than the cluster widths near their cores. Figure 14(a) shows an example of such a data set. The segmentation process used here is called the "binary watershed segmentation[20,21]".

Before discussing the watershed segmentation algorithm, it is necessary to introduce to the reader the concept of geodesic influence zones[1,2,3] "S_1" and "S_2" of sets "F_1" and "F_2" inside set "G" where $F_1 \cap F_2 = \emptyset$ and $F_1 \subset G$ and $F_2 \subset G$. In order to find these influence zones, they are first initialized as $S_1=F_1$ and $S_2=F_2$. In each step of the algorithm, conditional dilations of "S_1" and "S_2" are performed with neighborhood mask "B" inside the control set "G" to obtain sets "S_1'" and "S_2'" respectively. Thus

$$S_1' = S_1 (\oplus) (B, G) \text{ and } S_2' = S_2 (\oplus) (B, G) \quad (44)$$

The influence zones are updated as follows:

$$S''=S_1 \quad (45)$$

$$S_1=S_1'/(S_1' \cap S_2) \text{ and } S_2=S_2'/(S_2' \cap S''). \quad (46)$$

The process is continued until all the points of "G" belong to either "S_1" and/or "S_2", or in other words until $G=S_1 \cup S_2$. Physically, a point of "G" belongs to "S_1" if it is closer to its nearest point in set "F_1" than the nearest point in set "F_2". Thus the concept of geodesic influence zones do use the Euclidean distance measure in an indirect way. However, if the starting sets "F_1" and "F_2" are large and the separation between them is small, the effect of the distance norm on the overall clustering process is insignificant.

In order to perform the watershed segmentation, a sequence of sets is obtained by successive erosions with the nearest neighbor mask "B". For the given data set "S", let $S^1=S\ominus B$, $S^{k+1}=S\ominus B^{k+1}=S^k\ominus B$ for k=1,2,...n such that $S^{n+1}=\emptyset$. Each connected component of "S^n" is initialized as a different cluster. Let us call these clusters "C^k" for k=1,2,..N, where "N" is the total number of connected regions in "S^n". The connected component of a set can be determined from the segmentation procedure based on conditional dilation discussed in section 5.1.1. Now, let "C" be any connected component of "S^{n-1}". Three possibilities exist for the set "C":

(i) "C" does not contain any pre-assigned cluster, that is $C \cap C^k=\emptyset$ for all k.

(ii) "C" contains one and only one pre-assigned cluster, that is $C \cap C^k=\emptyset$ for $k \neq k'$ and $C^{k'} \subset C$.

(iii)"C" contains more than one pre-assigned clusters, that is $C \cap C^k \neq \emptyset$ or equivalently $C^k \subset C$, for more than one value of k.

In the first case "C" is assigned as a new cluster. In the second case "C" is assigned as cluster "$C^{k'}$". In the third case the clusters inside "C" are updated as the geodesic influence zones of the clusters inside "C". The clusters are further modified by the connected components of "S^{n-2}". The process is continued until the clusters are finally modified by the connected components of S. Figure 14 shows various steps of the watershed segmentation process. In figure 14(a), set "S" and structuring element "B" are drawn in a continuous space for the sake of simplicity. In this chapter, couple of more figures are also drawn in continuous spaces to illustrate the basic ideas. The actual algorithms in all these cases are applied in discrete spaces. In the example of figure 14, $S^7=S\ominus B^7=\emptyset$. Figure 14(b) shows S^6 which has only one connected region. Hence, according to case (i) above, this region is assigned as the first cluster. We will denote it "P". Figure 14(c) shows S^5 which also has one connected component. The connected component of 14(c) contains the cluster "P" in 14(b). Hence, according to case (ii) above, the cluster "P" is modified as the connected component in 14(c). Figure 14(d) shows S^4 which once again has only one connected component. The cluster "P" is reassigned to this component. Figure 14(e) shows S^3 which has two connected components. One of the connected components contains cluster "P" in 14(d) and thus updates this cluster. The other component does not contain any pre-defined cluster. Hence, this component is assigned as a new cluster "Q". Figure 14(f) shows S^2 which still has two connected components. One component contains clusters "P" in 14(e) while the other component contains cluster "Q" in 14(e). Hence, these regions update clusters "P" and "Q" respectively. Figure 14(g) shows S^1 which has only one connected component. This component contains both the clusters in 14(f). Hence, according to case (iii) above, clusters "P" and "Q" are updated as the geodesic influence zones of "P" and "Q" in 14(f) inside the connected component of 14(g). A partition is shown in 14(g) to separate the two influence zones. Figure 14(h) shows original set "S" which once again contains only one connected component. The final clusters are obtained as the influence zones of P and Q in 14(g) inside "S".

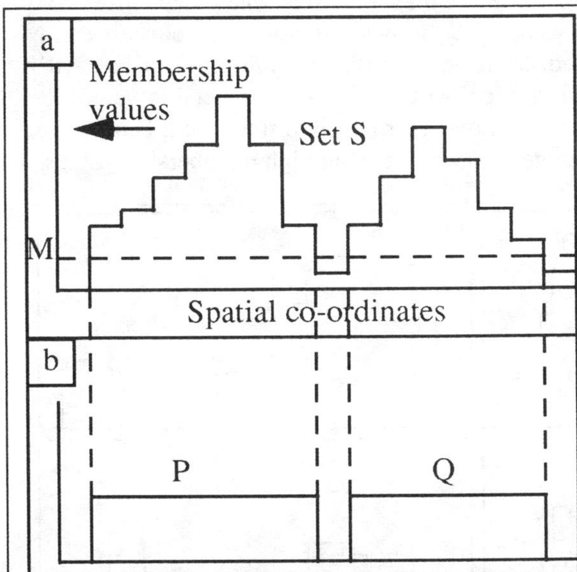

Figure 15: Segmentation of fuzzy data sets into clusters connected at membership values lower than a small number "M".
(a) Fuzzy set S. (b) Clusters "P" and "Q" of set "S".

In the binary watershed algorithm, new clusters formed from the connected components of S^m, where m is a small number, may be ignored. This modification prevents generation of small clusters which may arise from rough boundaries of larger clusters.

5.2 Gray scale morphology algorithms for the segmentation of fuzzy sets:

<u>5.2.1 Segmentation of disconnected clusters:</u>

Connectivity of regions of a fuzzy set may be defined in terms of membership values. Region A and B in figure 15(a) are connected at the membership value of "M".

We will describe an algorithm to segment a fuzzy data set "S" into regions connected at membership values greater than some threshold level "M". A binary set "S_b" is obtained by thresholding set "S" at membership value "M". "S_b" is set of all points with membership values greater than "M". The process of multiple controlled dilations, explained in section 5.1.1, is performed to segment "S_b" into disconnected regions. These regions represent clusters in original fuzzy set "S". The membership value of a point in a cluster may be assumed to be the degree to which the point belongs to the cluster. Thus the cluster cores are the regions with high membership values in the original fuzzy data set. Figure 15 illustrates the segmentation process discussed above.

The above approach of segmentation may be used when the clusters are connected at a nominal membership value "M". In these situations the clusters can be considered as practically disconnected. More involved algorithms need to be used if the value of "M" is significant, such that the data points with membership values less than "M" can not be ignored. The next section describes algorithms that can be used for segmentation of clusters that are in general connected at high membership values.

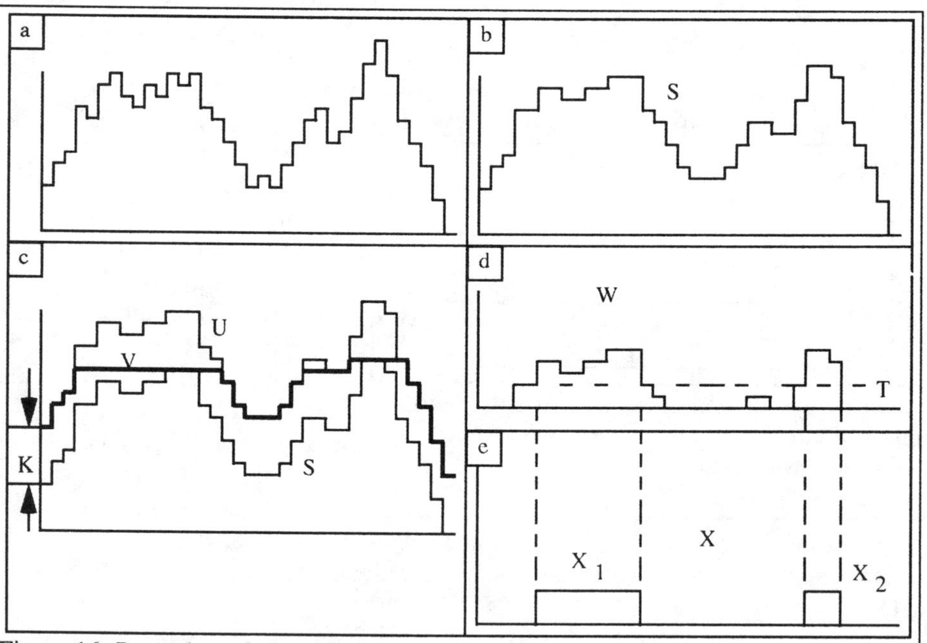

Figure 16: Detection of cluster cores of a fuzzy data sets.
(a) 1-D noisy data set, (b) Result of filtering, (c)U=S+K and $V=S^m=S(\oplus)(B,U)^m$ such that $S^m=S^{m+1}$, (d) W=U-V, (e) Binary set X obtained by thresholding W at membership value T. X_1 and X_2 are connected components of X.

5.2.2 Segmentation of connected clusters:

A fuzzy data set connected at a high membership value may be segmented either based on its shape or its membership values. The present section deals with the segmentation based on membership information. "Gray level watershed algorithm[22,23,24]" is the most popular technique[1] for this purpose. The clusters obtained by this technique may be further segmented using the "binary watershed algorithm" which uses the shape information. The "gray level watershed algorithm needs to know the cluster cores as a starting point. We will first mention a procedure to obtain cluster cores in the presence of noise.

5.2.2.1 Detection of cluster cores:

For simplicity we will show the processing for a 1-D data set. Figure 16(a) shows a 1-D fuzzy data set. The aim is to separate the two major clusters. Local maximas resulting from noise may be eliminated by filtering the data set prior to application of the segmentation algorithm. Alternating sequential filter may be used for this purpose. Figure 16(b) shows the data set after noise filtering. The two major humps still have local maximas which may be misunderstood as independent clusters. In order to eliminate the effect of these local maximas, a fuzzy set "U" is obtained by adding a value "K" to the membership values of set "S" such that

$$\mu_U = \mu_S + K \tag{47}$$

Next a set "V" is obtained by performing k^{th} order conditional dilation of "S" with the neighborhood mask "B" inside the control set "U". Thus

$$V=S(\oplus)(B,U)^k \tag{48}$$

The value k is adjusted such that

$$S(\oplus)(B,U)^{k-1} \neq F(\oplus)(B,G)^k = F(\oplus)(B,G)^{k+1} \tag{49}$$

$$\text{Let } W=U-V \tag{50}$$

Figure 16(c) shows sets "U" and "V", whereas figure 16(d) shows set "W". A binary set "X" is obtained by thresholding set "W" at membership value $T < K$ such that

$$X = \{x : \mu_W (x) > T\} \tag{51}$$

Disconnected components of "X" may be obtained by using the technique described in section 5.1.1. Notice that the set "X" in figure 16(e) shows only two disconnected components. The disconnected components of "X" define the cluster cores. We will denote these cluster cores by the notation "X_i" (i = 1,N) where "N" is the total number of disconnected regions of set "X". For the example in figure 16, N=2. The procedure illustrated in figures 16(c) and 16(d) eliminates the effects of shallow humps and dips and produces only one cluster core per cluster. If a cluster has several peaks of comparable heights the above process finds a cluster core which is a connected region containing all the peaks. The cluster on the left in figure 16(b) is an example of this situation. The differences between local maximas and minimas of such a cluster need to be less than (K-T). Another situation is when a cluster has one dominant peak such as the one in the left cluster of figure 16(b). The above process finds a cluster core which contains only the dominant peak. The difference between other local maximas and minimas need to be less than "T".

A more robust criterion called "dynamics" has been proposed[25] for the detection of dominant cluster cores of fuzzy data sets. This criterion has already been used[10,11] for the segmentation of multivariate histograms using "gray level watershed algorithm". We will not discuss this criterion in the present chapter.

5.2.2.2 Gray level watershed algorithm[22,23,24]*:*

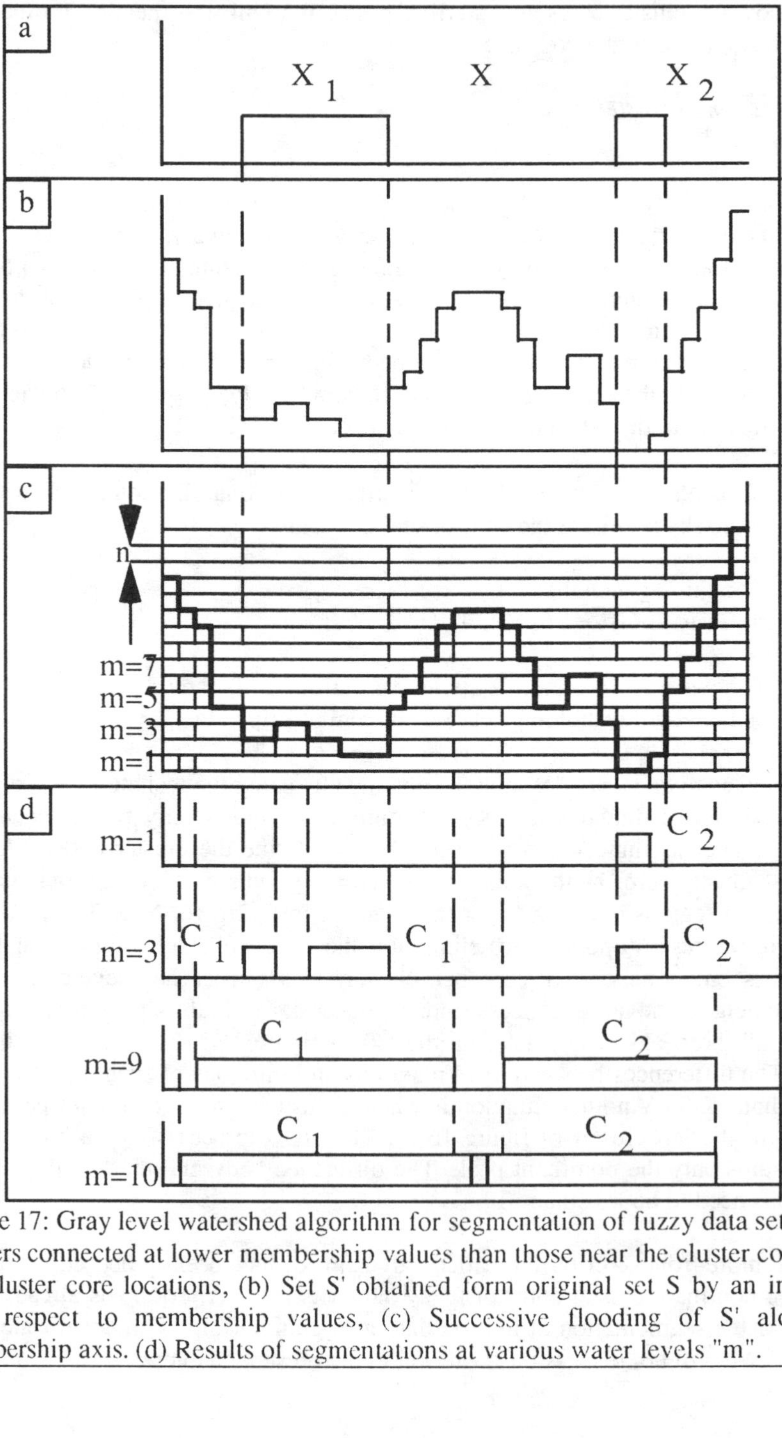

Figure 17: Gray level watershed algorithm for segmentation of fuzzy data sets into clusters connected at lower membership values than those near the cluster cores. (a) Cluster core locations, (b) Set S' obtained form original set S by an inversion with respect to membership values, (c) Successive flooding of S' along the membership axis. (d) Results of segmentations at various water levels "m".

Figure 17(a) shows the cluster cores "X_k " (k=1,2) from figure 16(e). Figure 17(b) shows the inverted form "S' " of the filtered data set "S" in figure 16(b). "S' " is defined by

$$\mu_{s'}(x) = \mu_M - \mu_s(x) \quad (52)$$

$$\text{where } \mu_M = \max(\mu_s(x)) \quad (53)$$

Set "S' " may be considered the set complement of the fuzzy set "S". Clusters "C_k " corresponding to cluster cores "X_k " (k=1,N) are initialized as

$$C_k = \varnothing \quad (k=1,N) \quad (54)$$

As shown in figure 17(c) a water level is imagined to rise along the membership axis in the steps of a value "n". Let "S_1 " be the binary set representing the set of points of "S' " the membership values of which are less than the water level. Thus, at step "m":

$$S_1 = \{\, x : \mu_{S'}(x) < mn \,\} \quad (55)$$

The connected components of "S_1" are obtained using the algorithm introduced in section 5.1.1. Three possibilities exist for a connected component "C" of set "S_1":

(i) "C" does not contain any portion of any cluster core, that is $C \cap X_k = \varnothing$ for k=1,N.

(ii) "C" contains a portion of only one cluster core, that is $C \cap X_k = \varnothing$ for $k \neq k'$ and $C \cap X_{k'} \neq \varnothing$.

(iii) "C" contains more than one cluster cores, that is $C \cap X_k \neq \varnothing$ for more than one value of k.

Nothing is done for the first case. For the second case the cluster "k" is updated as $C_k = C_k \cup C$. For the third case "C_k" s are updated by their geodesic influence zones inside "C". Here each "k" satisfies the condition in case (iii) above. Note that the cluster cores do not play an active role in the actual processing, but they provide a marker for each cluster. Figure 17(d) and 17(e) show the clustering for different values of water-level "m". Water flooding is continued until the entire set "S' " is under the water level. Alternately the flooding may be stopped when

$$\mu_M - mn < M \quad (56)$$

where "M" is a low value such that the regions in "S" with membership values lower than "M" may be ignored.

6. CONNECTIVITY PRESERVING FILTERS FOR MULTIDIMENSIONAL SETS

Figure 18 illustrates the need for connectivity-preserving filters. The closing operation on the data set in this figure, performed to remove the noisy holes in the data set, tends to connect different clusters. In the figure, neither the small noisy holes in the clusters nor the thin regions between the clusters can accommodate the structuring element. Hence both these regions are filled in the closing operation. One difference between the noisy holes and the thin region between the clusters is that the holes are confined in a small neighborhood whereas the regions between the clusters are long. Note that the notion of thin regions arises only in the case of multidimensional sets. The issue is irrelevant for one dimensional sets. Algebraic opening and closing operations are available to remove noisy regions confined in a small neighborhood, while preserving thin but long regions. These operations are called connectivity preserving closing and opening operations, or in general, connectivity-preserving filters[4]. A connectivity preserving opening preserves the connected regions of the data set under consideration whereas a connectivity preserving closing operation preserves

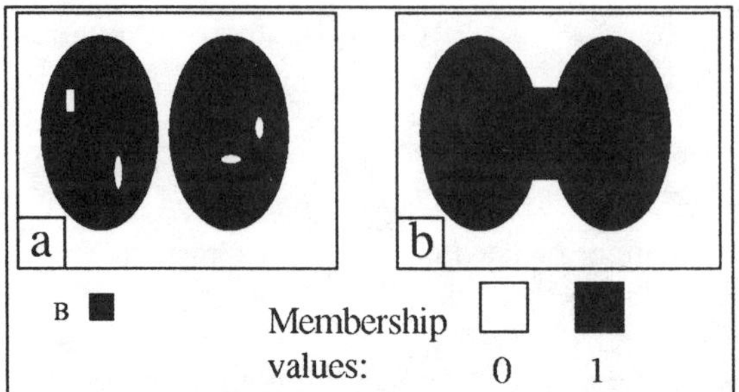

Figure 18: Side-effects of clustering.
(a) Noisy data set S, (b) $(S\bullet B)\circ B$

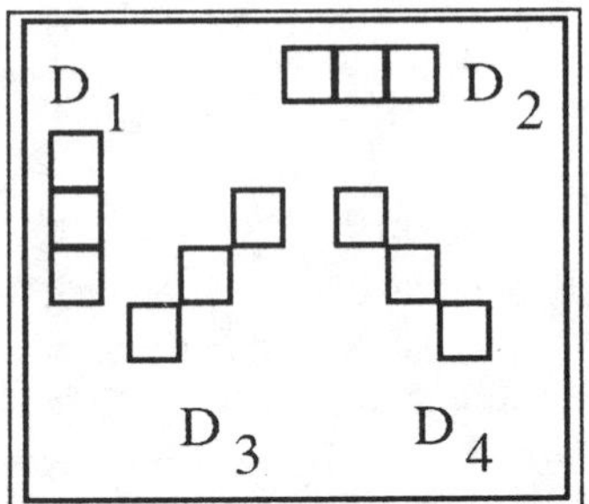

Figure 19:
Set $D = \{D_l : l=1,4\}$ of structuring elements useful for median closing and median opening operation.

connectivity of the complement of the set. Popular examples of connectivity preserving filters are "median closing" and "median opening" filters. Median closing of a binary or a fuzzy set A with a set of structuring elements $D=\{D_l : l=1,m\}$ is defined as:

A✚$D=\cap(A\bullet D_l)$: $l=1,m$. (57)

Median opening of A with D is given as (58)

A✣$D=\cup(A\circ D_l)$: $l=1,m$. (59)

The structuring elements "D_l" ($l=1,m$) are linear elements. Median closing satisfies the conditions of extensivity and idempotence in equations (36) and (37) respectively. Similarly, median opening satisfies the conditions of antiextensivity and idempotence in equations (34) and (35) respectively. Hence, median closing and opening operations are algebraic closing and algebraic opening respectively. Figure 19 shows an example of "D_l" for two dimensional space and m=4. In "median opening", small regions where none of the structuring elements "D_l" can fit, are removed. In "median closing," small holes which can not accommodate any of the structuring elements are filled. However, the thin regions which can accommodate at least one of the structuring elements are preserved. Median closing and median opening are popularly used for removing internal texture of clusters to obtain smooth regions. A limitation of these filters is that they preserve only those regions which can be probed by structuring elements in a limited number (m) of directions. In addition, these filters do not preserve the regions which change directions more than once within distances smaller than the length of the linear structuring elements. This limitation is particularly severe when the clusters of interest are thin or shell type clusters instead of solid clusters. Figure 20(a) shows an example of a shell type cluster

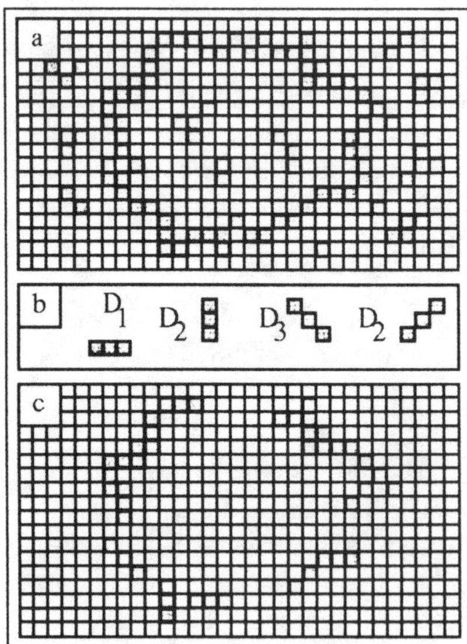

Figure 20: Destruction of shell type clusters by median closing and opening.
(a) Noisy shell type data set S.
(b) Set of linear structuring elements $D = \{D_1, D_2, D_3, D_4\}$
(c)(S✚D)✣D

in two dimensions. The median opening operation on this cluster using the linear structuring elements shown in figure 20(b) destroys a large portion of the shell structure. Hence, more robust filters which can preserve the connectivity completely are needed.

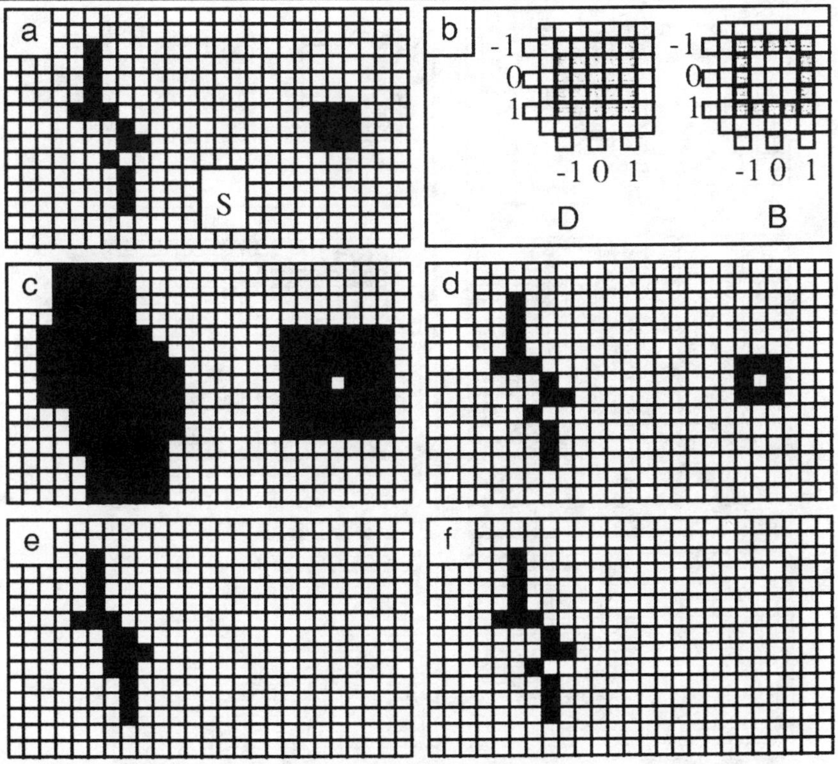

Figure 21: Comparative opening and bounded opening.
(a) Data set S containing a thin region and a noise spot. (b) Structuring elements (D,B), (c) $S \oplus B$, (d) Comparative opening: $S \bigcirc B = S \cap (S \oplus B)$, (e) $(S \oplus B) \ominus D$,
(f) Bounded opening $S \circledcirc (D,B) = S \cap ((S \oplus B) \ominus D)$

Two examples of opening/closing operations which preserve connectivity completely are comparative opening/closing[5,15] and bounded opening/closing[16]. Comparative closing and opening operations were introduced by Serra[5] and used by Safa and Flouzad[15] in their comparative filters to remove speckle noise from RADAR imagery. Bounded closing and opening operations were introduced by the present authors[16]. Let "D" be a solid structuring element with boundary "B". The comparative closing of "A" with "B" is defined as

$$A \bullet B = A \cup (A \ominus B) \tag{60}$$

Comparative opening is defined as

$$A \bigcirc B = A \cap (A \oplus B) \tag{61}$$

The bounded closing of A with {D,B} is defined as

$$A \odot (D,B) = A \cup ((A \ominus B) \oplus D) \tag{62}$$

The bounded opening is defined as

$$A \circledcirc (D,B) = A \cap ((A \oplus B) \ominus D) \tag{63}$$

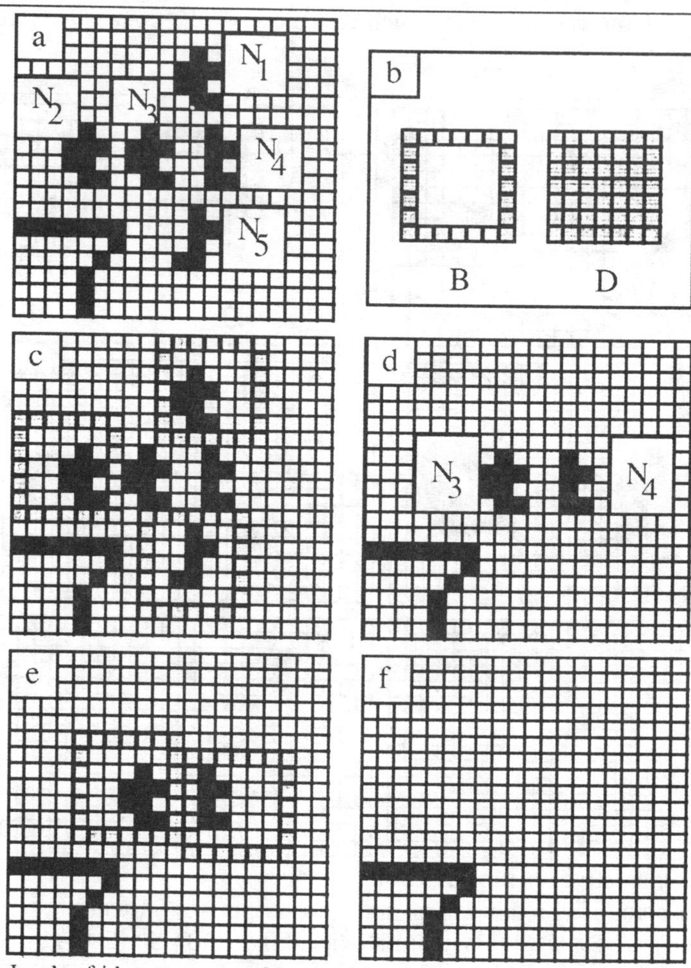

Figure 22: Lack of idempotence of bounded opening.
(a) Set "S" with noise spots N_1-N_5 and a thin region connected over larger space.
(b) Structuring elements (D,B) (c) Possible positions of B around N_1, N_2, and N_5.
(d) $S_1 = S \circledcirc (D,B)$ (e) Possible positions of B around N_3 and N_4. (f) $S_2 = S_1 \circledcirc (D,B)$

Figure 21 illustrates the processes of comparative opening and bounded opening on set "S". The aim is to retain the thin but connected region and remove the noise spot in "S". Both, comparative opening (figure 21(d)) and bounded opening (figure 21(f)) preserve the thin connected region. However, the comparative opening removes only part of the noisy spot, whereas the bounded opening removes the entire noise spot. Comparative filtering of a binary set with "B" of equation 61 removes certain points of the set such that when the center of "B" is placed at these points, "B" does not lie on any portion of the set. Notice that only the center of the noise spot in figure 21(a) satisfies this condition. Thus, in comparative opening shown in figure 21(d), only the center point of the noise spot is removed. On the other hand, bounded opening of a binary set removes those portions of the set which can lie inside the boundary "B" of the solid structuring element "D" while the boundary "B" does not fall on any portion of the set. The entire noise spot of figure 21(a) satisfies this condition. Thus, the bounded opening shown in figure 21(f) removes the entire noise spot. The above descriptions of comparative and bounded openings are more evident from alternate representations of these operators given by

$$A \bigcirc B=\{x : B+x \subset A^c \} \quad (64)$$

$$A \circledcirc (D,B)= \cup(D+x) : B+x \subset A^c \quad (65)$$

The ordering relation between morphological opening, median opening, comparative opening, and bounded opening may be stated as follows:

$$A \circ D \subset A \maltese C \subset A \circledcirc (D,B) \subset A \bigcirc B \subset A \quad (66)$$

Here "D" is a solid structuring element with boundary "B". "C" is set of linear structuring elements given by

$$C = \{ C_l : l=1,k\} \text{ such that } C_l \subset D \text{ for } l=1,k. \quad (67)$$

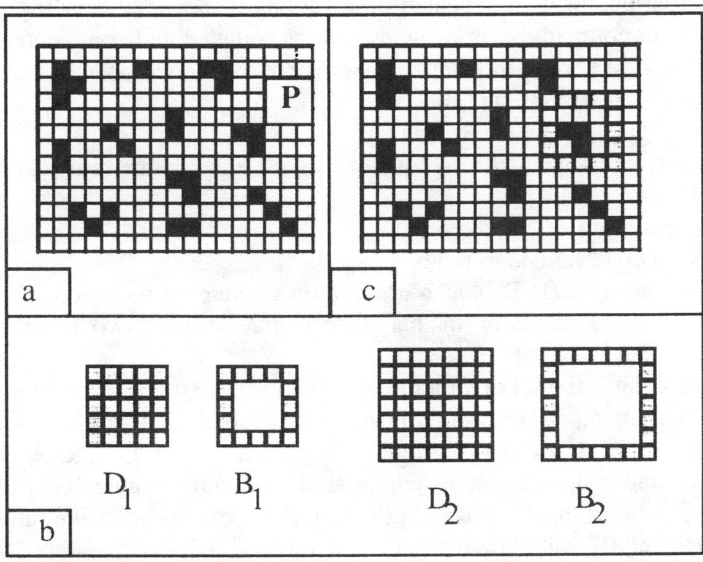

Figure 23: (a) Noise spots to be removed, (b) Structuring element pairs (D_1,B_1) and (D_2,B_2), (c)A possible position such that B_1 can fall around noise spot P, without hitting on any noise spot.

Equation (66) implies that the bounded opening removes more portion of the starting set "A" than the comparative opening. However, it can be shown[16] that the bounded opening preserves connected regions completely. Hence, bounded opening removes the noise more efficiently than the comparative opening. In fact, as illustrated in figure 21, if comparative filter can remove at least one point of a noise spot, the corresponding bounded opening will be able remove the noise spot completely. In bounded opening of a fuzzy set, the membership value of a point is unaltered if for any position of boundary "B" around the point, the minimum of the membership values of the points on "B' is always greater than or equal to the membership value of the point. Otherwise, the membership value of the point is changed to the maximum of all these minimum values. Bounded openings on binary and fuzzy sets satisfy the condition in equation 13 which is satisfied by all basic morphological operators. Comparative opening and median opening also satisfy the condition in equation 13. Closing plays the same role on the set complement, that opening plays on the set. Hence the above discussion for openings may be appropriately extended for closings.

Neither the comparative opening, nor the bounded opening satisfy the condition of idempotence in equation 35. Hence, neither of these opening operations are algebraic openings. Figure 22 illustrates this problem for bounded opening. The set in figure 22(a) contains five noisy spots and a connected region. As shown in figure 22(c) the boundary B of the solid structuring element D can lie around only three of these spots without lying on any region of the set. Hence, as shown in figure 22(d) bounded opening removes only these three spots from the set. However, once the above three spots are already removed from the set, "B" can lie around the other two spots without hitting any part of the set (22(e)). Hence, as shown in figure 22(f), in another application of bounded opening, these two spots are also removed. Note that bounded opening of the set in figure 22(f) will not change the set any more. The above example illustrates that bounded opening is not idempotent. However, a practically idempotent operation may be obtained by repeated applications of bounded opening. The same argument holds for comparative opening as well as bounded and comparative closings. Algebraic openings are obtained from comparative and bounded openings as follows:

$A(\bigcirc)B = (A\bigcirc B)^m = A\bigcirc B\bigcirc B$...m times. (68)

$A(\circledcirc)(D,B) = (A\circledcirc(D,B))^m = A\circledcirc(D,B)\circledcirc(D,B)$ ...m times. (69)

Similarly, algebraic closings are obtained from comparative and bounded closings as follows:

$A(\bullet)B = A\bullet B\bullet B$...m times. (70)

$A(\odot)(D,B) = A\odot(D,B)\odot(D,B)$...m times. (71)

In equations 68 through 71, D is a solid structuring element with boundary B. Also, "m" is a small integer. Practically "m" may take a value between two to five.

In figure 23(a) the set of structuring element (D_2,B_2) can not remove the noise spots by bounded opening, because their are no locations such that the boundary B_2 can lie around each of these spots without hitting another spot. On the other hand the set (D_1,B_1) can remove the same spots with bounded opening. Figure 20(b) illustrates a position for the structuring element such that a noise spot can lie in the interior of the structuring element B_1 while B_1 does not fall on any point of the set. Statistically, noise spots of a certain size may be best removed with structuring elements of size slightly larger than the size of the noise spot[16]. Hence, a practical connectivity preserving filter consists of ASF of equation 33 with algebraic closing and opening operations of equations (71) and (69) respectively. Thus a connectivity preserving ASF may be defined as

$ASF(A(\odot)(D,B)(\circledcirc)(D,B))^m = A(\odot)(D,B)(\circledcirc)(D,B)(\odot)(D^2,B^2)(\circledcirc)(D^2,B^2)$..m times (72)

where B^k is the boundary of the structuring element D^k.

The basic element "D" used in a connectivity preserving "ASF" of equation (72) may be a "N" dimensional cube of length three units, surrounding one cell in the given "N" dimensional digitized space. Erosions and dilations with increasing sizes of structuring elements (D^k,B^k) may be obtained using the fast algorithms outlined in section 2.4 and illustrated in figures 4,5, and 6. The filtering process may be further speeded up using sub-sampling of the data set[16].

Multidimensional data set "S" may be pre-processed to remove noise, prior to the application of the clustering techniques. A typical filtering procedure for multidimensional data may consist of a kth order "ASF' given by

$ASF(S \bullet C \circ C) = S \bullet C \circ C \bullet C^2 \circ C^2 .... \bullet C^k \circ C^k$ (73)
where "C" may be the nearest neighbor mask or a cube of length three units in the given discrete space. The filter in equation 73 may be operated for a small value of "k". The connectivity preserving filtering of equation 72 may then be performed for a large value of "m". The basic structuring element "D" used in equation 72 may be a cube of length three units.

7 CLUSTERING OF SCATTERED REGIONS

In segmentation algorithms we ignored the regions which formed small clusters. Some of these regions were omitted in filtering processes, while the others were ignored in the actual segmentation process. In many applications each of these data points may need to be assigned to one of the large clusters obtained in the segmentation process. We will describe a procedure to assign a point "P" of a binary or fuzzy data set to one of the clusters "C^i " (i=1,k) determined in the segmentation process.

For binary data set "S", let A(S) = the area of "S". (74)
Alternately for fuzzy set "S", let $A(s)=\Sigma\, \mu_S(x)$ (75)
Let $A^i=A((P\oplus B^m)\cap C^i)$ (76)
where "m" is the smallest integer such that "A^i " is non-zero for at least one value of "i". "B" is the neighborhood mask. Then, the point "P" is assigned to cluster "j" such that
$A^j=\max(A^i : i = 1,k)$ (77)
In figure 24, the point "P" is assigned to cluster "L". In this case m=2.

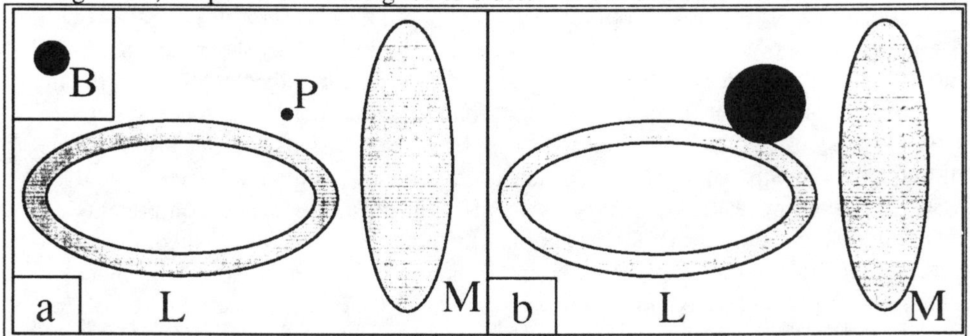

Figure 24: (a) Clusters L, M, and an unassigned region P, (b) "P" is classified as a member of cluster "L".

The same procedure may be extended for fuzzy data sets. For fuzzy sets, A(S) is the sum of membership values of all points of S. The present approach of assignment of scattered points to various clusters is not invariant with respect to scale changes. For larger values of m, the assignment may vary with a relative change of scale of different axes. In the lack of either shape information or membership-connectivity information, use of a distance measure is the only way to classify these scattered points. In section five, geodesic influence zones were often used in the clustering algorithms. The process of determining these zones does depend on the axes scaling. However, once again in spite of using the shape information or the membership value information, the distance norm may need to be used for segmentation of certain regions. The procedure described in this section uses the shortest distance of a scattered point from a given cluster instead of its distance from the cluster centroid. In thin clusters with shell structures, the notion of a single centroid is meaningless, and the present approach can perform better job of assignment of scattered points.

8. SUMMARY:

The tools of mathematical morphology, popular for the analysis of binary or gray level images may be applied for the structural analysis of the general class of multidimensional fuzzy data sets. Erosion, dilation, opening, and closing are the basic building blocks of morphological operations. Various properties of morphological operators may be exploited to develop fast algorithms which implement these operations. Alternating sequential filters (ASF) based on algebraic close-openings with increasing sizes of structuring elements comprise a very useful class of filters for the preprocessing of data sets. Morphological closing and opening operations are examples of algebraic closing and algebraic opening respectively. Since the membership functions are lower bounded, (the lower bound being zero), the close-open operation is a better filter than the open-close operation. Most of the numerical algorithms in mathematical morphology are developed for data sets in discrete space. Hence the data sets in continuous space need to be sampled to obtain corresponding data sets in discrete space. Binary sampling produces binary sets in discrete spaces whereas fuzzy sampling produces fuzzy sets in discrete spaces. Fuzzy sampling can encode the information more efficiently with a larger grid spacing and hence a smaller number of grid points than those needed for binary sampling. Many self learning clustering algorithms use distances from cluster centroids to classify the data points. These approaches are very sensitive to the scale changes of the axes of the co-ordinate systems. Determinations of appropriate scales or weighing factors may be a difficult and sometimes impossible task. Morphological techniques may be used to obtain reliable clustering which is stable with respect to the scale changes of the co-ordinate axes. Conditional dilation may be used to obtain various connected components of binary or fuzzy sets. The connectivity of a fuzzy set is defined in terms of the minimum membership value (M) at which each distinct component is connected. A simple segmentation task may consist of identifying these connected components. A more complex task is to segment overlapping clusters. The binary watershed algorithm uses the shape information to segment a binary data set into connected or disconnected clusters. The fuzzy watershed algorithm may be used to separate clusters based on the membership function information. These clusters may be further segmented based on their shape information using the binary watershed algorithm. The segmentation algorithms identify only the significant clusters. The regions which tend to form individual small clusters are made to merge with one of the dominant clusters. Morphological filtering based on simple close-opening operations could connect closely spaced clusters. The noise filtering may also cause damage to thin, shell type clusters. Connectivity preserving filters are used to remove noise while preserving thin regions which are connected over larger spaces than the sizes of the noise spots. Median closing and opening are popular connectivity preserving filters. These filters work well to preserve the fine details of the boundaries of solid clusters and at the same time smear the noisy texture of these regions. However, the median closing and opening operations perform poorly on thin, shell type clusters, destroying the structure of these regions. Comparative closing and opening and bounded closing and opening preserve the connectivity information completely. Bounded closing and opening remove noise more efficiently than comparative closing and opening. Hence, complex data sets with thin clusters or narrow cluster separations, may be preprocessed with ASF based on bounded close-opening. A practical filtering strategy may include a lower order ASF based on the simple close-opening followed by higher order ASF based on the bounded close-opening. The regions of the data set, eliminated in noise filtering or in clustering processes, may be assigned to various clusters using a morphological technique.

9. ACKNOWLEDGMENT

This work was supported by a grant from Northrop Co. under contract # HHH 880856K.

10 REFERENCES

1. Matheron G., Random Sets and Integral Geometry, John Wiley, New York, 1975.
2. Dougherty E., An Introduction to Morphological Image Processing, SPIE Optical Engineering Press, Washington, 1992.
3. Giardina C., Dougerty E., Morphological Methods in Image and Signal Processing, Prentice Hall, New Jersey, 1988.
4. Serra J., Image Analysis and Mathematical Morphology, Academic Press, New York, 1983.
5. Serra J., editor, Image Analysis and Mathematical Morphology, Vol. 2, Academic Press, New York, 1988.
6. Maragos P. and Shafer R., "Morphological Systems for Multidimensional Signal Processing," Proc. IEEE, Vol. 78, No. 4, pp. 690-710, April, 1990.
7. Serra J., and Vincent L., "An Overview of Morphological Filtering," Circuits Systems and Signal Processing, Vol. 11, No. 1, pp. 47-108, February, 1992.
8. Haralick R., Sternberg S., and Zuang, X., "Image Analysis Using Mathematical Morphology," IEEE Transactions on Pattern Analysis and Machine Intelligence, Vol. 9, No. 4, pp. 532-550, July, 1987.
9. Postaire J., Zhang R., and Lecocq-Botte C., "Clustering Analysis by Binary Morphology," IEEE Transactions on Pattern Analysis and Machine Intelligence, Vol. 15, No. 2, page 170, July, 1993.
10. Soille P., "Partitioning of Multispectral Images Using Mathematical Morphology," Vision and Voice Magazine, Vol. 6, No. 4, pp. 283-293, 1992.
11. Soille P. and Rivest J.-F., "Dimensionality of Morphological Operators and Cluster Analysis," Image Algebra and Morphological Image Processing IV, Proc. SPIE, vol. 2030, pp. 43-53, July 1993.
12. Maragos P., "Pattern Spectrum and Multi-Scale Shape Representation," IEEE Transactions on Pattern Analysis and Machine Intelligence, Vol. 11, pp. 701-716, July, 1989.
13. Kher A. and Mitra S., "Registration of noisy SAR imagery using morphological feature extractor and 2-D cepstrum," Applications of Digital Image Processing XV, Proc. SPIE, vol. 1771, 1992.
14. Kraske W., "Resolvable analysis and synthesis of synthetic aperture radar imagery," Intelligent Robots and Computer Vision XI, Proc. SPIE, vol. 1826, pp. 317-326, 1992.

15. Safa F. and Flouzat G., "Speckle Removal on Radar Imagery Based on Mathematical Morphology," Signal Processing, Vol. 16, No. 4, pp. 319-333, April, 1989.
16. Kher A. and Mitra S., "Optimum Morphological Filtering to Remove Speckle Noise From SAR Images," Image Algebra and Morphological Image Processing IV, Proc. SPIE, vol. 2030, pp. 97-108, July 1993.
17. Jain A. and Dubes R., Algorithms for Clustering Data, Prentice Hall, Englewood Cliffs, NJ, 1988.
18. Duda R. and Hart P., Pattern Classification and Scene Analysis, Wiley, New York, 1973.
19. Lippmann R., "An Introduction to Computing with Neural Nets," IEEE ASSP Magazine, April 1987.
20. Serra J. and Vincent L., Lecture Notes on Morphological Filtering, Ecole Nationale Superieure des Mines de Paris, Fontainbleau, 1989.
21. Vincent L., Algorithmes Morphologiques a Base de Files d'Attente et de Lacets. Extension aux Graphes, Ph.D. Theses, Ecole des Mines de Paris, May, 1990.
22. Beucher S., "Watersheds of functions and picture segmentation," IEEE Int. Conf. on Acoustic, Speech and Signal Processing, pp. 1928-1931, Paris, 1982.
23. Meyer F. and Beucher S., "Morphological Segmentation," Journal of Visual Communication and Image Representation, Vol. 1, No. 1, pp. 21-46, September 1990.
24. Vincent, L. and Soille, P., "Watersheds in Digital Spaces: An Efficient Algorithm Based on Immersion Simulations," IEEE Transactions on Pattern Analysis and Machine Intelligence, Vol. 13, No. 6, pp. 583-598, June, 1991.
25. Grimaud M., "New measure of contrast: dynamics," Image Algebra and Morphological Image Processing III, Proc. SPIE, Vol. 1769, pages 292-305, July 1992.